Real-Time Systems

Hermann Kopetz • Wilfried Steiner

Real-Time Systems

Design Principles for Distributed Embedded Applications

3rd ed. 2022

Hermann Kopetz
Vienna University of Technology
Vienna, Austria

Wilfried Steiner
TTTech Computertechnik AG
Vienna, Austria

ISBN 978-3-031-11994-1   ISBN 978-3-031-11992-7   (eBook)
https://doi.org/10.1007/978-3-031-11992-7

This Springer imprint is published by the registered company Springer Nature Switzerland AG
The registered company address is: Gewerbestrasse 11, 6330 Cham, Switzerland

Preface

The primary objective of this book is to serve as a textbook for students who take a senior undergraduate or a first-year graduate course on real-time embedded systems, also called *cyber-physical systems*. The book's structure—the material is organized into 14 chapters—maps to the 14 weeks of a semester. The book is also intended for practitioners in industry who want to learn about the state of the art in real-time embedded system design and need a reference book that explains the fundamental concepts of the field. More than a 1000 students used the first edition of this book, published about 25 years ago, as a text for the real-time systems course at the Vienna University of Technology. The feedback from these students and many new developments in this dynamic field of embedded real-time systems have been incorporated in the second edition. This third edition of the book includes learnings from industrial applications of its elaborated design principles and addresses the ongoing convergence of real-time embedded systems with IT (information technology) systems. It also covers emerging autonomous system developments.

The book's focus is on the design of distributed real-time systems at the architecture level. While a significant part of the established computer science literature abstracts from the progression of real time, real-time system designers cannot get away with such an abstraction. In this book, the progression of physical time is considered a first-order citizen that shapes many of the relevant concepts. The book explains the fundamental concepts related to the progression of time on a number of practical, insightful examples from industry. The conceptual model of a distributed real-time system has been extended, and precise definitions of important time-related concepts, such as *sparse time*, *state*, *temporal accuracy of real-time data*, and *determinism*, are given. This book emphasizes design principles that result in understandable systems, *irrespectively of a system's size*. Following these principles ensures that *simple models* faithfully represent an implemented system from different viewpoints and at different levels of abstraction.

Since the second edition of this book, about 11 years ago, cloud computing has revolutionized classical IT systems. Inspired by this development, strong market forces seek to replicate this success in the real-time embedded systems market by reusing native cloud technologies and principles. However, naively doing so will lead to disaster. Thus, this third edition includes a new chapter on *cloud and fog computing* that argues for the *fog architectural style* to decouple the *world of cloud* from the *world of real-time embedded systems*. The second major change is the rewritten chapter on *real-time networks* that now also covers IEEE 802.1 Time-Sensitive Networking (TSN), the incorporation of the time-triggered paradigm in the mainstream IT networking set of standards. All other chapters have been revised and updated. This third edition includes approximately 80 new references in total.

Since the publication of the first edition, a visible paradigm shift from the event-triggered to the time-triggered design methodology for dependable distributed real-time systems has taken place in a number of applications.

It is assumed that the reader of this book has a background in basic computer science or computer engineering or has some practical experience in the design or implementation of embedded systems.

The glossary, provided at the end, is an integral part of this book, providing definitions for many of the technical terms used throughout. If the reader is not sure about the meaning of a term, they are advised to refer to the glossary.

Acknowledgments

It is impossible to name all students, colleagues from industry, and fellow scientists who have contributed to this third edition of the book by asking intelligent questions or making constructive comments over the last decade—thanks to all of you.

Vienna, Austria   Hermann Kopetz
                                                                                          Wilfried Steiner

May 2022

Contents

Chapter 1
The Real-Time Environment

Overview

The purpose of this introductory chapter is to describe the environment of real-time
computer systems from a number of different perspectives. A solid understanding of
the technical and economic factors that characterize a real-time application helps to
interpret the demands that the system designer must cope with. The chapter starts
with the definition of a real-time system and with a discussion of its functional and
nonfunctional requirements. Particular emphasis is placed on the temporal requirements that are derived from the well-understood properties of control applications.
The objective of a control algorithm is to drive a process such that a performance
criterion is satisfied. Random disturbances occurring in the environment degrade
system performance and must be taken into account by the control algorithm. Any
additional uncertainty that is introduced into the control loop by the control system
itself, e.g., a non-predictable jitter of the control loop, results in a degradation of the
quality of control.

In Sects. 1.2, 1.3, 1.4 and 1.5, real-time applications are classified from a number
of viewpoints. Special emphasis is placed on the fundamental differences between
hard and *soft* real-time systems. Because soft real-time systems do not have severe
failure modes, a less rigorous approach to their design is often followed. Sometimes
resource-inadequate solutions that will not handle the rarely occurring peak-load
scenarios are accepted on economic arguments. In a hard real-time application, such
an approach is unacceptable because the safety of a design in all specified situations, even if they occur only very rarely, must be demonstrated vis-a-vis a certification agency. In Sect. 1.6, a brief analysis of the real-time system market is carried
out with emphasis on the field of embedded real-time systems. An embedded real-
time system is a part of a self-contained product, e.g., a television set or an automobile. Embedded real-time systems, also called *cyber-physical systems* (CPS), form
the most important market segment for real-time technology and the computer
industry in general.

© The Author(s), under exclusive license to Springer Nature Switzerland AG 2022   1
H. Kopetz, W. Steiner, *Real-Time Systems*,
https://doi.org/10.1007/978-3-031-11992-7_1

1.1   When Is a Computer System Real-Time?

A *real-time computer system* is a computer system where the correctness of the system behavior depends not only on the logical results of the computations but also on the physical time when these results are produced. By *system behavior* we mean the sequence of *outputs in time* of a system.

We model the flow of time by a directed timeline that extends from the past into the future. A cut of the timeline is called an *instant*. Any *ideal* occurrence that happens at an *instant* is called an *event*. Information that describes an event (see also Sect. 5.2.4 on event observation) is called *event information*. The present point in time, *now*, is a very special event that separates the *past* from the *future* (the presented model of time is based on Newtonian physics and disregards relativistic effects). An *interval* on the timeline, called a *duration*, is defined by two events, the *start event* and the *terminating event* of the interval. A digital clock partitions the timeline into a sequence of equally spaced durations, called the *granules* of the clock, which are delimited by special periodic events, the *ticks* of the clock.

A real-time computer system is always part of a larger system—this larger system is called a *real-time system* or a *cyber-physical system*. A real-time system changes as a function of physical time, e.g., a chemical reaction continues to change its state even after its controlling computer system has stopped. It is reasonable to decompose a real-time system into a set of self-contained subsystems called *clusters*. Examples of clusters are (Fig. 1.1) the physical plant or machine that is to be controlled (the *controlled cluster*), the real-time computer system (the *computational cluster*), and the *human operator* (the *operator cluster*). We refer to the *controlled cluster* and the *operator cluster* collectively as the *environment* of the *computational cluster* (the real-time computer system).

If the real-time computer system is *distributed* (and most of them are), it consists of a set of (computer) *nodes* interconnected by a real-time communication network.

The interface between the human operator and the real-time computer system is called the *man-machine interface*, and the interface between the controlled object and the real-time computer system is called the *instrumentation interface*. The man-machine interface consists of input devices (e.g., keyboard) and output devices (e.g., display) that interface to the human operator. The instrumentation interface consists

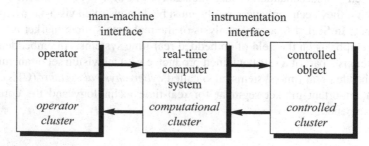

Fig. 1.1 Real-time system

of the sensors and actuators that transform the physical signals (e.g., voltages, currents) in the controlled cluster into a digital form and vice versa.

A real-time computer system must react to stimuli from its environment (the controlled cluster or the operator cluster) within time intervals dictated by its environment. The instant when a result must be produced is called a *deadline*. If a result has utility even after the deadline has passed, the deadline is classified as *soft*; otherwise it is *firm*. If *severe consequences* could result if a firm deadline is missed, the deadline is called *hard*.

> **Example**: Consider a traffic signal at a road before a railway crossing. If the traffic signal does not change to *red* before the train arrives, an accident could result.

A real-time computer system that must meet at least one hard deadline is called a *hard real-time computer system* or a *safety-critical real-time computer system*. If no hard deadline exists, then the system is called a *soft real-time computer system*.

The design of a hard real-time system is fundamentally different from the design of a soft real-time system. While a hard real-time computer system must sustain a guaranteed temporal behavior under all specified load and fault conditions, it is permissible for a soft real-time computer system to miss a deadline occasionally. The differences between soft and hard real-time systems will be discussed in detail in the following sections. The focus of this book is on the design of hard real-time systems.

1.2 Functional Requirements

The functional requirements of real-time systems are concerned with the functions that a real-time computer system must perform. They are grouped into data collection requirements, direct digital control requirements, and man-machine interaction requirements.

1.2.1 Data Collection

A controlled object, e.g., a car or an industrial plant, changes its state as a function of time (whenever we use the word *time* without a qualifier, we mean *physical time* as described in Sect. 3.1). If we freeze the time, we can describe the current state of the controlled object by recording the values of its state variables at that moment. Possible state variables of a controlled object *car* are the position of the car, the speed of the car, the position of switches on the dashboard, and the position of a piston in a cylinder. We are normally not interested in *all* state variables, but only in the *subset* of state variables that is *significant* for our purpose. A significant state variable is called a *real-time (RT) entity*.

Every RT entity is in the *sphere of control (SOC)* of a subsystem, i.e., it belongs to a subsystem that has the authority to change the value of this RT entity (see also Sect. 5.1.1). Outside its sphere of control, the value of an RT entity can be observed, but its *semantic content* (see Sect. 2.2.4) cannot be modified. For example, the current position of a piston in a cylinder of the engine is in the sphere of control of the engine. Outside the car engine, the current position of the piston can only be observed, but we are not allowed to modify the *semantic content* of this observation (the representation of the *semantic content* can be changed!).

The first functional requirement of a real-time computer system is the observation of the RT entities in a controlled cluster and the collection of these observations. An observation of an RT entity is represented by a *real-time (RT) image* in the computer system. Since the state of a *controlled object* in the *controlled cluster* is a function of real time, a given RT image is only *temporally accurate* for a limited time interval. The length of this time interval depends on the dynamics of the controlled object. If the state of the controlled object changes very quickly, the corresponding RT image has a very short *accuracy interval*.

Example: Consider the example of Fig. 1.2, where a car enters an intersection controlled by a traffic light. How long is the observation *the traffic light is green* temporally accurate? If the information *the traffic light is green* is used outside its accuracy interval, e.g., a car enters the intersection after the traffic light has switched to *red*, an accident may occur. In this example, an upper bound for the accuracy interval is given by the duration of the yellow phase of the traffic light. The car may only enter the intersection safely if the observation remains temporally accurate while the car passes through it.

The set of all temporally accurate real-time images of the controlled cluster is called the *real-time database*. The real-time database must be updated whenever an RT entity changes its value. These updates can be performed periodically, triggered by the progression of the real-time clock by a fixed period (*time-triggered (TT) observation*), or immediately after a change of state, which constitutes an event, occurs in the RT entity (*event-triggered (ET) observation*). A more detailed analysis of time-triggered and event-triggered observations will be presented in Chaps. 4 and 5.

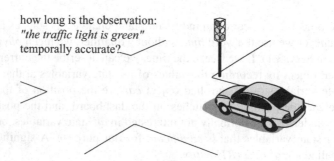

how long is the observation:
"the traffic light is green"
temporally accurate?

Fig. 1.2 Temporal accuracy of the traffic light information

Signal Conditioning A physical sensor, e.g., a thermocouple, produces a *raw data* element (e.g., a voltage). Often, a sequence of raw data elements is collected and an averaging algorithm is applied to reduce the measurement error. In the next step, the raw data must be calibrated and transformed to standard measurement units. The term *signal conditioning* is used to refer to all the processing steps that are necessary to obtain meaningful *measured data* of an RT entity from the raw sensor data. After signal conditioning, the measured data must be checked for plausibility and related to other measured data to detect a possible fault of the sensor. A data element that is judged to be a correct RT image of the corresponding RT entity is called an *agreed data element*.

Alarm Monitoring An important function of a real-time computer system is the continuous monitoring of the RT entities to detect abnormal process behaviors.

> **Example**: The rupture of a pipe, a *primary event*, in a chemical plant will cause many RT entities (diverse pressures, temperatures, liquid levels) to deviate from their normal operating ranges and to cross some preset alarm limits, thereby generating a set of correlated alarms, which is called an *alarm shower*.

The real-time computer system must detect and display these alarms and must assist the operator in identifying a *primary event* that was the initial cause of these alarms. For this purpose, alarms that are observed must be logged in a special alarm log with the exact instant when the alarm occurred. The exact temporal order of the alarms is helpful in identifying the *secondary alarms*, i.e., all alarms that can be a causal consequence of the primary event. In complex industrial plants, sophisticated knowledge-based systems are used to assist the operator in the alarm analysis.

> **Example**: In the final report on the August 14, 2003 power blackout in the USA and Canada, we find on [Tas03, p. 162] the following statement: *A valuable lesson from the August 14 blackout is the importance of having time-synchronized system data recorders. The Task Force's investigators labored over thousands of data items to determine the sequence of events much like putting together small pieces of a very large puzzle. That process would have been significantly faster and easier if there had been wider use of synchronized data recording devices.*

A situation that occurs infrequently but is of utmost concern when it does occur is called a *rare-event* situation. The validation of the performance of a real-time computer system in a rare-event situation is a challenging task and requires models of the physical environment (see Sect. 12.2.2).

> **Example**: The sole purpose of a nuclear power plant monitoring and shutdown system is reliable performance in a peak-load alarm situation (a *rare event*). Hopefully, this rare event will never occur during the operational life of the plant.

1.2.2   Direct Digital Control

Many real-time computer systems must calculate the *actuating variables* for the
actuators in order to control the controlled object directly (*direct digital control—
DDC*), i.e., without any underlying conventional control system.

Control applications are highly regular, consisting of an (infinite) sequence of
control cycles, each one starting with sampling (observing) of the RT entities, followed by the execution of the control algorithm to calculate a new actuating variable
and subsequently by the output of the actuating variable to the actuator. The design
of a proper control algorithm that achieves the desired control objective, and compensates for the random disturbances that perturb the controlled object, is the topic
of the field of control engineering. In the next section on temporal requirements,
some basic notions of control engineering will be introduced.

1.2.3   Man-Machine Interaction

A real-time computer system must inform the operator of the current state of the
controlled object and must assist the operator in controlling the machine or plant
object. This is accomplished via the man-machine interface, a critical subsystem of
major importance. Many severe computer-related accidents in safety-critical real-
time systems have been traced to mistakes made at the man-machine interface
[Lev95].

> **Example**: *Mode confusion* at the man-machine interface of an aircraft has been identified
> to be the cause of major aircraft accidents [Deg95].

Most process-control applications contain, as part of the man-machine interface,
an extensive data logging and data reporting subsystem that is designed according
to the demands of the particular industry.

> **Example**: In some countries, the pharmaceutical industry is required by law to record and
> store all relevant process parameters of every production batch in an archival storage in
> order that the process conditions prevailing at the time of a production run can be reexam-
> ined in case a defective product is identified on the market at a later time.

Man-machine interfacing has become such an important issue in the design of
computer-based systems that a number of courses dealing with this topic have been
developed. In the context of this book, we will introduce an abstract man-machine
interface in Sect. 4.5.2, but we will not cover its design in detail. The interested
reader is referred to standard textbooks on user interface design.

1.3 Temporal Requirements

1.3.1 Where Do Temporal Requirements Come From?

The most stringent temporal demands for real-time systems have their origin in the requirements of control loops, e.g., in the control of a fast process such as an automotive engine. The temporal requirements at the man-machine interface are, in comparison, less stringent because the human perception delay, in the range of *50–100* ms, is orders of magnitude larger than the latency requirements of fast control loops.

A Simple Control Loop Consider the simple control loop depicted in Fig. 1.3 consisting of a vessel with a liquid, a heat exchanger connected to a steam pipe, and a controlling computer system. The objective of the computer system is to control the valve (*control variable*) determining the flow of steam through the heat exchanger such that the temperature of the liquid in the vessel remains within a small range around the *set point* selected by the operator.

The focus of the following discussion is on the temporal properties of this simple control loop consisting of a *controlled object* and a *controlling computer system*.

The Controlled Object Assume that the system of Fig. 1.4 is in equilibrium. Whenever the steam flow is increased by a step function, the temperature of the liquid in the vessel will change according to Fig. 1.4 until a new equilibrium is reached. This *response function* of the temperature in the vessel depends on the environmental conditions, e.g., the amount of liquid in the vessel, and the flow of steam through the heat exchanger, i.e., on the dynamics of the controlled object. (In the following section, we will use *d* to denote a *duration* and *t* to denote an *instant*, i.e., a point in time.)

There are two important temporal parameters characterizing this elementary step-response function, the *object delay* d^{object} (sometimes called the *lag time* or *lag*) after which the *measured variable* temperature begins to rise (caused by the initial inertia of the process and the instrumentation, called the *process lag*) and the *rise*

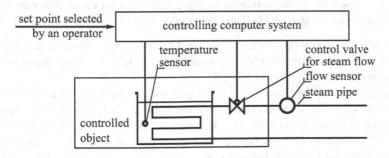

Fig. 1.3 A simple control loop

Fig. 1.4 Delay and rise time of the step response

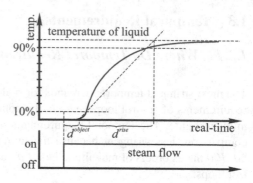

time d^{rise} of the temperature until the new equilibrium state has been reached. To determine the *object delay* d^{object} and the *rise time* d^{rise} from a given experimentally recorded shape of the step-response function, one finds the two points in time where the response function has reached *10%* and *90%* of the difference between the two stationary equilibrium values. These two points are connected by a straight line (Fig. 1.4). The significant points in time that characterize the *object delay* d^{object} and the *rise time* d^{rise} of the step-response function are constructed by finding the intersection of this straight line with the two horizontal lines that denote the two liquid temperatures that correspond to the stable equilibrium states before and after the application of the step function.

Controlling Computer System The controlling computer system must sample the temperature of the vessel periodically to detect any deviation between the intended value and the actual value of the controlled variable *temperature*. The constant duration between two sampling points is called the sampling period d^{sample}, and the reciprocal $1/d^{sample}$ is the sampling frequency, f^{sample}. A rule of thumb says that, in a digital system which is expected to behave like a quasi-continuous system, the sampling period should be less than one-tenth of the rise time d^{rise} of the step-response function of the controlled object, i.e., $d^{sample} < (d^{rise}/10)$. The computer compares the *measured temperature* to the *temperature set point* selected by the operator and calculates the *error term*. This error term forms the basis for the calculation of a new value of the control variable by a *control algorithm*. Given a time interval after each sampling point, called the *computer delay* $d^{computer}$, the controlling computer will output this new value of the actuating variable to the control valve, thus closing the control loop. The delay $d^{computer}$ should be smaller than the sampling period d^{sample}.

The difference between the maximum and the minimum values of the delay of the computer is called the *jitter* of the computer delay, $\Delta d^{computer}$. This jitter is a sensitive parameter for the quality of control.

The *dead time* of the control loop is the time interval between the observation of the RT entity and the start of a reaction of the controlled object due to a computer action based on this observation. The dead time is the sum of the controlled object delay d^{object}, which is in the sphere of control of the controlled object and is thus determined by the controlled object's dynamics, and the computer delay $d^{computer}$,

Fig. 1.5 Delay and delay jitter

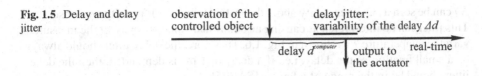

observation of the controlled object

delay jitter: variability of the delay Δd

delay $d^{computer}$

output to the acutator

real-time

Table 1.1 Parameters of an elementary control loop

Symbol	Parameter	Sphere of control	Relationships
d^{object}	Controlled object delay	Controlled object	Physical process
d^{rise}	Rise time of step response	Controlled object	Physical process
d^{sample}	Sampling period	Computer	$d^{sample} \ll d^{rise}$
$d^{computer}$	Computer delay	Computer	$d^{computer} < d^{sample}$
$\Delta d^{computer}$	Jitter of the computer delay	Computer	$\Delta d^{computer} \ll d^{computer}$
$d^{deadtime}$	Dead time	Computer and controlled object	$d^{computer} + d^{object}$

which is determined by the computer implementation. To reduce the dead time in a control loop and to improve the stability of the control loop, these delays should be as small as possible. The computer delay $d^{computer}$ is defined by the time interval between the *sampling points*, i.e., the observation of the controlled object, and the *use* of this information (see Fig. 1.5), i.e., the output of the corresponding actuator signal, the *actuating variable*, to the controlled object. Apart from the necessary time for performing the calculations, the computer delay is determined by the time required for communication and the reaction time of the actuator.

Parameters of a Control Loop Table 1.1 summarizes the temporal parameters that characterize the elementary control loop depicted in Fig. 1.3. In the first two columns, we denote the symbol and the name of the parameter. The third column denotes the sphere of control in which the parameter is located, i.e., what subsystem determines the value of the parameter. Finally, the fourth column indicates the relationships between these temporal parameters.

1.3.2 Minimal Latency Jitter

The data items in control applications are state-based, i.e., they contain images of the RT entities. The computational actions in control applications are mostly time-triggered, e.g., the control signal for obtaining a sample is derived from the progression of time within the computer system. This control signal is thus in the sphere of control of the computer system. It is known in advance when the next control action must take place. Many control algorithms are based on the assumption that the delay jitter $\Delta d^{computer}$ is very small compared to the *delay* $d^{computer}$, i.e., the delay is close to constant. This assumption is made because control algorithms can be designed to compensate a *known* constant delay. Delay jitter brings an additional uncertainty into the control loop that has an adverse effect on the quality of control. The jitter

Δd can be seen as an uncertainty about the instant when the RT entity was observed. This jitter can be interpreted as causing an additional value error ΔT of the measured variable temperature T as shown in Fig. 1.6. Therefore, the delay jitter should always be a small fraction of the delay, i.e., if a delay of 1 ms is demanded, then the delay jitter should be in the range of a few µs [SAE95].

1.3.3   Minimal Error-Detection Latency

Fig. 1.6 The effect of jitter on the measured variable T

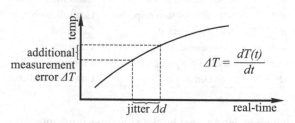

$$\Delta T = \frac{dT(t)}{dt}$$

Hard real-time applications are, by definition, safety-critical. It is therefore important that any error within the control system, e.g., the loss or corruption of a message or the failure of a node, is detected within a short time with a very high probability. The required *error-detection latency* must be in the same order of magnitude as the sampling period of the fastest critical control loop. It is then possible to perform some corrective action, or to bring the system into a safe state, before the consequences of an error can cause any severe system failure. Almost-no-jitter systems will have shorter guaranteed error-detection latencies than systems that allow for jitter.

1.4   Dependability Requirements

The notion of dependability covers the nonfunctional attributes of a computer system that relate to the quality of service a system delivers to its users during an extended interval of time. (A user could be a human or another technical system.) The following measures of dependability attributes are of importance [Avi04].

1.4.1   Reliability

The *reliability* $R(t)$ of a system is the probability that a system will provide the specified service until time t, given that the system was operational at the beginning, i.e., $t = t_0$. The probability that a system will fail in a given interval of time is expressed by the *failure rate*, measured *in FITs* (failure in time). A failure rate of *1 FIT* means that the mean time to a failure *(MTTF)* of a device is 10^9 hours, i.e., one failure occurs in about *115,000 years*. If a system has a constant *failure rate* of λ *failures/hour*, then the reliability at time t is given by

$$R(t) = \exp(-\lambda(t-t_0)),$$

where $t-t_0$ is given in hours. The inverse of the failure rate $1/\lambda = MTTF$ is called the *mean time to failure (MTTF)* (in hours). If the failure rate of a system is required to be in the order of 10^{-9} failures/h or lower, then we speak of a system with an *ultrahigh reliability* requirement.

1.4.2   Safety

Safety is reliability regarding *critical failure modes*. A critical failure mode is said to be *malign*, in contrast with a noncritical failure, which is *benign*. In a malign failure mode, the cost of a failure can be orders of magnitude higher than the utility of the system during normal operation. Examples of malign failures are an airplane crash due to a failure in the flight-control system and an automobile accident due to a failure of a computer-controlled intelligent brake in the automobile. Safety-critical (hard) real-time systems must have a failure rate with regard to critical failure modes that conforms to the *ultrahigh reliability* requirement.

> **Example**: Consider the example of a computer-controlled brake in an automobile. The failure rate of a computer-caused critical brake failure must be lower than the failure rate of a conventional braking system. Under the assumption that a car is operated about 1 hour per day on the average, one safety-critical failure per million cars per year translates into a failure rate in the order of 10^{-9} failures/h.

Similarly low failure rates are required in flight-control systems, train-signaling systems, and nuclear power plant monitoring systems.

Certification   In many cases the design of a safety-critical real-time system must be approved by an independent certification agency. The certification process can be simplified if the certification agency can be convinced that:

(i)   The subsystems that are critical for the safe operation of the system are protected by fault-containment mechanisms that eliminate the possibility of error propagation from the rest of the system into these safety-critical subsystems.

(ii) From the point of view of design, all scenarios that are covered by the given load and fault hypothesis can be handled according to the specification without reference to probabilistic arguments. This makes a resource-adequate design necessary.

(iii) The architecture supports a constructive modular certification process where the certification of subsystems can be done independently of each other. At the system level, only the *emergent properties* must be validated.

[Joh92] specifies the required properties for a system that is *designed for validation*:

(i) A complete and accurate reliability model can be constructed. All parameters of the model that cannot be deduced analytically must be measurable in feasible time under test.

(ii) The reliability model does not include state transitions representing design faults; analytical arguments must be presented to show that design faults will not cause system failure.

(iii) Design tradeoffs are made in favor of designs that minimize the number of parameters that must be measured.

1.4.3   Maintainability

Maintainability is a measure of the time interval required to repair a system after the occurrence of a benign failure. Maintainability is measured by the probability $M(d)$ that the system is restored within a time interval d after the failure. In keeping with the reliability formalism, a constant repair rate μ (repairs per hour) and a *mean time to repair (MTTR)* are introduced to define a quantitative maintainability measure.

There is a fundamental conflict between reliability and maintainability. A maintainable design requires the partitioning of a system into a set of *field-replaceable units* (*FRUs*) connected by serviceable interfaces that can be easily disconnected and reconnected to replace a faulty *FRU* in case of a failure. A serviceable interface, e.g., a plug connection, has a significantly higher physical failure rate than a non-serviceable interface. Furthermore, a serviceable interface is more expensive to produce.

In the field of *ambient intelligence*, automatic diagnosis and *maintainability by an untrained end user* are important system properties that are critical for the market success of a product.

1.4.4   Availability

Availability is a measure of the delivery of correct service with respect to the alternation of correct and incorrect service. It is measured by the fraction of time that the system is ready to provide the service.

Example: Whenever a user picks up the phone, the telephone switching system should be ready to provide the telephone service with a very high probability. A telephone exchange is allowed to be out of service for only a few minutes per year.

In systems with constant failure and repair rates, the reliability (*MTTF*), maintainability (*MTTR*), and availability (*A*) measures are related by

$$A = MTTF / (MTTF + MTTR).$$

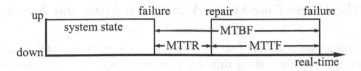

Fig. 1.7 Relationship between *MTTF*, *MTBF*, and *MTTR*

The sum *MTTF* + *MTTR* is sometimes called the *mean time between failures* (*MTBF*). Figure 1.7 shows the relationship between *MTTF*, *MTTR*, and *MTBF*.

A high availability can be achieved either by a long *MTTF* or by a short *MTTR*. The designer has thus some freedom in the selection of her/his approach to the construction of a high-availability system.

1.4.5 Security

A dependable system must also be secure. *Security* is concerned with the authenticity and integrity of information and the ability of a system to prevent unauthorized access to information or services (see also Sect. 6.2). There are difficulties in defining a quantitative security measure, e.g., the specification of a *standard burglar* that takes a certain time to intrude a system. Traditionally, security issues have been associated with large databases, where the concerns are confidentiality, privacy, and authenticity of information. During the last few years, security issues have also become important in real-time systems, e.g., a cryptographic theft-avoidance system that locks the ignition of a car if the user cannot present the specified access code. In the *Internet of Things (IoT)*, where the endpoints of the Internet are embedded systems that bridge the gap between the cyber world and physical world, security concerns are of crucial importance, since an intruder cannot only corrupt a data structure in a computer, but can cause harm in the physical environment.

1.5 Classification of Real-Time Systems

In this section we classify real-time systems from different perspectives. The first two classifications, *hard real-time* versus *soft real-time* (online) and *fail-safe* versus *fail-operational*, depend on the characteristics of the application, i.e., on factors *outside* the computer system. The second three classifications, *guaranteed timeliness* versus *best effort*, *resource-adequate* versus *resource-inadequate*, and *event-triggered* versus *time-triggered*, depend on the design and implementation of the computer application, i.e., on factors *inside* the computer system.

1.5.1 Hard Real-Time System Versus Soft Real-Time System

The design of a *hard real-time system*, which always must produce the results at the correct instant, is fundamentally different from the design of a *soft real-time* or an *online system*, such as a transaction processing system. In this section we will elaborate on these differences. Table 1.2 compares the characteristics of hard real-time systems versus soft real-time systems.

Response Time The demanding response time requirements of hard real-time applications, often in the order of milliseconds or less, preclude direct human intervention during normal operation or in critical situations. A hard real-time system must be highly autonomous to maintain safe operation of the process. In contrast, the response time requirements of soft real-time and online systems are often in the order of seconds. Furthermore, if a deadline is missed in a soft real-time system, no catastrophe can result.

Peak-Load Performance In a hard real-time system, the peak-load scenario must be well defined. It must be guaranteed by design that the computer system meets the specified deadlines in all situations, since the utility of many hard real-time applications depends on their predictable performance during *rare-event scenarios* leading to a peak load. This is in contrast to the situation in a soft real-time system, where the *average* performance is important, and a degraded operation in a rarely occurring peak-load case is tolerated for economic reasons.

Control of Pace A hard real-time computer system is often paced by the state changes occurring in the environment. It must keep up with the state of the environment (the controlled object and the human operator) under all circumstances. This is in contrast to an online system, which can exercise some control over the environment in case it cannot process the offered load.

> **Example:** Consider the case of a transaction processing system, such as an airline reservation system. If the computer cannot keep up with the demands of the users, it just extends the response time and forces the users to slow down.

Table 1.2 Hard real-time versus soft real-time systems

Characteristic	Hard real-time	Soft real-time (online)
Response time	Hard-required	Soft-desired
Peak-load performance	Predictable	Degraded
Control of pace	Environment	Computer
Safety	Often critical	Noncritical
Size of data files	Small/medium	Large
Redundancy type	Active	Checkpoint-recovery
Data integrity	Short-term	Long-term
Error detection	Autonomous	User assisted

Safety The safety criticality of many real-time applications has a number of consequences for the system designer. In particular, error detection and recovery must be autonomous such that the system can initiate appropriate recovery actions and arrive at a safe state within the time intervals dictated by the application without human intervention.

Size of Data Files The *real-time database* that is composed of the temporally accurate images of the RT entities is normally of small size. The key concern in hard real-time systems is on the *short-term* temporal accuracy of the real-time database that is invalidated by the flow of real time. In contrast, in online transaction processing systems, the maintenance of the *long-term* integrity and availability of large data files is the key issue. Large data files may also be present in hard real-time applications, like real-time camera data for self-driving cars. However, these large files are typically only part of the state during a computation that contracts them into small-sized RT images of the RT entities (see Sect. 5.3 for RT images and Sect. 4.2.2 for state expansion and contraction). For example, although the cameras of self-driving cars produce large data files, the resulting information of the position, speed, and direction of surrounding objects is only of small size.

Redundancy Type After an error has been detected in an online system, the computation is rolled back to a previously established checkpoint to initiate a recovery action. In hard real-time systems, rollback/recovery is of limited utility for the following reasons:

(i) It is difficult to guarantee the deadline after the occurrence of an error, since the rollback/recovery action can take an unpredictable amount of time.

(ii) An action can be irrevocable because its effect on the environment cannot be undone.

(iii) The temporal accuracy of the checkpoint data may be invalidated by the time difference between the checkpoint time and the instant *now*.

The topic of temporal accuracy of real-time data is discussed at length in Sect. 5.4, while the issues of error detection and types of redundancy are dealt with in Chap. 6.

1.5.2 Fail-Safe Versus Fail-Operational

In many hard real-time systems, one or more safe states, which can be reached in case of a system failure, can be identified. If such a safe state can be identified and quickly reached upon the occurrence of a failure, then we call the system *fail-safe*. Fail safeness is a characteristic of the controlled object, not the computer system. In fail-safe applications, the computer system must have a *high error-detection coverage*, i.e., the probability that an error is detected, provided it has occurred, must be close to one.

Example: In case a failure is detected in a railway signaling system, it is possible to set all
signals to red and thus stop all the trains in order to bring the system to a safe state.

In many real-time computer systems, a special external device, a *watchdog*, is
provided to monitor the operation of the computer system. The computer system
must send a periodic life sign (e.g., a digital output of predefined form) to the watchdog. If this life sign fails to arrive at the watchdog within the specified time interval,
the watchdog assumes that the computer system has failed and forces the controlled
object into a safe state. In such a system, timeliness is needed only to achieve high
availability, but is not needed to maintain safety since the watchdog forces the controlled object into a safe state in case of a timing violation.

There are, however, applications where a safe state that can be reached quickly
cannot be identified, e.g., a flight-control system aboard an airplane. In such an
application, the computer system must remain operational and provide a minimal
level of service even in the case of a failure to avoid a catastrophe. This is why these
applications are called *fail-operational*.

1.5.3   Guaranteed Response Versus Best Effort

If we start out with a specified *fault* and *load hypothesis* and deliver a design that
makes it possible to reason about the adequacy of the design without reference to
probabilistic arguments (even in the case of a peak-load and fault scenario), then we
can speak of a system with a *guaranteed response*. The probability of failure of a
perfect system with guaranteed response is reduced to the probability that the
assumptions about the peak load and the number and types of faults do not hold in
reality. This probability is called *assumption coverage* [Pow95]. Guaranteed
response systems require careful planning and extensive analysis during the
design phase.

If such an analytic response guarantee cannot be given, we speak of a *best-effort*
design. Best-effort systems do not require a rigorous specification of the load and
fault hypothesis. The design proceeds according to the principle *best possible effort
taken*, and the sufficiency of the design is established during the test and integration
phases. It is difficult to establish that a best-effort design operates correctly in a rare-
event scenario. At present, many non-safety-critical real-time systems are designed
according to the best-effort paradigm.

1.5.4   Resource-Adequate Versus Resource-Inadequate

Guaranteed response systems are based on the principle of resource adequacy, i.e.,
there are enough computing resources available to handle the specified peak-load
and the fault scenario. Many non-safety-critical real-time system designs are based

on the principle of resource inadequacy. It is assumed that the provision of sufficient resources to handle every possible situation is not economically viable and that a dynamic resource allocation strategy based on resource sharing and probabilistic arguments about the expected load and fault scenarios is acceptable.

It is expected that, in the future, there will be a paradigm shift to resource-adequate designs in many applications. The use of computers in important volume-based applications, e.g., in cars, raises both the public awareness and concerns about computer-related incidents and forces the designer to provide convincing arguments that the design functions properly under *all* stated conditions. Hard real-time systems must be designed according to the guaranteed response paradigm that requires the availability of adequate resources.

1.5.5 Event-Triggered Versus Time-Triggered

The distinction between *event-triggered* and *time-triggered* depends on the type of internal *triggers* and not the external behavior of a real-time system. A *trigger* is an event that causes the start of some action in the computer, e.g., the execution of a task or the transmission of a message. Depending on the triggering mechanisms for the start of communication and processing actions in each node of a computer system, two distinctly different approaches to the design of the control mechanisms of real-time computer applications can be identified, *event-triggered control* and *time-triggered control*.

In *event-triggered (ET)* control, all communication and processing activities are initiated whenever a significant event other than the regular event of a clock tick occurs. In an ET system, the signaling of significant events to the central processing unit (CPU) of a computer is realized by the well-known interrupt mechanism. ET systems require a dynamic scheduling strategy to activate the appropriate software task that services the event.

In a *time-triggered (TT) system*, all activities are initiated by the progression of real time. There is only one interrupt in each node of a distributed TT system, the periodic real-time clock interrupt. Every communication or processing activity is initiated at a periodically occurring predetermined tick of a clock. In a distributed TT real-time system, it is assumed that the clocks of all nodes are synchronized to form a *global time* that is available at every node. Every observation of the controlled object is time-stamped with this global time. The granularity of the global time must be chosen such that the time order of any two observations made anywhere in a distributed TT system can be established from their time-stamps with adequate faithfulness [Kop09]. The topics of global time and clock synchronization will be discussed at length in Chap. 3.

> **Example:** The distinction between *event-triggered* and *time-triggered* can be explained by an example of an elevator control system. When you push a *call button* in the event-triggered implementation, the event is immediately relayed to the interrupt system of the computer in order to start the action of calling the elevator. In a time-triggered system, the

button push is stored locally, and periodically, e.g., every second, the computer asks to get the state of all push buttons. The flow of control in a time-triggered system is managed by the progression of time, while in an event-triggered system, the flow of control is determined by the events that happen in the environment or the computer system.

1.6   The Real-Time System Market

In a market economy, the cost/performance relation is a decisive parameter for the market success of any product. There are only a few scenarios where cost arguments are not the major concern. The total life-cycle cost of a product can be broken down into three rough categories: non-recurring development cost, production cost, and operation and maintenance cost. Depending on the product type, the distribution of the total life-cycle cost over these three cost categories can vary significantly. We will examine this life-cycle cost distribution by looking at two important examples of real-time systems, embedded real-time systems and plant automation systems.

1.6.1   Embedded Real-Time Systems

The ever-decreasing price/performance ratio of microcontrollers makes it economically attractive to replace conventional mechanical or electronic control system within many products by an embedded real-time computer system. There are numerous examples of products with embedded computer systems: cellular phones, engine controllers in cars, heart pacemakers, computer printers, television sets, washing machines, and even some electric razors contain a microcontroller with some thousand instructions of software code. Because the external interfaces (particularly the man-machine interface) of the product often remain unchanged relative to the previous product generation, it is often not visible from the outside that a real-time computer system is controlling the product behavior.

Characteristics An embedded real-time computer system is always part of a well-specified larger system, which we call an *intelligent product*. An intelligent product consists of a physical (mechanical) subsystem: the controlling embedded computer and, most often, a man-machine interface. The ultimate success of any intelligent product depends on the relevance and quality of service it can provide to its users. A focus on the genuine user needs is thus of utmost importance.

Embedded systems have a number of distinctive characteristics that influence the system development process:

(i) Mass Production: many embedded systems are designed for a mass market and consequently for mass production in highly automated assembly plants. This implies that the production cost of a single unit must be as low as possible, i.e., efficient memory and processor utilization are of concern.

(ii) Static Structure: the computer system is embedded in an intelligent product of given functionality and rigid structure. The known a priori static environment can be analyzed at design time to simplify the software, to increase the robustness, and to improve the efficiency of the embedded computer system. In many embedded systems, there is no need for flexible dynamic software mechanisms. These mechanisms increase the resource requirements and lead to an unnecessary complexity of the implementation.

(iii) Man-Machine Interface: if an embedded system has a man-machine interface, it must be specifically designed for the stated purpose and must be easy to operate. Ideally, the use of the intelligent product should be self-explanatory and not require any training or reference to an operating manual.

(iv) Minimization of the Mechanical Subsystem: to reduce the manufacturing cost and to increase the reliability of the intelligent product, the complexity of the mechanical subsystem is minimized.

(v) Functionality Determined by Software in Read-Only Memory (ROM): the integrated software that often resides in ROM determines the functionality of many intelligent products. Since it is not possible to modify the software in a ROM after its release, the quality standards for this software are high.

(vi) Maintenance Strategy: many intelligent products are designed to be non-maintainable, because the partitioning of the product into replaceable units is too expensive. If, however, a product is designed to be maintained in the field, the provision of an excellent diagnostic interface and a self-evident maintenance strategy is of importance.

(vii) Ability to Communicate: many intelligent products are required to interconnect with some larger system or the Internet. Whenever a connection to the Internet is supported, the topic of security is of utmost concern.

(viii) Limited Amount of Energy: many mobile embedded devices are powered by a battery. The lifetime of a battery load is a critical parameter for the utility of a system.

A large fraction of the life-cycle cost of many intelligent products is in the production, i.e., in the hardware. The known a priori static configuration of the intelligent product can be used to reduce the resource requirements, and thus the production cost, and also to increase the robustness of the embedded computer system. Maintenance cost can become significant, particularly if an undetected design fault (software fault) requires a recall of the product and the replacement of a complete production series.

Example: In [Neu96] we find the following laconic one-liner: *General Motors recalls almost 300 K cars for engine software flaw.*

Future Trends   During the last few years, the variety and number of embedded computer applications have grown to the point that, by now, this segment is by far the most important one in the computer market. The embedded system market is driven by the continuing improvements in the cost/performance ratio of the semiconductor industry that makes computer-based control systems cost-competitive relative to their mechanical, hydraulic, and electronic counterparts. Among the key mass markets are the domains of consumer electronics and automotive electronics. The automotive electronics market is of particular interest, because of stringent timing, dependability, and cost requirements that act as *technology catalysts*.

Automotive manufacturers view the proper exploitation of computer technology as a key competitive element in the never-ending quest for increased vehicle performance and reduced manufacturing cost. While some years ago, the computer applications on board a car focused on noncritical body electronics or comfort functions, there is now a substantial growth in the computer control of core vehicle functions, e.g., engine control, brake control, transmission control, and suspension control. We observe an integration of many of these functions with the goal of increasing the vehicle stability in critical driving maneuvers. Obviously, an error in any of these core vehicle functions has severe safety implications.

At present the topic of computer safety in cars is approached at two levels. At the basic level, a mechanical system provides the proven safety level that is considered sufficient to operate the car. The computer system provides optimized performance on top of the basic mechanical system. In case the computer system fails cleanly, the mechanical system takes over. Consider, for example, an *Electronic Stability Program (ESP)*. If the computer fails, the conventional mechanical brake system is still operational. Soon, this approach to safety may reach its limits for three reasons:

 (i) If the computer-controlled system is further improved, the magnitude of the difference between the performance of the computer-controlled system and the performance of the basic mechanical system is further increased. A driver who is used to the high performance of the computer-controlled system might consider the fallback to the inferior performance of the mechanical system a safety risk.
 (ii) The improved price/performance of the microelectronic devices will make the implementation of fault-tolerant computer systems cheaper than the implementation of mixed computer/mechanical systems. Thus, there will be an economic pressure to eliminate the redundant mechanical system and to replace it with a computer system using active redundancy.
(iii) Self-driving cars cannot rely on a human to exercise the mechanical backup. Therefore, the real-time computer system that controls the self-driving car must be designed according to the ultrahigh reliability requirement.

The embedded system market is expected to continue steep growth during the next 10 years. Today many embedded systems already connect to the Internet, forming the Internet of Things (see Chap. 13). We expect this trend to even intensify by fog and cloud computing (see Chap. 14).

1.6.2   Plant Automation Systems

Characteristics Historically, industrial plant automation was the first field for the application of real-time digital computer control. This is understandable since the benefits that can be gained by the computerization of a sizable plant are much larger than the cost of even an expensive process-control computer of the late 1960s. In the early days, human operators controlled the industrial plants locally. With the refinement of industrial plant instrumentation and the availability of remote automatic controllers, plant monitoring and command facilities were concentrated into a central control room, thus reducing the number of operators required to run the plant. In the 1970s, the next logical step was the introduction of central process-control computers to monitor the plant and assist the operator in her/his routine functions, e.g., data logging and operator guidance. In the beginning, the computer was considered an *add-on* facility that was not fully trusted. It was the duty of the operator to judge whether a set point calculated by a computer made sense and could be applied to the process (*open-loop control*). In the next phase, *Supervisory Control and Data Acquisition* (SCADA) systems calculated the set points for the *programmable logic controllers* (PLCs) in the plant. With the improvement of the process models and the growth of the reliability of the computer, control functions have been increasingly allocated to the computer, and gradually the operator has been taken out of the control loop (*closed-loop control*). Sophisticated nonlinear control techniques, which have response time requirements beyond human capabilities, have been implemented.

Usually, every plant automation system is unique. There is an extensive amount of engineering and software effort required to adapt the computer system to the physical layout, the operating strategy, the rules and regulations, and the reporting system of a particular plant. To reduce these engineering and software efforts, many process-control companies have developed a set of modular building blocks, which can be configured individually to meet the requirements of a customer. Compared to the development cost, the production cost (hardware cost) is of minor importance. Maintenance cost can be an issue if a maintenance technician must be on-site for 24 h in order to minimize the downtime of a plant.

Future Trends The market of industrial plant automation systems is limited by the number of plants that are newly constructed or are refurbished to install a computer-control system. During the last 30 years, many plants have already been automated. This investment must pay off before a new generation of computers and control equipment is installed.

Furthermore, the installation of a new generation of control equipment in a production plant causes disruption in the operation of the plant with a costly loss of production that must be justified economically. This is difficult if the plant's efficiency is already high, and the margin for further improvement by refined computer control is limited.

The size of the plant automation market is too small to support the mass production of special application-specific components. This is the reason why many VLSI components that are developed for other application domains, such as automotive electronics, are taken up by this market to reduce the system cost. Examples of such components are sensors, actuators, real-time local area networks, and processing nodes. Already several process-control companies have announced a new generation of process-control equipment that takes advantage of the low-priced mass-produced components that have been developed for the automotive market, such as the chips developed for the Controller Area Network (CAN—see Sect. 7.3.2). On the other hand, the plant automation market can use cross-industry solutions, such as IEEE 802.1 Time-Sensitive Networking (TSN—see Sect. 7.5.3).

We also expect a steep increase in the connectivity of industrial automation systems to fog and cloud computing systems (for use cases, see Sect. 14.5) to realize Industry 4.0 and the Industrial Internet of Things (Industrial IoT).

1.6.3 Multimedia Systems

Characteristics The multimedia market is a mass market for specially designed soft and firm real-time systems. Although the deadlines for many multimedia tasks, such as the synchronization of audio and video streams, are firm, they are not hard deadlines. An occasional failure to meet a deadline results in a degradation of the *quality of the user experience*, but will not cause a catastrophe. The processing power required to transport and render a continuous video stream is large and difficult to estimate, because it is possible to improve a good picture even further. The resource allocation strategy in multimedia applications is thus quite different from that of hard real-time applications; it is not determined by the given application requirements, but by the amount of available resources. A fraction of the given computational resources (processing power, memory, bandwidth) is allocated to a user domain. *Quality of experience* considerations at the end user determine the detailed resource allocation strategy. For example, if a user reduces the size of a window and enlarges the size of another window on his multimedia terminal, then the system can reduce the bandwidth and the processing allocated to the first window to free the resources for the other window that has been enlarged. Other users of the system should not be affected by this local reallocation of resources.

Future Trends The marriage of the Internet with smartphones and multimedia personal computers leads to many new volume applications. The focus of this book is not on multimedia systems, because these systems belong to the class of soft and firm real-time applications.

1.7 Examples of Real-Time Systems

In this section, three typical examples of real-time systems are introduced. These examples will be used throughout the book to explain the evolving concepts. We start with an example of a very simple system for flow control to demonstrate the need for *end-to-end protocols* in process input/output.

1.7.1 Controlling the Flow in a Pipe

It is the objective of the simple control system depicted in Fig. 1.8 to control the flow of a liquid in a pipe. A given flow set point determined by a client should be maintained despite changing environmental conditions. Examples for such changing conditions are the varying level of the liquid in the vessel or the temperature-sensitive viscosity of the liquid. The computer interacts with the controlled object by setting the position of the control valve. It then observes the reaction of the controlled object by reading the flow sensor F to determine whether the desired effect, the intended change of flow, has been achieved. This is a typical example of the necessary *end-to-end protocol* [Sal84] that must be put in place between the computer and the controlled object (see also Sect. 7.2.5.1). In a well-engineered system, the effect of any control action of the computer must be monitored by one or more independent sensors. For this purpose, many actuators contain a number of sensors in the same physical housing. For example, the control valve in Fig. 1.8 might contain a sensor, which measures the mechanical position of the valve in the pipe, and two limit switches, which indicate the firmly closed and the completely open positions of the valve. A rule of thumb is that there are about three to seven sensors for every actuator.

The dynamics of the system in Fig. 1.8 is essentially determined by the speed of the control valve. Assume that the control valve takes *10* s to open or close from *0%* to *100%* and that the flow sensor F has a precision of *1%*. If a sampling interval of *100* ms is chosen, the maximum change of the valve position within one sampling interval is *1%*, the same as the precision of the flow sensor. Because of this finite speed of the control valve, an output action taken by the computer at a given time will lead to an effect in the environment at some later time. The observation of this effect by the computer will be further delayed by the given latency of the sensor. All these latencies must either be derived analytically or measured experimentally, before the temporal control structure for a stable control system can be designed.

Fig. 1.8 Flow of liquid in a pipe

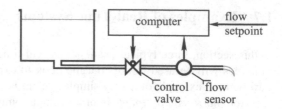

1.7.2   Engine Control

The task of an engine controller in an automobile engine is the calculation of the proper amount of fuel and the exact moment at which the fuel must be injected into the combustion chamber of each cylinder. The amount of fuel and the timing depend on a multitude of parameters: the intentions of the driver, articulated by the position of the accelerator pedal, the current load on the engine, the temperature of the engine, the condition of the cylinder, and many more. A modern engine controller is a complex piece of equipment. Up to 100 concurrently executing software tasks must cooperate in tight synchronization to achieve the desired goal, a smoothly running and efficient engine with a minimal output of pollutants.

The up- and downward moving piston in each cylinder of a combustion engine is connected to a rotating axle, the *crankshaft*. The intended start point of fuel injection is relative to the position of the piston in the cylinder and must be precise within an accuracy of about *0.1* degree of the measured angular position of the crankshaft. The precise angular position of the crankshaft is measured by a number of digital sensors that generate a rising edge of a signal at the instant when the crankshaft passes these defined positions. Consider an engine that turns with *6000* rpm (revolutions per minute), i.e., the crankshaft takes *10* ms for a *360* degree rotation. If the required precision of *0.1* degree is transformed into the time domain, then a temporal accuracy of *3* μs is required. The fuel injection is realized by opening a solenoid valve or a piezoelectric actuator that controls the fuel flow from a high-pressure reservoir into the cylinder. The latency between giving an *open* command to the valve and the actual point in time when the valve opens can be in the order of hundreds of μsec and changes considerably depending on environmental conditions (e.g., temperature). To be able to compensate for this latency jitter, a sensor signal indicates the point in time when the valve has actually opened. The duration between the execution of the output command by the computer and the start of opening of the valve is measured during every engine cycle. The measured latency is used to determine when the output command must be executed during the next cycle so that the intended effect, the start of fuel injection, happens at the proper point in time.

This example of an engine controller has been chosen because it demonstrates convincingly the need for extremely precise temporal control. For example, if the processing of the signal that measures the exact position of the crankshaft in the engine is delayed by a few μsec, the quality of control of the whole system is compromised. It can even happen that the engine is mechanically damaged if a valve is opened at an incorrect moment.

1.7.3   Rolling Mill

A typical example of a distributed plant automation system is the computer control
of a rolling mill. In this application a slab of steel (or some other material, such as
paper) is rolled to a strip and coiled. The rolling mill of Fig. 1.9 has three drives and
some instrumentation to measure the quality of the rolled product. The distributed
computer-control system of this rolling mill consists of seven nodes connected by a
real-time communication system. The most important sequence of actions—we call
this a *real-time (RT) transaction*—in this application starts with the reading of the
sensor values by the sensor computer. Then, the RT transaction passes through the
model computer that calculates new set points for the three drives and finally reaches
the control computers to achieve the desired action by readjusting the rolls of the
mill. The RT transaction thus consists of three processing actions connected by two
communication actions.

The total duration of the RT transaction (bold line in Fig. 1.9) is an important
parameter for the quality of control. The shorter the duration of this transaction, the
better the control quality and the stability of the control loop, since this transaction
contributes to the *dead time* of the critical control loop. The other important term of
the dead time is the time it takes for the strip to travel from the drive to the sensor.
A jitter in the dead time that is not compensated for will reduce the quality of control significantly. It is evident from Fig. 1.9 that the latency jitter in an event-
triggered system is the sum of the jitter of all processing and communication actions
that form the critical RT transaction.

Note that the communication pattern among the nodes of this control system is
multicast, not *point-to-point*. This is typical for most distributed real-time control
systems. Furthermore, the communication between the model node and the drive
nodes has an *atomicity requirement*. Either all of the drives are changed according
to the output of the model or none of them is changed. The loss of a message, which
may result in the failure of a drive to readjust to a new position, may cause mechanical damage to the drive.

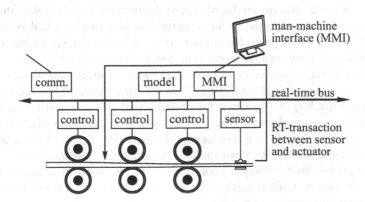

Fig. 1.9 An RT transaction

Points to Remember

- A real-time computer system must react to stimuli from the controlled object (or the operator) within time intervals *dictated* by its environment. If a catastrophe could result in case a firm deadline is missed, the deadline is called *hard*.
- In a hard real-time computer system, it must be guaranteed by design that the computer system will meet the specified deadlines in all situations because the utility of many hard real-time applications can depend on predictable performance during a peak-load scenario.
- A hard real-time system must maintain synchrony with the state of the environment (the controlled object and the human operator) in all operational scenarios. It is thus paced by the state changes occurring in the environment.
- Because the state of the controlled object changes as a function of real time, an observation is *temporally accurate* only for a limited time interval.
- A *trigger* is an event that causes the start of some action, e.g., the execution of a task or the transmission of a message.
- Real-time systems have only small data files, the *real-time database* that is formed by the temporally accurate images of the RT entities. The key concern is on the *short-term* temporal accuracy of the real-time database that is invalidated by the flow of real time.
- The real-time database must be updated whenever an RT entity changes its value. This update can be performed periodically, triggered by the progression of the real-time clock by a fixed period (*time-triggered observation*) or immediately after the occurrence of an event in the RT entity (*event-triggered observation*).
- The most stringent temporal demands for real-time systems have their origin in the requirements of the control loops.
- The temporal behavior of a simple controlled object can be characterized by *process lag* and *rise time* of the *step-response function*.
- The *dead time* of a control loop is the time interval between the observation of the RT entity and the start of a reaction of the controlled object as a consequence of a computer action based on this observation.
- Many control algorithms are based on the assumption that the delay jitter is a very small fraction of the *delay* since control algorithms are designed to compensate a known constant delay. Delay jitter brings an additional uncertainty into the control loop that has an adverse effect on the quality of control.
- The term *signal conditioning* is used to refer to all processing steps that are needed to get a meaningful RT image of an RT entity from the raw sensor data.
- The *reliability* $R(t)$ of a system is the probability that a system will provide the specified service until time t, given that the system was operational at $t = t_0$.
- If the failure rate of a system is required to be about 10^{-9} failures/h or lower, then we are dealing with a system with an *ultrahigh reliability* requirement.
- Safety is reliability regarding *malign (critical) failure modes*. In a malign failure mode, the cost of a failure can be orders of magnitude higher than the utility of the system during normal operation.

- *Maintainability* is a measure of the time it takes to repair a system after the last experienced benign failure and is measured by the probability $M(d)$ that the system is restored within a time interval d after the failure.
- *Availability* is a measure for the correct service delivery regarding the alternation of correct and incorrect service and is measured by the probability $A(t)$ that the system is ready to provide the service at time t.
- The main security concerns in real-time systems are the *authenticity, integrity,* and *timeliness* of the real-time information.
- The probability of failure of a perfect system with guaranteed response is reduced to the probability that the assumptions concerning the peak load and the number and types of faults are valid in reality.
- If we start out from a specified fault and load hypothesis and deliver a design that makes it possible to reason about the adequacy of the design without reference to probabilistic arguments (even in the case of the extreme load and fault scenarios), we speak of a system with a *guaranteed response*.
- An embedded real-time computer system is part of a well-specified larger system, an *intelligent product*. An intelligent product normally consists of a mechanical subsystem, the controlling embedded computer, and a man-machine interface.
- The static configuration, known a priori, of the intelligent product can be used to reduce the resource requirements and increase the robustness of the embedded computer system.
- Usually, every plant automation system is unique. Compared to development cost, the production cost (hardware cost) of a plant automation system is less important.
- The embedded system market continues to grow significantly during the next 10 years. Automated vehicles like self-driving cars and Industry 4.0 are example drivers. Compared with other information technology markets, this market will offer the best employment opportunities for the computer engineers of the future.

Bibliographic Notes

There exist a number of textbooks on real-time and embedded systems, such as *Introduction to Embedded Systems – A Cyber-Physical Systems Approach* [Lee10] by Ed Lee and Seshia, *Real-Time Systems* by Jane Liu [Liu00], *Hard Real-Time Computing Systems: Predictable Scheduling Algorithms and Applications* by Giorgio Buttazzo [But04], and *Real-Time Systems and Programming Languages: Ada, Real-Time Java and C/Real-Time POSIX* by Burns and Wellings [Bur09]. *Principles of Cyber-Physical Systems* by Rajeev Alur [Alu15] emphasizes cyber-physical systems' formal and theoretical aspects. The Journal *Real-Time Systems* by Springer publishes archival research articles on the topic.

Review Questions and Problems

1.1 What makes a computer system a *real-time* computer system?

1.2 What are typical functions that a real-time computer system must perform?

1.3 Where do the *temporal requirements* come from? What are the parameters that describe the temporal characteristics of a controlled object?

1.4 Give a "rule of thumb" that relates the *sampling period* in a quasi-continuous system to the *rise time* of the step-response function of the controlled object.

1.5 What are the effects of delay and delay jitter on the quality of control? Compare the error-detection latency in systems with and without jitter.

1.6 What does *signal conditioning* mean?

1.7 Consider an RT entity that changes its value periodically according to $v(t) = A_0 \sin (2\pi t/T)$ where T, the period of the oscillation, is *100* ms. What is the maximum change of value of this RT entity within a time interval of *1* ms? (Express the result in percentage of the amplitude A_0.)

1.8 Consider an engine that rotates with *3000* rpm. By how many degrees will the crankshaft turn within *1* ms?

1.9 Give some examples where the predictable rare-event performance determines the utility of a hard real-time system.

1.10 Consider a fail-safe application. Is it necessary that the computer system provides guaranteed timeliness to maintain the safety of the application? What is the level of error-detection coverage required in an ultrahigh dependability application?

1.11 What is the difference between availability and reliability? What is the relationship between maintainability and reliability?

1.12 When is there a simple relation between the MTTF and the failure rate?

1.13 Assume you are asked to certify a safety-critical control system. How would you proceed?

1.14 What are the main differences between a soft real-time system and a hard real-time system?

1.15 Why is an *end-to-end protocol* required at the interface between the computer system and the controlled object?

1.16 What is the fraction *development cost/production cost* in embedded systems and in plant automation systems? How does this relation influence the system design?

1.17 Assume that an automotive company produces *2,000,000* electronic engine controllers of a special type. The following design alternatives are discussed:

   (i) Construct the engine control unit as a single SRU with the application software in read-only memory (ROM).The production cost of such a unit is *250* \$. In case of an error, the complete unit has to be replaced.

   (ii) Construct the engine control unit such that the software is contained in a ROM that is placed on a socket and can be replaced in case of a software error. The production cost of the unit without the ROM is *248* \$. The cost of the ROM is *5* \$.

(iii) Construct the engine control unit as a single SRU where the software is loaded in a flash EPROM that can be reloaded. The production cost of such a unit is *255* $.

The labor cost of repair is assumed to be *50* $ for each vehicle. (It is assumed to be the same for each one of the three alternatives.) Calculate the cost of a software error for each one of the three alternative designs if *300,000* cars have to be recalled because of the software error (example in Sect. 1.6.1). Which one is the lowest-cost alternative if only *1000* cars are affected by a recall?

1.18 Estimate the relation (development cost)/(production cost) in an embedded consumer application and in a plant automation system.

1.19 Compare the peak load (number of messages, number of task activations inside the computer) that can be generated in an event-triggered and a time-triggered implementation of an elevator control system!

Chapter 2
Simplicity

Overview
The report on *Software for Dependable Systems: Sufficient Evidence?* [Jac07] by
the National Academies contains as one of its central recommendations: *One key to
achieving dependability at reasonable cost is a serious and sustained commitment
to simplicity, including simplicity of critical functions and simplicity in system interactions. This commitment is often the mark of true expertise.* We consider *simplicity*
to be the antonym of *cognitive complexity* (in the rest of this book, we mean *cognitive complexity* whenever we use the word *complexity*). In everyday life, many
embedded systems seem to move in the opposite direction. The ever-increasing
demands on the functionality and the nonfunctional constraints (such as safety,
security, or energy consumption) that must be satisfied by embedded systems lead
to a growth in system complexity.

In this chapter we investigate the notion of cognitive complexity and develop
guidelines for building *understandable* computer systems. We ask the question:
What does it mean when we say we understand a scenario? We argue that it is not
the embedded system but the *models* of the embedded system that must be simple
and understandable relative to the *background knowledge* of the observer. The models must be based on clear concepts that capture the relevant properties of the scenario under investigation. The *semantic content* of a program variable is one of
these concepts that we investigate in some detail. The major challenge of design is
the building of an *artifact* that can be modeled at different levels of abstraction by
models of adequate simplicity.

This chapter is structured as follows. Section 2.1 focuses on the topic of cognition and problem-solving and an elaboration of the two different human problem-
solving subsystems, the *intuitive-experiential* subsystem and the *analytic-rational*
subsystem. *Concept formation* and the *conceptual landscape*, that is, the private
knowledge base that a human develops over his lifetime, are discussed in Sect. 2.2.

31
H. Kopetz, W. Steiner, *Real-Time Systems*,
https://doi.org/10.1007/978-3-031-11992-7_2

Section 2.3 looks at the essence of model building and investigates what makes a task difficult. Section 2.4 deals with the important topic of emergence in large systems and system complexity. Finally, Sect. 2.5 defines seven principles that guide system design.

2.1 Cognition

Cognition deals with the study of thought processes and the interpretation and binding of sensory inputs to the existing knowledge base of an individual [Rei10]. It is an interdisciplinary effort that stands between the humanities, i.e., philosophy, language studies, and social science on one side and the natural sciences, such as neural science, logic, and computer science, on the other side. The study of model building, problem-solving, and knowledge representation forms an important part of the cognitive sciences.

2.1.1 Problem-Solving

Humans have two quite different mental subsystems for solving problems: the *intuitive-experiential subsystem* and the *analytic-rational subsystem* [Eps08]. Neuroimaging studies have shown that these two subsystems are executed in two different regions of the human brain [Ami01]. Table 2.1 compares some of the distinguishing characteristics of these two subsystems.

Table 2.1 Intuitive-experiential versus analytic-rational problem-solving strategy

Intuitive experiential	Analytic rational
Holistic	Analytic
Emotional (what feels good)	Logical reason oriented (what is sensible)
Unreflective associative connections	Cause-and-effect connections, causal chains
Outcome oriented	Process oriented
Behavior mediated by vibes from past experience	Behavior mediated by conscious appraisal of events
Encodes reality in concrete images, metaphors, and narratives	Encodes reality in abstract symbols, words, and numbers
More rapid processing, immediate action	Slower processing, delayed action
Slow to change the fundamental structure: changes with repetitive or intense experience	Changes more rapidly, changes with the speed of thought
Experience processed passively and preconsciously, seized by our emotions	Experience processed actively and consciously, in control of our thoughts
Self-evidently valid: *seeing is believing*	Requires justification via logic and evidence

Adapted from [Eps08, p. 26]

Example: A typical task for the *intuitive-experiential subsystem* is *face recognition*, a demanding task that a baby at the age of 6 months can accomplish. A typical task for the *analytic-rational subsystem* is the *confirmation* of a *proof of a mathematical theorem.*

The experiential subsystem is a preconscious emotionally based subsystem that operates holistically, automatically, and rapidly and demands minimal cognitive resources for its execution. Since it is nearly effortless, it is used most of the time. It is assumed that the experiential subsystem has access to a large coherent knowledge base that represents an implicit model of the world. This subjective knowledge base, which is one part of what we call the *conceptual landscape* of an individual, is mainly built up and maintained by experience and emotional events that are accumulated over the lifetime of an individual. Although this knowledge base is continually adapted and extended, its core structure is rather rigid and cannot be changed easily. *Experiential reasoning* is holistic and has the tendency to use limited information for general and broad classifications of scenarios and subjects (e.g., this is a *good* or *bad* person). The experiential system does assimilate the data about reality in a coherent stable conceptual framework. The concepts in this framework are mostly linked by *unconscious associative connections*, where the *source* of an association is often *unknown*.

The rational subsystem is a conscious analytic subsystem that operates according to the laws of causality and logic. Bunge [Bun08, p. 48] defines a *causality relationship* between a *cause C* and an *event E* as follows: *If C happens, then (and only then) E is always produced by it.* We try to get an understanding of a dynamic scenario by isolating a *primary cause*, suppressing seemingly irrelevant detail, and establishing a *unidirectional causal chain* between this primary cause and an observed *effect*. If cause and effect cannot be cleanly isolated, such as is the case in a feedback scenario, or if the relationship between cause and effect is *nondeterministic* (see also Sect. 5.6.1 on the definition of *determinism*), then it is more difficult to understand a scenario.

Example: Consider the analysis of a car accident that is caused by *the skidding* of a car. There are a number of conditions that must hold for skidding to occur: the speed of the car, the conditions of the road (e.g., icy road), the conditions of the tires, abrupt maneuver by the driver, the nonoptimal functioning of the computer-based skid-control system, etc. In order to *simplify* the model of the situation (the reality is not simplified), we often *isolate* a primary cause, e.g., the *speed*, and consider the other conditions as secondary.

The rational subsystem is a verbal and symbolic reasoning system, driven by a controlled and noticeable mental effort to investigate a scenario. Adult humans have a conscious *explicit model* of reality in their rational subsystem, in addition to their *implicit model* of reality in the experiential subsystem. These two models of reality coincide to different degrees and form jointly the *conceptual landscape* of an individual. There seem to be a nearly unlimited set of resources in the experiential subsystem, whereas the cognitive resources that are available to the rational subsystem are limited [Rei10].

There are many subtle interrelationships between these two problem-solving subsystems, which form the extremes of a continuum of problem-solving strategies where both systems cooperate to arrive at a solution. It is not infrequent that, after unsuccessful tries by the rational subsystem, at first a solution to a problem is produced unconsciously by the experiential subsystem. Afterward this solution is justified by analytical and logical arguments that are constructed by the rational subsystem.

Similarly, the significance of a new scenario is often recognized at first by the experiential subsystem. At a later stage, it is investigated and analyzed by the rational subsystem, and rational problem-solving strategies are developed. Repeated encounters of similar problems—the accumulation of *experience*—effortful learning, and drill move the problem-solving process gradually from the rational subsystem to the experiential subsystem, thus freeing the cognitive resources that have previously been allocated to the problem-solving process in the limited rational subsystem. There exist many practical examples that demonstrate this phenomenon: learning a foreign language, learning a new sport, or learning how to drive a car. It is characteristic for a *domain expert* that she/he has mastered this transition in her/his domain and mainly operates in the effortless experiential mode, where a fast, holistic, and intuitive approach to problem-solving dominates.

> **Example**: A brain-imaging study of the chess-playing strategy of amateurs vs. grandmasters investigated the activity in different sections of the brain immediately after a chess move by the partner. The amateurs displayed the highest activity in the *medial temporal lobe* of the brain, which is consistent with the interpretation that their mental activity is focused on the rational analysis of the new move. The highly skilled grandmasters showed more activity in the *frontal and parietal cortices*, indicating that they are retrieving stored information about previous games from expert memory in order to develop an *understanding of the scenario* [Ami01].

2.1.2 Definition of a Concept

In a changing world, *knowledge* about permanent and characteristic properties of objects and situations must be identified and maintained since such knowledge is of critical importance for survival. This knowledge is acquired by the process of *abstraction*, by which the particular is subordinated to the general, so that what is known about the general is applicable to many particulars. Abstraction is a fundamental task of the human cognitive system.

> **Example**: Face recognition is an example for the powerful process of *abstraction*. Out of many particular images of the face of a person—varying angles of observation, varying distance, and changing lighting conditions—characteristic permanent features of the face are identified and stored in order that they can be used in the future to recognize the face again. This demanding abstraction process is executed unconsciously, seemingly without effort, in the experiential subsystem. Only its results are delivered to the rational subsystem.

Abstraction forms categories, where a *category* is a set of elements that share common *characteristic features*. The notion of category is *recursive*: the *elements of a category* can themselves be *categories*. We thus arrive at a hierarchy of categories, going from the concrete to the abstract. At the lowest level, we find immediate sensory experiences.

A *concept* is a category that is augmented by a *set of beliefs* about its relations to other categories [Rei01, pp. 261–300]. The set of beliefs relates a *new concept* to already *existing concepts* and provides for an *implicit theory* (a subjective mental model). As a new domain is penetrated, new concepts are formed and linked to the concepts that are already present in the conceptual landscape. A concept is a mental construct of the *generalizable aspects* of a known entity. It has an intension (*What is the essence?*) and an extension, answering the question as to which *things* and *mental constructs* are exemplars of the concept. A concept can also be considered as a *unit of thought* [Vig62].

2.1.3 Cognitive Complexity

What do we mean when we say *an observer understands a scenario*? It means that the concepts and relationships that are employed in the representation of the scenario have been adequately linked with the conceptual landscape and the methods of reasoning of the observer. *The tighter the links are, the better is the understanding. Understanding* (and therefore *simplicity*) is thus a *relation* between an observer and a scenario, *not* a *property* of the scenario.

We take the view of Edmonds [Edm00] that *complexity* can only be assigned to *models of physical systems*, but not to the physical systems themselves, no matter whether these physical systems are natural or man-made. A physical system has a nearly infinite number of properties—every single transistor of a billion-transistor *system-on-chip* consists of a huge number of atoms that are placed at distinct positions in space. We need to abstract, to build *models* that leave out the seemingly irrelevant detail of the micro-level, in order to be able to reason about properties of interest to us at the macro-level.

What then is a good measure for *the cognitive complexity* of a *model*? We are looking for a quantity that measures the cognitive effort needed to understand the model by a human observer. *We consider the elapsed time needed to understand a model by a given observer a reasonable measure for the cognitive effort and thus for the complexity of a model relative to the observer.* We assume that the *given observer* is representative for the intended user group of the model.

According to the *scientific tradition*, it would be desirable to introduce an objective notion of cognitive complexity without reference to the subjective human experience. However, this does not seem to be possible, since cognitive complexity refers to a *relation* between an objective external scenario and the subjective internal conceptual landscape of the observer.

The perceived complexity of a model depends on the relationship between the existing subjective conceptual landscape and the problem-solving capability of the observer and the concepts deployed in the representation of the model, the interrelations among these concepts, and the notation used to represent these concepts. If the observer is an expert, such as the chess grandmaster in the previous example, the experiential subsystem provides an understanding of the scenario within a short time and without any real effort. According to our metric, the scenario will be judged as *simple*. An amateur has to go through a tedious cause-and-effect analysis of every move employing the rational subsystem that takes time and explicit cognitive effort. According to the above metric, the same chess scenario will be judged as *complex*.

There are models of behavior and tasks that are *intrinsically difficult to comprehend* under any kind of representation. The right column of Table 2.3 in Sect. 2.5 lists some characteristics of intrinsically difficult tasks. It may take a long time, even for an expert in the field, to gain an understanding of a model that requires the comprehension of the behavior of difficult tasks—if at all possible. According to the introduced metric, these models are classified as exceedingly complex.

In order to gain an understanding of a large system, we have to understand many models that describe the system from different viewpoints at different abstraction levels (see also Sect. 2.3.1). The cognitive complexity of a large system depends on the number and complexity of the different models that must be comprehended in order to understand the complete system. The time it takes to understand all these models can be considered as a measure for the *cognitive complexity of a large system*.

Case studies about the understanding of the behavior of large systems have shown that the *perceptually available information* plays an important role for developing an understanding of a system [Hme04]. *Invisible information flows* between *identified subsystems* pose a considerable barrier to understanding.

If every embedded system is one of its kind and no relationships between different instances of systems can be established, then there is hardly a chance that experience-based expert knowledge can be developed, and the transition from the tedious and effortful rational subsystem to the effortless experiential subsystem can take place.

One route to simplification is thus the development of a *generic model of an embedded system* that can be successfully deployed in many different domains at a proper level of abstraction. This model should contain few orthogonal mechanisms that are used recursively. The model must support simplification strategies and make public the internal information flow between identified subsystems, such that the process of gaining an understanding of the behavior is supported. By getting intimately acquainted with this model and gaining experience by using this model over and over again, the engineer can incorporate this model in the experiential subsystem and become an expert. It is one stated goal of this book to develop such a generic cross-domain model of embedded systems.

2.1.4 Simplification Strategies

The resources in the rational problem-solving subsystem of humans, both in storage and processing capacity, are limited. The seminal work of Miller [Mil56] introduced a limit of five to seven chunks of information that can be stored in short-term memory at a given instant. Processing limitations are established by the *relational complexity theory* of Halford [Hal96]. Relational complexity is considered to correspond to the *arity* (number of arguments) of a relation. For example, binary relations have two arguments as in LARGER-THAN (elephant, mouse). The relational complexity theory states that the upper limits of adult cognition seem to be relations at the quaternary level.

If a scenario requires cognitive resources that are beyond the given limits, then humans tend to apply simplification strategies to reduce the problem size and complexity in order that the problem can be tackled (possibly well, possibly inadequately) with the limited cognitive resources at hand. We know of four strategies to simplify a complex scenario in order that it can be processed by the limited cognitive capabilities of humans: *abstraction, partitioning, isolation,* and *segmentation:*

- *Abstraction* refers to the formation of a higher-level concept that captures the essence of the problem at hand and reduces the complexity of the scenario by omitting irrelevant detail that is not needed, given the purpose of the abstraction. Abstraction is applied *recursively*.
- *Partitioning* (also known as *separation of concerns*) refers to the *division* of the problem scenario into nearly independent parts that can be studied successfully in isolation. Partitioning is at the core of *reductionism*, the preferred simplification strategy in the natural sciences over the past 300 years. Partitioning is not always possible. It has its limits when *emergent properties* are at stake.
- *Isolation* refers to the suppression of seemingly irrelevant detail when trying to find a *primary cause*. The primary cause forms the starting point of the causal chain that links a sequence of events between this primary cause and the observed effect. There is a danger that the simplification strategy of *isolation* leads to a too simplistic model of reality (see the example on skidding of a car in Sect. 2.1.1).
- *Segmentation* refers to the *temporal decomposition* of intricate behavior into smaller parts that can be processed *sequentially*, one after the other. Segmentation reduces the amount of information that must be processed in parallel at any particular instant. Segmentation is difficult or impossible if the behavior is formed by highly concurrent processes, depends on many interdependent variables, and is strongly nonlinear, caused by positive or negative feedback loops.

2.2 The Conceptual Landscape

The notion of *conceptual landscape*, or the *image* [Bou61], refers to the *personal knowledge base* that is built up and maintained by an individual in the experiential and rational subsystem of the mind. The knowledge base in the experiential subsystem is *implicit*, while the knowledge base in the rational subsystem is *explicit*. The conceptual landscape can be thought of as a structured network of interrelated *concepts* that defines the *world model*, the *personality*, and the *intentions* of an individual. It is built up over the lifetime of an individual, starting from pre-wired structures that are established during the development of the *genotype* to the *phenotype*, and continually augmented as the individual interacts with its environment by exchanging messages via the sensory systems.

2.2.1 Concept Formation

The formation of concepts is governed by the following two principles [And01]:

- The *principle of utility* states that a new concept should encompass those properties of a scenario that are of utility in achieving a stated purpose. The purpose is determined by the human desire to fulfill basic or advanced needs.
- The *principle of parsimony* (also called *Occam's razor*) states that out of a set of alternative conceptualizations that are of comparable utility, the one that requires the least amount of mental effort is selected.

There seems to be a *natural level of categorization*, neither too specific nor too general, that is used in human communication and thinking about a domain. We call the concepts at this natural level of categorization *basic-level concepts* [Rei01, p. 276].

> **Example**: The basic-level concept *temperature* is more fundamental than the sub-concept *oil-temperature* or the encompassing concept *sensor data*.

Studies with children have shown that basic-level concepts are *acquired earlier* than *sub-concepts* or *encompassing* concepts. As a child grows up, it continually builds and adds to its *conceptual landscape* by observing regularities in the perceptions and utility in grouping properties of perceptions into new categories [Vig62]. These new categories must be interlinked with the already existing concepts in the child's mind to form a consistent *conceptual landscape*. By abstracting not only over perceptions but also over already existing concepts, new concepts are formed.

A new concept requires for its formation a number of experiences that have something in common and form the basis for the abstraction. *Concept acquisition* is normally a bottom-up process, where sensory experiences or basic concepts are the starting point. Examples, prototypes, and feature specification play an important role in concept formation. A more *abstract concept* is understood best *bottom up* by generalizations from a set of a suitable collection of examples of already acquired concepts. Abstract analysis and concrete interpretation and explanation should be intertwined frequently. If one remains only at a *low level of abstraction*, then the amount of nonessential detail is overwhelming. If one remains only at a *high level of abstraction*, then relationships to the world as it is experienced are difficult to form.

In the *real world* (in contrast to an *artificial world*), a *precise definition* of a concept is often not possible, since many concepts become fuzzy at their boundaries [Rei10, p. 272].

> **Example**: How do you define the concept of *dog*? What are its characteristic features? Is a dog, which has lost a leg, still a dog?

Understanding a new concept is a matter of establishing connections between the new concept and already familiar concepts that are well embedded in the conceptual landscape.

> **Example**: In order to understand the new concept of *counterfeit money*, one must relate this new concept to the following already familiar concepts: (i) the concept of *money*, (ii) the concept of a *legal system*, (iii) the concept of a *national bank* that is *legalized* to print money, and (iv) the concept of *cheating*. A *counterfeit money bill* looks like an *authentic money bill*. In this situation, *examples* and *prototypes* are of limited utility.

In the course of cognitive development and language acquisition, *words (names)* are associated with concepts. The *essence of a concept* associated with a word can be assumed to be the *same* within a natural language community (*denotation*), but different individuals may associate different *shades of meaning* with a concept (*connotation*), dependent on their *individual existing conceptual landscape* and the differing personal emotional experiences in the acquisition of the concept.

> **Example**: If communicating partners refer to different concepts when using a word or if the concept behind a word is not well established in the (scientific) language community (i.e., does not have a well-defined denotation), then effective communication among partners becomes difficult to impossible.

If we change the language community, the names of concepts will be changed, although the *essence of the concept, its semantic content*, remains the same. The names of concepts are thus relative to the context of discourse, while the *semantic content* remains invariant.

> **Example**: The semantic *content* of the concept *speed* is precisely defined in the realm of physics. Different language communities give different names to the same concept: in German *Geschwindigkeit*, in French *vitesse*, and in Spanish *velocidad*.

2.2.2   Scientific Concepts

In the world of science, new concepts are introduced in many publications in order to be able to express new *units of thought*. Often these concepts are named by a *mnemonic*, leading to, what is often called, *scientific jargon*. In order to make an exposition *understandable*, new concepts should be introduced sparingly and with utmost care. A new scientific concept should have the following properties [Kop08]:

* *Utility*: The new concept should serve a useful well-defined purpose.
* *Abstraction and Refinement*: The new concept should abstract from lower-level properties of the scenario under investigation. It should be clear which properties

are not parts of the concept. In the case of refinement of a *basic-level concept*, it should be clearly stated what additional aspects are considered in the refined concept.

- *Precision*: The characteristic properties of the new concept must be precisely defined.
- *Identity*: The new concept should have a distinct identity and should be significantly different from other concepts in the domain.
- *Stability*: The new concept should be usable uniformly in many different contexts without any qualification or modification.
- *Analogy*: If there is any concept in the existing *conceptual landscape* that is, in some respects, analogous to the new concept, this similarity should be pointed out. The analogy helps to establish links to the *existing conceptual landscape* of a user and facilitates understanding. According to [Hal96, p. 5]: *Analogical reasoning mechanisms are important to virtually every area of higher cognition, including language comprehension, reasoning and creativity. Human reasoning appears to be based less on an application of formal laws of logic than on memory retrieval and analogy.*

The availability of a useful, well-defined, and stable set of concepts and associated terms that are generally accepted and employed by the scientific community is a mark for the maturity of a scientific domain. An ontology is a shared taxonomy that classifies terms in a way useful to a specific application domain in which all participants share similar levels of understanding of the meaning of the terms [Fis06, p. 23]. Progress in a field of science is intimately connected with concept formation and the establishment of a well-defined ontology.

> **Example:** The main contributions of Newton in the field of mechanics are not only in the formulation of the laws that bear his name but also in the isolation and conceptualization of the abstract notions *power*, *mass*, *acceleration*, and *energy* out of an unstructured reality.

Clear concept formation is an essential prerequisite for any formal analysis or formal verification of a given scenario. The mere replacement of fuzzy concepts by formal symbols will not improve the understanding.

2.2.3   The Concept of a Message

We consider a *message* as a *basic concept* in the realm of communication. A message is an *atomic unit* that captures the value domain and the temporal domain of a unidirectional information transport at a level of abstraction that is applicable in many diverse scenarios of human communication [Bou61] and machine communication. A basic message-transport service (BMTS) transports a message from a sender to one or a set of receivers. The BMTS can be realized by different means, e.g., biological or electrical.

For example, the message concept can be used to express the information flow from the human sensory system to the conceptual landscape of an individual. The message concept can also model the indirect high-level interactions of a human with his environment that are based on the use of language.

> **Example**: We can model the sensory perception, e.g., of temperature, by saying that a message containing the sensed variable (temperature) is sent to the *conceptual landscape*. A message could also contain verbal information about the temperature at a location that is outside the realm of direct sensory experience.

The message concept is also a *basic concept* in the domain of distributed embedded computer systems at the architecture level. If the BMTS between encapsulated subsystems is based on *unidirectional temporally predictable multicast messages*, then the data aspect, the timing aspect, the synchronization aspect, and the publication aspect are integrated in a single mechanism. The BMTS can be refined at a lower level of abstraction by explaining the transport mechanism. The transport mechanism could be wired or wireless. The information can be coded by different signals. These refinements are relevant when studying the implementation of the message mechanism at the physical level, but are irrelevant at a level where the only concern is the timely arrival of the information sent by one partner to another partner.

A *protocol* is an abstraction over a sequence of *rule-based* message exchanges between communicating partners. A protocol can provide additional services, such as flow control or error detection. A protocol can be understood by breaking it down to the involved messages without the need to elaborate on the concrete transport mechanisms that are used.

2.2.4   Semantic Content of a Variable

The concept of a *variable*, a fundamental concept in the domain of computing, is of such importance for the rest of the book that it justifies some special elaboration. A variable can be considered as a *language construct* that assigns an *attribute* to a *concept*. If the point in real time, the *instant*, when this assignment is valid, is of relevance, then we call the *variable* a *state variable*. As time progresses, the attribute of a state variable may change, while the concept remains the same. A variable thus consists of two parts, a *fixed part*, the *variable name* (or the *identifier*), and a *variable part* called the *value of the variable* that is assigned to the variable. The variable name designates the concept that determines *what we are talking about*. In a given context, the variable name—which is analogous to the name of a concept in a natural language community—must be unique and point to the same concept at all communicating partners. The meaning that is conveyed by a variable is called the *semantic content* of the variable. As we will show in the latter part of this section, the semantic content of a variable is *invariant* to a change in representation. The requirement of *semantic precision* demands that the concept that is associated with a variable name and the domain of values of the variable are unambiguously defined in the model of the given application.

Example: Consider the variable name *engine-temperature* that is used in an automotive application. This concept is too abstract to be meaningful to an automotive engineer, since there are different temperatures in an automotive engine: the temperature of the oil, the temperature of the water, or the temperature in the combustion chamber of the engine.

The unambiguous definition of a concept does not only relate to the meaning of the concept associated with the variable but also to the specification of the *domain of values* of the variable. In many computer languages, the *type of a variable*, which is introduced as an attribute of the variable name, specifies primitive attributes of the value domain of the variable. These primitive attributes, like *integer* or *floating point number*, are often not sufficient to properly describe all relevant attributes of the value domain. An extension of the type system will alleviate the problem.

Example: If we declare the *value domain* of the variable *temperature* to be *floating point*, we still have not specified whether the temperature is measured in units of *Celsius, Kelvin*, or *Fahrenheit*.

Example: The Mars Climate Orbiter crash occurred because the ground-based software used different system units than the flight software. The first of the recommendations in the report of the mishap investigation board was *that the MPL (Mars Polar Lander) project verify the consistent use of units throughout the MPL spacecraft design and operation* [NAS99].

In different language communities, different variable names may be used to point to the same concept. For example, in an English-speaking language community, *the temperature of the air* may be abbreviated by *t-air*, while a German-speaking community may call it *t-luft*. If we change the representation of the value domain of a variable, e.g., if we replace the units for measuring the temperature from *Celsius* to *Fahrenheit* and adapt the value of the variable accordingly, the *semantic content* expressed by the variable remains the same.

Example: On the *surface* the two variables *t-air* = 86 and *t-luft* = 30 are completely different since they have different names and different values. If, however, *t-air* and *t-luft* refer to the same concept, i.e., the temperature of the air, and the value of *t-air* is expressed in degrees Fahrenheit and that of *t-luft* in degrees Celsius, then it becomes evident that the *semantic contents* of these two variables are the same.

These differences in the representations of the *semantic content* of a variable become important when we look at *gateway components* which link two subsystems of a *system of systems* that have been developed by two different organizations according to two different *architectural styles*. The term *architectural style* refers to all *explicit* and *implicit* principles, rules, and conventions that are followed by an organization in the design of a system, e.g., the representation of data, protocols, syntax, naming, semantics, etc. The gateway component must translate the variable names and representations from one architectural style to the other architectural style, while keeping the semantic content *invariant*.

Data that describes the properties of *(object) data* is sometimes called *meta-data*. In our model of a variable, data that describes the properties of the *fixed parts of a variable* is *meta-data*, while the *variable part* of a variable, *the value set*, is *(object) data*. *Meta-data* thus describes the properties of the concept that is referred to by the variable name. Since *meta-data* can become *object data* of another level, the distinction between *data* and *meta-data* is relative to the viewpoint of the observer.

> **Example**: The price of a product is *data*, while the currency used to denote the price, the time interval, and the location where this price is applicable are *meta-data*.

2.3   The Essence of Model Building

Given the rather limited cognitive capabilities of the rational subsystem of the human mind, we can only develop a rational understanding of the world around us if we build *simple models* of those properties that are of relevance and interest to us and disregard (abstract from) detail that proves to be irrelevant for the given purpose. *A model is thus a deliberate simplification of reality with the objective of explaining a chosen property of reality that is relevant for a particular purpose.*

> **Example**: The purpose of a model in *celestial mechanics* is the explanation of the movements of the heavenly bodies in the universe. For this purpose it makes sense to introduce the abstract concept of a *mass point* and to reduce the whole diversity of the world to a single *mass point in space* in order that the interactions with other mass points (heavenly bodies) can be studied without any distraction by unnecessary detail.

When a new level of abstraction (a new model) is introduced that successfully conceptualizes the properties relevant for the given purpose and disregards the rest, *simplicity* emerges. Such simplicity, made possible by the formation of proper concepts, gives rise to new insights that are at the roots of the *laws of nature*. As Popper [Pop68] points out, due to the inherent imperfection of the *abstraction* and *induction* process, laws of nature can only be falsified, but never be proven to be absolutely correct.

2.3.1   Purpose and Viewpoint

At the start of any modeling activity, a clear purpose of the model must be established. Formulating the precise questions the model must address helps to concretize the purpose of the model. If the purpose of a model is not crystal clear or if there are multiple divergent purposes to satisfy, then it is not possible to develop a *simple* model.

> **Example**: The purpose of a *model of behavior* of a real-time computer system is to provide answers to the question at *what points in real time* will the computer system produce *what kind of outputs*. If our computer system is a *system-on-chip (SoC)* with a billion transistors, then we must find a hierarchy of behavioral models to meet our purpose.

The recursive application of the *principles of abstraction* leads to such a hierarchy of models that Hayakawa [Hay90] calls the *abstraction ladder*. Starting with *basic-level concepts* that are essential for understanding a domain, more general concepts can be formed by *abstraction*, and more concrete concepts can be formed by *refinement*. At the lowest level of the abstraction ladder are the direct sensory experiences.

> **Example**: The *four-universe model* of Avizienis [Avi82] introduces a hierarchy of models in order to simplify the description of the behavior of a computer system. At the lowest level of the hierarchy, the *physical level*, the analog signals of the circuits are observed, such as the rise time of the voltages as a transistor performs a switching operation. The analysis of a circuit behavior at the physical (analog) level becomes difficult as soon as more and more transistors get involved (*emerging complexity*). The next higher level, the *digital logic level*, abstracts from the physical analog quantities and the dense time and introduces binary logic values (*high* or *low*) of signals at discrete instants, resulting in a much simpler representation of the behavior of an elementary circuit, e.g., an AND gate (*emerging simplicity*). Complexity creeps in again as we combine more and more logic circuits. The next higher level, the *information level*, lumps a (possible large) sequence of binary values into a meaningful data structure (e.g., a pointer, a real-valued variable, or a complete picture) and introduces powerful high-level operations on these data structures. Finally, at the *external level*, only the services of the computer system to the environment, as seen by an outside user, are of relevance.

A posed question about a distinct property of a *real system* gives rise to the construction of a hierarchy of models of that system that are intended to answer the posed question. Figure 2.1 depicts two hierarchies of models that are introduced to serve two purposes, purpose A and purpose B. Purpose A could refer to a hierarchy of behavioral models, while purpose B could refer to a hierarchy of dependability models of the same *real system*. At the top of each hierarchy is the stated purpose, i.e., the questions that must be answered. The different levels of the hierarchy—the abstraction levels—are introduced to support a stepwise refinement of the stated question considering more detail, where each step takes consideration of the limited cognitive capabilities of the human mind. At the low end of the hierarchy is the real system. The analysis is substantially simplified if the structure of the model corresponds with the structure of the system. Otherwise we have to resolve a structure clash that complicates the issues.

> **Example**: The model for predicting the temporal properties of the behavior of a real-time computer system is straightforward if there is a predictable sequence of computational and communication actions between the start of a computation and the termination of a computation. Conversely, if the actual durations of the computational and communication actions depend on global system activity (e.g., arbitration for access to shared resources such as caches, communication links, etc.), then it will not be possible to construct a *simple model* for predicting the temporal properties of the behavior.

In a computer system, given rules relate the different levels of abstraction (like the levels discussed in Avizienis' four-universe model above) to each other. These rules even allow the specification of an abstract model and to *synthesize* a lower level of abstraction. However, the resource-optimized refinement from one abstraction level to the next lower one can require sophisticated tooling.

> **Example**: In chip design two succeeding levels of abstraction on the abstraction ladder are the *register-transfer level* (e.g., a VHDL or Verilog model) and the *netlist level* (the concrete

Fig. 2.1 Purpose and abstraction level of a model

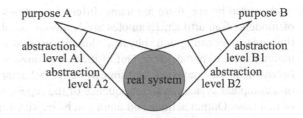

list of electronic components and their connections in a circuit). There are rules that determine the relation between the register-transfer and the netlist level. Although these rules are given, the chip industry relies on sophisticated logic synthesis tools to produce a resource-optimized netlist from the register-transfer level.

Simple Verification Procedure Although the synthesis of lower-level models from higher-level models can become computationally intense, verifying that an existing lower-level model is a correct refinement of an existing higher-level model is sometimes simple.

> **Example**: In a puzzle game, the picture on the box that contains the puzzle pieces serves as a model. Assembling the puzzle is computationally intense and requires frequent trial and error. However, once the puzzle has been completed, it is simple to verify that the puzzle equals the picture on the box.

An abstraction ladder with simple verification procedures on each step (i.e., between every two succeeding models of abstraction) is of exceptionally high value: such an abstraction ladder simplifies verification and validation (see Chap. 12). The static scheduling of tasks (see Sect. 10.3) and the static scheduling of messages (see Sect. 7.5) have *simple verification procedures*.

2.3.2 The Grand Challenge

Whereas the natural scientist must uncover the regularities in a given reality and find appropriate concepts at a suitable level of abstraction in order to formulate models and theories that explain the observed phenomena, the computer scientist is—*at least theoretically*—in a much better position: The computer scientist has the freedom to design the system—*an artifact*—which is the subject of his modeling. The requirement to build artifacts, the properties of which can be analyzed by *simple models*, should thus be an explicit design driver. In many areas of computer science, this principle of building artifacts that can be modeled by *simple models* is violated. For example, the temporal behavior of a modern pipelined microprocessor with multiple caches cannot be captured in a *simple model*.

The major challenge of design is the building of a software/hardware artifact (an embedded computer system) that provides the intended behavior (i.e. the service) under given constraints and where relevant properties of this artifact (e.g., the behavior) can be modeled at different levels of abstraction by models of adequate simplicity.

As stated before, there are many different purposes that give rise to a hierarchy of models of an artifact. Examples are behavior, reliability, man-machine interaction, energy consumption, physical dimension, cost of manufacturing, or cost of maintenance, to name a few. Out of these, the most important one is the *model of behavior*. In the context of real-time systems, behavior specifies the output actions of a computer system as a consequence of the inputs, the state and the progression of real time. Output actions and input can be captured in the concepts of *input messages* and *output messages*. In Chap. 4 of this book, we present a cross-domain model for the behavior of a real-time computer system using these concepts.

2.4 Emergence

We speak of *emergence* when the interactions of subsystems give rise to unique global properties at the system level that are not present at the level of the subsystems [Mor07]. Nonlinear behavior of the subsystems, feedback and feedforward mechanisms, and time delays are of relevance for the appearance of emergent properties. Up to now, the phenomenon of emergence is not fully understood and a topic of intense study.

2.4.1 Irreducibility

Emergent properties are irreducible, holistic, and novel—they disappear when the system is partitioned into its subsystem. Emergent properties can appear unexpectedly or they are planned. In many situations, the first appearance of the emergent properties is unforeseen and unpredictable. Often a fundamental revision of state-of-the-art models is required to get a better understanding of the conditions that lead to the intended emergence. In some cases, the emergent properties can be captured in a new conceptualization (model) at a higher level of abstraction resulting in an *abrupt simplification* of the scenario.

> **Example**: The emergent properties of a *diamond*, such as *brilliance* and *hardness*, which are caused by the *coherent* alignment of the carbon atoms, are substantially different from the properties of graphite (which consists of the *same atoms*). We can consider the diamond with its characteristic properties a new concept, a *new unit of thought*, and forget about its composition and internal structure. Simplicity comes out as a result of the intricate interactions among the elements that help to generate a *new whole* with its new emergent properties.

2.4.2 Prior and Derived Properties

When dealing with emergence, it is helpful to distinguish between the *prior properties* of the components and the new *derived properties* that come about by the interactions of the components.

Example: The high reliability of the services of a fault-tolerant system (*derived property*) that is the result of the interactions of many unreliable components (*prior property*) is an emergent property.

In many cases the prior properties and the derived properties can be of a completely different kind. It often happens that the *derived properties* open a completely new domain of science and engineering. This new domain requires the formation of novel concepts that capture essential properties of this new domain.

Example: The property of *being able to fly* which comes about by the proper interaction of the subsystems of an airplane, such as the wings, the fuselage, the engines, and the controls, is only present in the airplane as a whole but not in any of the isolated subsystems. *Being able to fly* has opened the domain of the air transportation industry with its own rules and regulations. For example, the subject of *air traffic control* is far removed from the prior properties of the components that make up an airplane.

Prior properties and derived properties are relative to the viewpoint of the observer. When climbing up the abstraction ladder, the derived properties at one level of abstraction become the prior properties at the next higher level of abstraction and so on, since a new form of emergence can appear at higher levels.

Example: In the evolution of the universe, two very significant stages of emergence are the *appearance of life* and at a further stage the *appearance of consciousness* that forms the basis for the development of human culture. The realm of human culture has developed its own system of concepts in the arts, sciences, etc. that are far removed from the biological prior properties that are characterizing the human brain.

Emergent behavior cannot be predicted analytically, but must be detected in an operating system. Thus control elements must incorporate hooks for monitoring system performance in real time [Par97, p. 7]. The multicast message concept, discussed in Sect. 2.2.3, provides the basis for the nonintrusive observation of system behavior.

2.4.3   Complex Systems

In 2003, Joseph Sussman compiled a working paper on *Collected Views on Complexity in Systems* [Sus03]. Out of this paper, we have extracted the following eight quotes on the topic of complexity to address different perspectives: what makes systems complex, what are insights about complex systems, what are potential threats of complex systems, and how can we manage complex systems?

What Makes Systems Complex?

1.   "Complexity has to do with interconnections between parts of a system, and it has to do with the nature of these interconnections (their intricateness)." [Mos]

In a cyber-physical system, the interconnections among the parts of a system are realized by the exchange of real-time messages. Let us look at the unidirectional transport of a message that contains *time-sensitive data elements*.

Table 2.2 Simple and complex transmission of real-time data

Time-triggered protocol	Event-triggered protocol
Communication delay fixed and known	Communication delay not known
Error detection at the receiver of a multicast message	Error detection at the sender of a multicast message
Independent observation of the message contents possible	Independent observation of the message contents not possible
Error propagation from a faulty receiver to the sender eliminated by design	Error propagation from a faulty receiver to the sender possible

From Table 2.2 it can be concluded that the characteristics of a time-triggered protocol are in better agreements with the needs of a real-time connection than the characteristic of an event-triggered protocol (see also Chap. 7). It is thus simpler to understand the behavior of a real-time control system if time-triggered protocols are used to connect the parts.

2.   "A system is complex when it is composed of a group of related units (subsystems), for which the degree and nature of the relationships is imperfectly known." [Sus00]
3.   "Complex systems are characterized by unfamiliar or unintended feedback loops (often closed by stigmergic communication channels)." [Per99]

Complexity, thus, arises in systems consisting of many parts (components, subsystems, etc.) interacting in various ways. Moreover, both the parts and their interactions are likely only partially known and characterized.

> **Example**: Earth's weather is a complex system. Some parts of the weather system are position and altitude on earth, local population density, spatial proximity to significant water sources, intensity and angle of solar irradiation, air and water streams, etc. Unfortunately, we do not know the precise interaction of all these parts sufficiently, and there are likely many further, yet unknown, parts and interactions as well.

> **Example**: Large enterprise IT systems can be complex systems. Besides the imperfection and shortcomings of current IT solutions, many of these systems have grown over time and give rise to unknown interactions between different generations of equipment. Furthermore, the system includes human users, and human behavior is often irrational.

> **Example**: Power grids can be complex systems. Power generation and distribution may fail because of system design faults or natural phenomena (e.g., earthquakes). On the other hand, a large portion of power consumption depends on human behavior. Emerging consumer trends like electric vehicles have a tremendous impact on the power grid. Furthermore, many consumers also become producers via photovoltaic installations.

> **Example**: The human body is a complex system, and so is the behavior of social groups.

In this book, we classify a system as *complex*, if we are not in the position to develop a set of models of *adequate simplicity*—commensurate to the rational capabilities of the human mind—to explain the structure and behavior of the system. Further examples of complex systems are *life* and *consciousness*, the earth's climate, the global economy, living organisms, and many large computer systems, to name a few.

What Are Insights About Complex Systems?

4.   "The world abounds with complex systems that have successfully evolved—organisms, economies, our legal system. We should begin to ask, What kinds of complex systems can be assembled by an evolutionary process? I should stress that no general answer is known, but that systems with some kinds of redundancy are almost certainly far more readily evolved than those without redundancy. Unfortunately, we only roughly understand what redundancy actually means in evolving systems." [Kau93]

We do not comprehend complex systems but can observe common patterns in successful ones. For example, *redundancy* is almost always present in purposeful complex systems.

> **Example**: Redundancy is inherent in the human body. For example, we have two eyes, two ears, and two kidneys and remain mobile even if we lose some extremities.

> **Example**: Large and complex IT systems typically implement redundancy, for example, in form of backup servers.

What Are Potential Threats of Complex Systems?

5.   "Its overall emergent behavior is difficult to predict, even when subsystem behavior is readily predictable" [Sus00].
6.   "Familiar computer programs are precisely the kind of complex systems that do not have the property that small changes in structure yield small changes in behavior. Almost all small changes in structure lead to catastrophic changes in behavior." [Kau93]

Since we do not understand complex systems, we cannot predict their future behavior with adequate certainty. As a consequence, *black swan events*, i.e., unanticipated events with major effect, often on a global scale, may erupt [Tal08].

> **Example**: Power outages (*blackouts*) occur commonly. In 2020–2021 alone, several hundred thousand households in the USA have been subject to power outages.

> **Example**: Cybersecurity is a growing concern in IT enterprises, but all kinds of organizations are under attack more recently. For example, hackers gained control over the IT systems in hotels, hospitals, and factories.

> **Example**: Globalization and free travel significantly increased the spread of infectious diseases. As a result, in 2020, Covid-19 evolved into a global pandemic.

Although we lack complex system understanding, we still aim to achieve a particular future system behavior by a modification of its components. Often, this modification does not have the desired outcome.

> **Example**: Software updates, including bug fixes, are common in IT systems. However, frequently, updates result in unexpected emergent phenomena. For example, software updates in banking IT systems have caused millions of inaccessible bank accounts. On the other hand, the operating system update in mobile phones regularly causes phones to malfunction, i.e., these phones are said to be *bricked*.

> **Example**: Medical drugs may have severe side effects on the human body. For example, pregnant women consuming the tranquilizer *Contergan* gave birth to infants with congenital disabilities.

How Can We Manage Complex Systems?

7.   "Scientists have broken down many kinds of systems. They think they know most of the elements and forces. The next task is to reassemble them, at least in mathematical models that capture the key properties of the entire ensembles. Success in this enterprise [i.e., the accurate and complete description of complex systems] will be measured by the power researchers acquire to predict emergent phenomena when passing from general to more specific levels of organization." [Wil98]

8.   "In evolving systems, bursts of simplicity often cut through growing complexity and establish new bases upon which complexity can then grow." [Art94]

> **Example**: The Avizienis *four-universe model* (see Sect. 2.3.1) is an example of a *burst of simplicity* that forms the basis for modern chip development.

We hold the opinion that a fundamental understanding of a complex system can only be achieved by a *proper conceptualization* and not by the execution of elaborate computer simulations. This view is also shared by *Mesarovic* et al. [Mes04, p. 19] when he speaks about biology: *We further argue that for a deeper understanding in systems biology investigations should go beyond building numerical mathematical or computer models—important as they are ... Such a categorical perspective led us to propose that the core of understanding in systems biology depends on the search for organizing principles rather than solely on construction of predictive descriptions (i.e. models) that exactly outline the evolution of systems in space and time. The search for organizing principles requires an identification/ discovery of new concepts and hypotheses.*

Maybe, sometimes in the future, we will form appropriate concepts that will lead to an abrupt simplification of some of today's complex systems. If this happens, the system will not be classified as complex anymore.

Whereas system biology deals with a natural system, a large computer system is an *artifact* developed by humans. When designing such an artifact, we should take consideration of the limited rational problem-solving capability of humans in order that we can describe the behavior of the artifact by models of adequate simplicity. These models should guide the design process, such that a structure clash between the model and the artifact is avoided.

> **Example**: Let us look at the technical example of designing the on-chip communication infrastructure for the communication among IP cores on a system-on-chip. There are basically two technical alternatives, the provision of a *shared memory* that can be accessed by all IP cores or the provision of *local memory to each one of the IP cores* and the design of a message-passing subsystem that enables the exchange of messages among IP cores [Pol07, Lev08]. The message-passing subsystem isolates and makes explicit the global communication among subsystems and thus supports the introduction of a new level in the hierarchy where a distinction is made between the *intra-IP core* interactions and the *inter-IP core* interactions. The common memory intermixes global intra-IP core and local inter-IP core interactions and makes it very difficult to separate global and local concerns, leading to a more complex system model.

> **Example**: The electrical/electronic vehicle system of modern cars is complicated. The push toward self-driving cars moves it toward the edge to become a complex system (i.e., per our definition, a system that exceeds the rational capabilities of the human mind). A conceptual architecture that defines an abstraction ladder of simple models is the essential enabler of

safe self-driving cars. On the top level of this ladder, the simple model should distinguish a *critical* and a *noncritical subsystem* with minimal information exchange. While the purpose of the critical subsystem is the safe maneuvering of the car, the purpose of the noncritical subsystem is passenger comfort (including infotainment). A conceptual architecture to realize the critical subsystem is given by Kopetz [Kop21] (see example in Sect. 6.5.1).

2.5   How Can We Achieve Simplicity?

Cognitive scientists have studied how students learn and understand different tasks [Fel04]. They have identified a set of task characteristics that require a disproportional mental effort for understanding the task. Table 2.3 compares the characteristics of *simple* tasks versus *difficult* tasks. We thus need to design a generic model for expressing the behavior of an embedded system that avoids the characteristics of difficult tasks. It should be possible to apply the model *recursively*, such that large systems can be modeled at different levels of abstraction using the same modeling mechanisms.

Table 2.3 Characteristics of simple versus difficult tasks

Characteristics of a simple task	Characteristics of a difficult task
Static: The properties of the task do not change over time	Dynamic: The properties of the task are time dependent
Discrete: The variables that characterize the task can only take values from discrete sets	Continuous: The domain of the variables is continuous
Separable: Different subtasks are nearly independent. There is only a weak interaction among tasks	Non-separable: Different subtasks are highly interactive. It is difficult to isolate the behavior of a single task
Sequential: Behavior can be understood by a sequential step-by-step analysis	Simultaneous: Many concurrent processes interact in generating visible behavior. Step-by-step analysis is difficult
Homogeneous: Components, explanatory schemes, and representations are alike	Heterogeneous: Many different components, explanatory schemes, and representations
Mechanism: Cause-and-effect relations dominate	Organicism: Behavior characterized by a multitude of feedback mechanisms
Linear: Functional relationships are linear	Nonlinear: Functional relationships are nonlinear
Universal: Explanatory principles do not depend on context	Conditional: Explanatory principles are context dependent
Regular: Domain characterized by a high regularity of principles and rules	Irregular: Many different context-dependent rules
Surface: Important principles and rules are apparent by looking at observable surface properties	Deep: Important principles are covert and abstract and not detectable when looking at surface properties

Adapted from [Fel04], p. 91

The real-time system model, presented in Chap. 4, is such a generic model that gives guidance on building understandable systems. Simplicity is achieved by adhering to the following seven design principles:

(i) *Principle of Abstraction:* The introduction of a component (a hardware/software unit) as a basic structural and computational unit makes it possible to use the component on the basis of its precise interface specifications without any need to understand the internals of the component operation. In order to maintain the abstraction of a component even in the case that faults are occurring, a component should be a fault-containment unit (Sect. 6.1.1). If components stand in a hierarchical relationship to each other, different *levels of abstraction* can be distinguished. At a high level of abstraction, the behavior of a complete autonomous *constituent system* (consisting of many clusters of components) of a system of systems (SoS) is captured in the precise linking interface specification of its gateway component (see Sects. 4.6 and 4.7.3).

(ii) *Principle of Separation of Concerns:* This principle helps to build simple systems by disentangling functions that are separable in order that they can be grouped in self-contained architectural units, thus generating *stable intermediate forms* [Sim81]. This principle is sometimes called *principle of partitioning* [Ses08]. An example is the strict separation of computational activities from communication activities such that the communication system and the computational components can be developed independently (Sect. 4.1.1).

(iii) *Principle of Causality:* The analytical-rational problem-solving subsystem of humans excels in reasoning along causal chains. The *deterministic behavior* of basic mechanisms makes it possible that a causal chain between a cause and the consequent effect can be established without a doubt (Sect. 5.6).

(iv) *Principle of Segmentation:* This principle suggests that hard-to-understand behavior should be decomposed, wherever possible, into a serial behavioral structure such that a sequential step-by-step analysis of the behavior becomes possible. Each step requires only the investigation of the limited context that is of relevance at this step (Sect. 5.4).

(v) *Principle of Independence:* This principle suggests that the interdependence of architectural units (*components* or *clusters*; see Sect. 1.1) should be reduced to the necessary minimum that is required by the application. An example is the provision of a single unidirectional primitive for the communication among components such that any low-level dependency of the sender of a message on the correct operation of the receiver is eliminated by design. This principle is of paramount importance in the design of fault-tolerant systems to ensure that backpropagation of failures is avoided and the independence of failures of fault-containment units can be assumed (see fault-containment unit in Sect. 6.1.1 and flow control in Sect. 7.2.3).

(vi) *Principle of Observability:* Non-visible communication channels among architectural units pose a severe impediment for the understanding of system behavior. This can be avoided by supporting a multicast topology in the basic message-passing primitive. It is then possible to observe the external behavior of any component without a *probe* effect (Sect. 12.2).

(vii) *Principle of a Consistent Time:* The progression of real time is an important independent variable in any behavioral model of the physical subsystem of an embedded system. This principle suggests that *a global time base* should be introduced in the distributed computer system such that system-wide consistent temporal relations (e.g., simultaneity) and temporal distances among events can be established on the basis of global time-stamps (Sect. 3.3). The availability of a global time simplifies the solution of many problems in distributed systems (see Sect. 11.7).

Bibliographic Notes

The textbook by Reisberg [Rei10] gives a good overview of the state of the art in the field of cognition and introduces many of the terms that have been used in this chapter. Epstein [Eps08] discusses the characteristics of the *intuitive-experiential subsystem* and the *analytic-rational subsystem* of problem-solving. Boulding [Bou61] elaborates extensively on the notion of *conceptual landscape* (which he calls the *image*) and the role of the *message metaphor* in all types of communication. The hierarchy of concepts, the *abstraction ladder*, is taken from Hayakawa [Hay90]. The relational complexity theory of Halford [Hal96] establishes limits for the rational reasoning capability of humans, while Miller [Mil56] elaborates on the limits of the human short-term memory capacity. Popper [Pop68] and Edmonds [Edm00] discuss the relevance and limitations of model building for understanding physical systems. The book by Bedau [Bed08] is devoted to the topic of emergence. The comparison of simple versus difficult tasks is taken from [Fel04]. The PhD thesis by Rumpler [Rum08] deals with *design comprehension* of embedded real-time systems. Roger Session's book [Ses08] *Simple Architectures for Complex Enterprises* contains practical guidelines for designing understandable enterprise information architectures.

Points to Remember

- Humans have two quite different mental subsystems for solving problems: the *intuitive-experiential subsystem* and the *analytic-rational subsystem*.
- The *experiential subsystem* is a preconscious emotionally based subsystem that operates holistically, automatically, and rapidly and demands minimal cognitive resources for its execution.
- The *rational subsystem* is a conscious analytic subsystem that operates according to the laws of logic. It is well equipped to handle deterministic relations and *causality*.
- Adult humans have a conscious *explicit model* of reality in their rational subsystem, in addition to their *implicit model* of reality in the experiential subsystem.

These two models of reality coincide to different degrees and form jointly the *conceptual landscape* of an individual.

- Knowledge is acquired by the process of *abstraction*, by which the particular is subordinated to the general, so that what is known about the general is applicable to many particulars.
- A *concept* is a category that is augmented by a *set of beliefs* about its relations to other categories. The set of beliefs relates a *new concept* to already *existing concepts* and provides for an *implicit theory* (a subjective mental model).
- *Understanding* means that the concepts and relationships that are employed in the representation of a scenario have been adequately linked with the conceptual landscape and the methods of reasoning of the observer. *The tighter the links are, the better is the understanding. Understanding* (and therefore *simplicity*) is thus a *relation* between an observer and a scenario, *not* a *property* of the scenario.
- The elapsed time needed to understand a model by an intended observer is a reasonable measure for the cognitive effort and thus for the *complexity* of a *model relative to the observer*.
- *Complexity* can only be assigned to models of *physical systems*, but not to the physical systems themselves, no matter whether these physical systems are natural or man-made.
- The complexity of a large system depends on the number and complexity of the models that must be comprehended in order to understand the complete system. The time it takes to understand all these models can be considered as a measure for the *cognitive complexity of a large system*.
- Invisible information flows between *identified subsystems* pose a considerable barrier for understanding.
- The resources in the rational problem-solving subsystem of humans, both in storage and processing capacity, are limited.
- The four strategies to simplify a complex scenario in order that it can be processed by the limited cognitive capabilities of humans are *abstraction, partitioning, isolation,* and *segmentation*.
- The formation of concepts is governed by the following two principles the *principle of utility* and the *principle of parsimony* (also called *Occam's razor*).
- The *essence of a concept, i.e., the semantic content of a concept*, associated with a *name*, can be assumed to be the *same* within a natural language community (*denotation*), but different individuals may associate different *shades of meaning* with a concept (*connotation*).
- A *variable* is a *language construct* that assigns an *attribute* to a *concept* at the given *instant*. A variable thus consists of two parts, a *fixed part*, the *variable name*, and a *variable part* called the *value of the variable* that is assigned to the variable at a particular instant.
- Differences in the representations of the *semantic content* of a variable become important when we look at *gateway components* which link two subsystems that have been developed by two different organizations according to two different *architectural styles*.

- A *model* is a deliberate simplification of reality with the objective of explaining a chosen property of reality that is relevant for a particular purpose.
- If the purpose of a model is not crystal clear or if there are multiple divergent purposes to satisfy, it is not possible to develop a *simple* model.
- The recursive application of the *principles of abstraction* leads to such a hierarchy of models. More general models can be formed by *abstraction* and more concrete models can be formed by *refinement*.
- The major challenge of design is the building of a software/hardware artifact (an embedded computer system) that provides the intended behavior (i.e., the service) under given constraints and where relevant properties of this artifact (e.g., the behavior) can be modeled at different levels of abstraction by models of adequate simplicity.
- We talk about *emergence* when the interactions of subsystems give rise to unique global properties at the system level that are not present at the level of the subsystems. Emergent properties are irreducible, holistic, and novel—they disappear when the system is partitioned into its subsystems.
- We classify a system as *complex* if we are not in the position to develop a set of models of *adequate simplicity*—commensurate to the rational capabilities of the human mind—to explain the structure and behavior of the system.

Review Questions and Problems

2.1. What are the distinguishing characteristics of the *intuitive-experiential* and the *analytic-rational* subsystems for human problem-solving?

2.2. Give concrete examples for typical tasks for the *intuitive-experiential* and the *analytic-rational* problem-solving subsystems!

2.3. How is a *concept* defined? What are the principles that guide *concept formation*? What is the *conceptual landscape*? What are basic-level concepts?

2.4. What is characteristic for a *domain expert*?

2.5. What do we mean when we say we *understand a scenario*? How is *cognitive complexity* defined? Give an example of a *barrier to understanding*!

2.6. Which are known *simplification strategies*?

2.7. What are the characteristics of scientific concepts?

2.8. What is the concept of a message? What is a *protocol*?

2.9. What is the *semantic content* of a variable? What is the relationship between *the representation of a variable* and its *semantic content*?

2.10. What is the essence of model building?

2.11. Explain the *four-universe model* of a computer system!

2.12. What makes a task *simple* or *complex*?

2.13. What do we mean by *emergence*? What are *prior* and *derivative properties*?

2.14. What are the advantages and disadvantages of message-based inter-IP core communication on an MP-SoC (multiprocessor system-on-chip)?

Chapter 3
Global Time

Overview

This chapter starts in Sect. 3.1 with a general discussion on time and order. The notions of causal order, temporal order, and delivery order and their interrelationships are elaborated. The parameters that characterize the behavior and the quality of a digital clock are investigated. Section 3.2 proceeds along the positivist tradition by introducing an omniscient external observer with an absolute reference clock that can generate precise time-stamps for all relevant events. These absolute time-stamps are used to reason about the precision and accuracy of a global time base and to expose the fundamental limits of time measurement in a distributed real-time system.

In Sect. 3.3, the model of a *sparse time base* is introduced to establish a consistent view of the order of computer-generated events in a distributed real-time system without having to execute an agreement protocol. The *cyclic model of time* presented in this section is well suited to deal with the progression of time in cyclic systems, such as in many control and multimedia systems.

The topic of internal clock synchronization is covered in Sect. 3.4. First, the notions of convergence function and drift offset are introduced to express the synchronization condition that must be satisfied by any synchronization algorithm. Then, the simple central master algorithm for clock synchronization is presented, and the precision of this algorithm is analyzed. Section 3.4.3 deals with the more complex issue of fault-tolerant distributed clock synchronization. The jitter of the communication system is a major limiting factor that determines the precision of the global time base.

The topic of external synchronization is studied in Sect. 3.5. The role of a time gateway and the problem of faults in external synchronization are discussed. Finally, the network time protocol (NTP) of the Internet, the time format of the IEEE 1588 clock synchronization protocol, and the time format of the TTA are presented.

3.1   Time and Order

Applying the principles of *utility* and *parsimony* (Sect. 2.2.1), we base our model of time on Newtonian physics, because the models of *Newtonian physics* are simpler than the models of *relativistic physics* and sufficient to deal with most temporal phenomena in embedded systems. In many engineering disciplines (e.g., Newtonian mechanics), time is introduced as an independent variable that determines the sequence of states of a system. The basic constants of physics are defined in relation to the standard of time, the physical second. This is why the global time base in a cyber-physical real-time system should be based on the metric of the physical second.

In a typical real-time application, the distributed computer system performs a multitude of different functions concurrently, e.g., the monitoring of real-time (RT) entities (both their value and rate of change), the detection of alarm conditions, the display of the observations to the operator, and the execution of control algorithms to find new set points for many distinct control loops. These diverse functions are normally executed at different nodes. In addition, replicated nodes are introduced to provide fault tolerance by active redundancy. To guarantee a consistent behavior of the entire distributed system, it must be ensured that all nodes process all events in the same consistent order, preferably in the same temporal order in which the events occurred (see also the example in Sect. 5.5) in the controlled object. A proper global time base helps to establish such a consistent temporal order on the basis of the time-stamps of the events.

3.1.1   Different Orders

Temporal Order The continuum of Newtonian real time can be modeled by a *directed timeline* consisting of an infinite set {T} of *instants* (or *points in time*) with the following properties [Wit90, p. 208]:

(i) {T} is an ordered set, that is, if *p* and *q* are any two instants, then either *p* is simultaneous with *q*, or *p* precedes *q*, or *q* precedes *p*, where these relations are mutually exclusive. We call the order of instants on the timeline the *temporal order*.
(ii) {T} is a dense set. This means that there is at least one *q* between *p* and *r* *iff p is not the same instant as r*, where *p*, *q*, and *r* are instants.

A section of the timeline between two different instants is called a *duration*. In our model, an *event* takes place at an instant of time and does not have a duration. If two events occur at the *same* instant, then the two events are said to occur *simultaneously*. Instants are totally ordered; however, events are only partially ordered, since simultaneous events are not in the order relation. Events can be totally ordered if another criterion is introduced to order events that occur simultaneously, e.g., in a distributed computer system, the number of the node at which the event occurred can be used to order events that occur simultaneously [Lam78].

Causal Order In many real-time applications, the causal dependencies among events are of interest. The computer system must assist the operator in identifying the *primary event* of an *alarm shower* (see Sect. 1.2.1). Knowledge of the exact temporal order of the events is helpful in identifying this primary event. If an event *e1* occurs after an event *e2*, then *e1* cannot be the cause of *e2*. If, however, *e1* occurs before *e2*, then it is possible, but not certain, that *e1* is the cause of *e2*. The *temporal* order of two events is necessary, but not sufficient, for their *causal* order. Causal order is *more* than temporal order.

Reichenbach [Rei57, p. 145] defined causality by a mark method without reference to time: If event *e1* is a cause of event *e2*, then a small variation (a mark) in *e1* is associated with small variation in *e2*, whereas small variations in *e2* are not necessarily associated with small variations in *e1*.

> **Example**: Suppose there are two events *e1* and *e2*:
>
> *e1* Somebody enters a room.
> *e2* The microwave beeps.
> Consider the following two cases:
> (i) *e2* occurs after *e1*.
> (ii) *e1* occurs after *e2*.
> In both cases the two events are temporally ordered. However, while it is unlikely that there is a causal order between the two events of case (i), it is likely that such a causal order exists between the two events of case (ii), since the person might enter the room to enjoy her lunch.

If the (partial) temporal order between alarm events has been established, it is possible to exclude an event from being the *primary event* if it *definitely occurred later* than another alarm event. Subsequently, we will show that a precise global time base helps to determine the event set that is in this *definitely-occurred-later-than* relation (see also the example in 1.2.1).

Delivery Order A weaker order relation that is often provided by distributed communication systems is a consistent *delivery order*. The communication system guarantees that all nodes see a defined *set of related events* in the same delivery order. This delivery order is not necessarily related to the temporal order of event occurrences or the causal relationship between events. Some distributed algorithms, e.g., *atomic broadcast algorithms*, establish a consistent delivery order.

3.1.2 Clocks

In ancient history, the measurement of durations between events was mainly based on subjective judgment. With the advent of modern science, objective methods for measuring the progression of time by using *physical clocks* have been devised.

Digital Physical Clock   A *(digital physical) clock* is a device for measuring time. It contains a *counter* and a *physical oscillation mechanism* that periodically generates an event to increase the counter. The periodic event is called the *microtick* of the clock. (The term *tick* is introduced in Sect. 3.2.1 to denote the events generated by the global time.)

Granularity   The duration between two consecutive microticks of a digital physical clock is called a *granule* of the clock. The granularity of a given clock can be measured only if there is a clock with a finer granularity available. The granularity of any digital clock leads to a digitalization error in time measurement.

There also exist analog physical clocks, e.g., *sundials* that do not have granularity. In the following, we only consider digital physical clocks.

In subsequent definitions, we use the following notation: clocks are identified by natural numbers $1, 2, ..., n$. If we express properties of clocks, the property is identified by the clock number as a superscript with the microtick or tick number as a subscript. For example, microtick i of clock k is denoted by $microtick_i^k$.

Reference Clock   Assume an *omniscient external observer* who can observe all events that are of interest in a given context (remember that relativistic effects are disregarded). This observer possesses a *unique reference clock z* with frequency f^z, which is in perfect agreement with the international standard of time. The counter of the reference clock is always the same as that of the international time standard. We call $1/f^z$ the *granularity* g^z of clock z. Let us assume that f^z is very large, say 10^{15} microticks/second, so that the granularity g^z is 1 femtosecond (10^{-15} s). Since the granularity of the reference clock is so small, the digitalization error of the reference clock is considered a second order effect and disregarded in the following analysis.

Absolute Time-Stamp   Whenever the omniscient observer perceives the occurrence of an event e, she/he will instantaneously record the current state of the reference clock as the time of occurrence of this event e and will generate a *time-stamp* for e. *Clock*(e) denotes the time-stamp generated by the use of a given *clock* to time-stamp an event e. Because z is the single reference clock in the system, $z(e)$ is called the *absolute time-stamp* of the event e.

The *duration* between two events is measured by counting the microticks of the reference clock that occur in the interval between these two events. The *granularity* g^k of a given clock k can now be measured and is given by the nominal number n^k of microticks of the reference clock z between two microticks of this clock k.

The temporal order of events that occur between any two consecutive microticks of the reference clock, i.e., within the granularity g^z, cannot be reestablished from their absolute time-stamps. This is a fundamental limit in time measurement.

Clock Drift   The *drift* of a physical clock k between microtick i and microtick $i + 1$ is the frequency ratio between this clock k and the reference clock, at the instant of microtick i. The drift is determined by measuring the duration of a granule of clock

k with the reference clock z and dividing it by the nominal number n^k of reference clock microticks in a granule:

$$\text{drift}_i^k = \frac{z\left(\text{microtick}_{i+1}^k\right) - z\left(\text{microtick}_i^k\right)}{n^k}$$

Because a good clock has a drift that is very close to 1, for notational convenience the notion of a *drift rate* ρ_i^k is introduced as

$$\rho_i^k = \left| \frac{z\left(\text{microtick}_{i+1}^k\right) - z\left(\text{microtick}_i^k\right)}{n^k} - 1 \right|$$

A perfect clock will have a drift rate of 0. Real clocks have a varying drift rate that is influenced by environmental conditions, e.g., a change in the ambient temperature, a change in the voltage level that is applied to a crystal oscillator, or aging of the crystal. Within specified environmental parameters, the drift rate of an oscillator is bounded by the *maximum drift rate* ρ_{max}^k, which is documented in the data sheet of the resonator. Typical maximum drift rates ρ_{max}^k are in the range of 10^{-2}–10^{-7} sec/sec, or better, depending on the quality (and price) of the oscillator. Because every clock has a nonzero drift rate, *free-running* clocks, i.e., clocks that are never resynchronized, leave any bounded relative time interval after a finite time, even if they are fully synchronized at startup.

> **Example**: During the Gulf War on February 25, 1991, a Patriot missile defense system failed to intercept an incoming Scud rocket. The clock drift over a 100-h period (which resulted in a tracking error of 678 m) was blamed for the Patriot missing the Scud missile that hit an American military barrack in Dhahran, killing 29 and injuring 97. The original requirement was a 14-h mission. The clock drift during a 14-h mission could be handled [Neu95, p. 34].

Failure Modes of a Clock   A physical digital clock can exhibit two types of failures. The counter could be mutilated by a fault so that the counter value becomes erroneous, or the drift rate of the clock could depart from the specified drift rate (the shaded area of Fig. 3.1) because the clock starts ticking faster (or slower) than specified.

3.1.3   Precision and Accuracy

Offset   The offset at microtick i between two clocks j and k with the same granularity is defined as

Fig. 3.1 Failure modes of
a physical clock

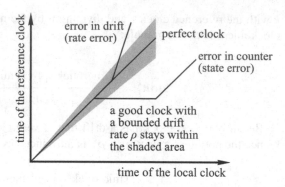

$$\text{offset}_i^{jk} = \left| z\left(\text{microtick}_i^j\right) - z\left(\text{microtick}_i^k\right) \right|$$

The offset denotes the time difference between the respective microticks of the two clocks, measured in the number of microticks of the reference clock.

Precision Given an ensemble of n clocks $\{1, 2, ..., n\}$, the maximum offset between any two clocks of the ensemble

$$\Pi_i = \max_{\forall j,k:1 \leq j,k \leq n} \left\{\text{offset}_i^{jk}\right\}$$

is called the *precision* Π_i of the ensemble at microtick i. The maximum of Π_i over an *interval of interest* is called the *precision* Π of the ensemble. The precision denotes the *maximum offset* of respective microticks of any two clocks of the ensemble during a duration of interest. The precision is expressed in the number of microticks of the reference clock.

Because of the drift rate of any physical clock, the clocks of an ensemble will drift apart if they are not resynchronized periodically (i.e., brought closer together). The process of mutual resynchronization of an ensemble of clocks to maintain a bounded precision is called *internal synchronization*.

Accuracy The offset of clock k with respect to the reference clock z at microtick i is called the accuracy$_i^k$. The maximum offset over all microticks i that is *of interest* *is* called the accuracy^k of clock k. The accuracy denotes the maximum offset of a given clock from the external time reference during a duration of interest.

To keep a clock within a bounded interval of the reference clock, it must be periodically resynchronized with an external time reference. This process of resynchronization of a clock with an external time reference is called *external synchronization*.

If all clocks of an ensemble are externally synchronized with an accuracy A, then the ensemble is also internally synchronized with a precision of at most $2A$. The converse is not true. An ensemble of internally synchronized clocks will drift from the external time if the clocks are never resynchronized with the external time base.

3.1.4   Time Standards

In the last decades, a number of different time standards have been proposed to measure the time difference between any two events and to establish the position of an event relative to some commonly agreed origin of a time base, the *epoch*. Two of these time bases are relevant for the designer of a distributed real-time computer system, the *International Atomic Time* (TAI) and the *Universal Time Coordinated* (UTC).

International Atomic Time (TAI—Temps Atomique Internationale) The need for a time standard that can be generated in a laboratory gave birth to the International Atomic Time (TAI). TAI defines the second as the duration of 9,192,631,770 periods of the radiation of a specified transition of the cesium atom 133. The intention was to define the duration of the TAI second so that it agrees with the second derived from astronomical observations. TAI is a *chronoscopic* timescale, i.e., a timescale without any discontinuities (e.g., leap seconds). The epoch of TAI starts on January 1, 1958, 00:00 hours Greenwich Mean Time (GMT). The time base of the global positioning system (GPS) is based on TAI with the epoch starting on January 6, 1980, at 00:00 hours.

Universal Time Coordinated (UTC) UTC is a time standard that has been derived from astronomical observations of the rotation of the earth relative to the sun. It is the basis for the time on the *wall clock*. However, there is a known offset between the local wall-clock time and UTC determined by the time zone and by the political decisions about when daylight savings time must be used. The UTC time standard was introduced in 1972, replacing the Greenwich Mean Time (GMT) as an international time standard. Because the rotation of the earth is not smooth, but slightly irregular, the duration of the GMT second changes slightly over time. In 1972, it was internationally agreed that the duration of the second should conform to the TAI standard and that the number of seconds in an hour would have to be modified occasionally by inserting a *leap second* into the UTC to maintain synchrony between the UTC (wall-clock time) and astronomical phenomena, like day and night. Because of this leap second, the UTC is not a chronoscopic timescale, i.e., it is not free of discontinuities. It was agreed that on January 1, 1958, at midnight, both the UTC and the TAI had the same value. Since then the UTC has deviated from TAI by about *30* s. The point in time when a leap second is inserted into the UTC is determined by the Bureau International de l'Heure and publicly announced, so that the current offset between the UTC and the TAI is always known.

Example: In *Software Engineering Notes* of March 1996 [Pet96, p. 16] was the following story:

Ivan Peterson reported on a problem that occurred when a leap second was added at midnight on New Year's Eve 1995. The leap second was added, but the date inadvertently advanced to Jan. 2. Ivars heard from a source at AP radio that the synchronization of their broadcast networks depends on the official time signal, and this glitch affected their operation for several hours until the problem was corrected. You can't even count on the national timekeepers to get it right all the time.

Bob Huey responded that making corrections at midnight is obviously risky: (1) The day increments to January 1, 1996, 00:00:00. (2) You reset the clock to 23:59:59, back one second. (3) The clock continues running. (4) The day changes again, and it is suddenly January 2, 1996, 00:00:00. No wonder they had problems.

3.2   Time Measurement

If the real-time clocks of all nodes of a distributed system were perfectly synchronized with the reference clock z and all events were time-stamped with this reference time, then it would be easy to measure the interval between any two events or to reconstruct the temporal order of events, even if variable communication delays generated differing delivery orders. In a loosely coupled distributed system where every node has its own local oscillator, such a tight synchronization of clocks is not possible. A weaker notion of a universal time reference, the concept of *global time*, is therefore introduced into a distributed system.

3.2.1   Global Time

Suppose a set of nodes exists, each one with its own local physical clock c^k that ticks with granularity g^k. Assume that all of the clocks are internally synchronized with a precision Π, i.e., for any two clocks j, k, and all microticks i:

$$\left| z\left(\text{microtick}_i^j\right) - z\left(\text{microtick}_i^k\right) \right| < \Pi.$$

(In Sect. 3.4, methods for the internal synchronization of the clocks are presented.) It is then possible to select a *subset of the microticks* of each local clock k for the generation of the local implementation of a global notion of time. We call such a selected local microtick i a *macrotick* (or a *tick*) of the global time. For example, every tenth microtick of a local clock k may be interpreted as the global tick, the *macrotick* t_i^k, of this clock (see Fig. 3.2). If it does not matter at which clock k the (macro)tick occurs, we denote the tick t_i without a superscript. A global time is thus an *abstract notion* that is *approximated* by properly selected microticks from the synchronized local physical clocks of an ensemble.

Reasonableness Condition The global time t is called *reasonable*, if all local implementations of the global time satisfy the condition.

$$g > \Pi$$

the *reasonableness condition* for the global granularity g. This reasonableness condition ensures that the synchronization error is *bounded* to less than one *macrogranule*, i.e., the duration between two (macro)ticks. If this reasonableness condition is satisfied, then for a single event e, that is observed by any two different clocks of the ensemble,

$$\left| t^j(e) - t^k(e) \right| \le 1,$$

i.e., the global time-stamps for a single event can differ by at most one tick. *This is the best we can achieve.* Because of the impossibility of synchronizing the clocks perfectly and the granularity of any digital time, there is always the possibility of the following sequence of events: clock j ticks, event e occurs, and clock k ticks. In such a situation, the single event e is time-stamped by the two clocks j and k with a difference of one tick (Fig. 3.2).

One-Tick Difference—What Does It Mean? What can we learn about the temporal order of two events, observed by different nodes of a distributed system with a reasonable global time, given that the global time-stamps of these two events differ by one tick?

In Fig. 3.3, four events are depicted, *event 17*, *event 42*, *event 67*, and *event 69* (time-stamps from the reference clock). Although the duration between *event 17* and *event 42* is *25* microticks and the duration between *event 67* and *event 69* is only 2 microticks, both durations lead to the *same* measured difference of one macrogranule. The global time-stamp for *event 69* is *smaller* than the global time-stamp for *event 67*, although *event 69* occurred *after event 67*. Because of the accumulation of the synchronization error and the digitalization error, it is not possible to reconstruct the temporal order of two events from the knowledge that the global time-stamps differ by one tick. However, if the time-stamps of two events differ by two ticks, then the temporal order can be reconstructed because the sum of the synchronization and digitalization error is always less than *two* granules in a clocking system with a reasonable global time base.

This fundamental limitation in time measurement limits the *faithfulness* of the digital computer model of a controlled physical subsystem. The time base in the

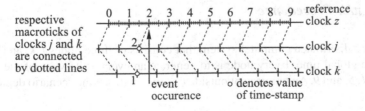

Fig. 3.2 Time-stamps of a single event

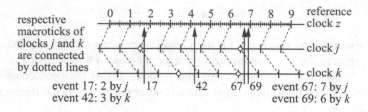

Fig. 3.3 Temporal order of two events with a difference of one tick

physical part of a cyber-physical system is dense, while the time base in the computer system is discrete. Whenever two events in the physical subsystem occur close together, compared to the granularity of the global time, it is not possible to reconstruct the physical temporal order of the events in the computer system faithfully. The only way out of this dilemma is the provision of a global time base with a smaller granularity, such that temporal errors are reduced [Kop09].

3.2.2   Interval Measurement

An interval is delimited by two events, the start event of the interval and the terminating event of the interval. The measurement of these two events relative to each other can be affected by the synchronization error and the digitalization error. The sum of these two errors is *less than 2g* because of the reasonableness condition, where g is the granularity of the global time. It follows that the true duration d_{true} of an interval is bounded by

$$(d_{obs} - 2g) < d_{true} < (d_{obs} + 2g)$$

where d_{obs} is the observed difference between the start event and the terminating event of the interval. Figure 3.4 depicts how the observed duration of an interval of length 25 microticks can differ, depending on which node observes the start event and the terminating event. The global tick, assigned by an observing node to an event delimiting the interval, is marked by a small circle in Fig. 3.4.

3.2.3   π/Δ-Precedence

Consider a distributed computer system that consists of three nodes j, k, and m that support a global time. Every node is to generate an event at its view of the global instants *1*, *5*, and *9*. An omniscient outside observer will see the scenario depicted in Fig. 3.5.

Fig. 3.4 Errors in interval measurement

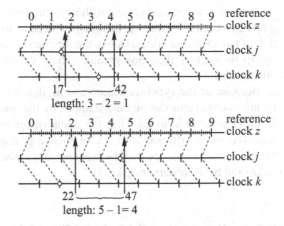

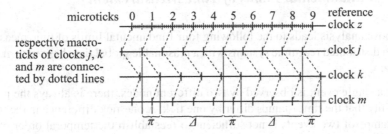

Fig. 3.5 π/Δ precedence

All events that are generated locally at the same global clock tick will occur within a small interval π, where $\pi \leq \Pi$, the precision of the ensemble (because of the reasonableness condition). Events that occur at different ticks will be at least Δ apart (Fig. 3.5). The outside observer should not order the events that occur within π, because these events are *supposed* to occur at the same instant. Events that occur at different ticks should be ordered. How many granules of silence must exist between the event subsets such that an outside observer or another cluster will always recover the temporal order intended by the sending cluster? Before we can answer this question (in Sect. 3.3.2), we must introduce the notion of π/Δ precedence.

Given a set of events {E} and two durations π and Δ where $\pi < < \Delta$, such that for any two elements e_i and e_j of this set, the following condition holds:

$$\left[\left|z(e_i)-z(e_j)\right| \leq \pi\right] \vee \left[\left|z(e_i)-z(e_j)\right| > \Delta\right]$$

where z is the reference clock. Such an event set is called π/Δ-*precedent.* π/Δ-*Precedence* means that a subset of the events that happen at about the same time (and that are therefore close together within π) is separated by a substantial interval (at least Δ) from the elements in another subset. If π is zero, then any two events of the *0/Δ-precedent* event set occur either at the same instant or are at least a duration Δ apart.

Assume a distributed system with a reasonable global time base with granularity g and two events, $e1$ and $e2$, that are produced at the same locally generated global tick of two different nodes. Due to the synchronization error, these events can differ by up to but less than one granule. These events are observed by some of the other nodes.

Because of the synchronization and digitalization error, the two (simultaneous by intention) events can be time-stamped by the observers with two-tick difference. In order to be able to establish the intended temporal order of events from their time-stamps, a sufficient duration of silence is needed before the next event may occur in order to ensure that the intended simultaneity of the events can always be recovered by all observers [Ver94].

3.2.4   Fundamental Limits of Time Measurement

The above analysis leads to the following four fundamental limits of time measurement in distributed real-time systems with a reasonable global time base with granularity g:

 (i) If a single event is observed by two different nodes, there is always the possibility that the time-stamps differ by one tick. A one-tick difference in the time-stamps of two events is not sufficient to reestablish the temporal order of the events from their time-stamps.
 (ii) If the observed duration of an interval is d_{obs}, then the true duration d_{true} is bounded by

$$\left(d_{obs} - 2g\right) < d_{true} < \left(d_{obs} + 2g\right)$$

 (iii) The temporal order of events can be recovered from their time-stamps if the difference between their time-stamps is equal to or greater than *two* ticks.
 (iv) The temporal order of events can *always* be recovered from their time-stamps, if the event set is at least *0/3g* precedent.

These fundamental limits of time measurement are also the *fundamental limits to the faithfulness* of a digital model of a physical system.

3.3   Dense Time Versus Sparse Time

Example: It is known a priori that a particular train will arrive at a train station every hour. If the train is always on time and all clocks are synchronized, it is possible to uniquely identify each train by its time of arrival. Even if the train is slightly off, say, by 5 min, and the clocks are slightly out of synchronization, say, by 1 min, there will be no problem in uniquely identifying a train by its time of arrival. What are the limits within which a train can still be uniquely identified by its time of arrival?

Assume a set {E} of events that are of interest in a particular context. This set {E} could be the ticks of all clocks, or the events of sending and receiving messages. If these events are allowed to occur at any instant of the timeline, then we call the time base *dense*. If the occurrence of these events is restricted to some *active intervals* of duration ε, with an interval of silence of duration Δ between any two active intervals, then we call the time base ε/Δ-*sparse*, or simply *sparse* for short (Fig. 3.6). If a system is based on a sparse time base, there are time intervals during which no significant event is allowed to occur. Events that occur only in the active intervals are called *sparse events*.

It is evident that the occurrences of events can only be restricted if the given system has the authority to control these events, i.e., these events are in the sphere of control of the computer system [Dav79]. The occurrence of events outside the sphere of control of the computer system cannot be restricted. These external events are based on a dense time base and cannot be forced to be *sparse events*.

Example: Within a distributed computing system, the sending of messages can be restricted to some intervals of the timeline and can be forbidden at some other intervals—they can be designed to be *sparse events*.

3.3.1 Dense Time Base

Suppose that we are given two events *e1* and *e2* that occur on a dense time base. If these two events are closer together than *3g*, where *g* is the granularity of the global time, then it is not always possible to establish the temporal order, or even a consistent order of these two events on the basis of the time-stamps generated by the different nodes if no *agreement protocol* (see below) is applied.

Example: Consider the scenario of Fig. 3.7 with two events, *e1* and *e2* which are 2.5 granules apart. Event *e1* is observed by node *j* at time 2 and by node *m* at time *1*, while *e2* is only

Fig. 3.6 Sparse time base

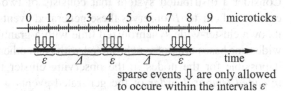

sparse events ⇩ are only allowed
to occure within the intervals ε

Fig. 3.7 Different observed order of two events *e1* and *e2*

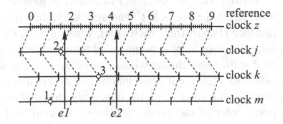

observed by node k that reports its observation "*e2* occurred at *3*" to node j and node m. Node j calculates a time-stamp difference of one tick and concludes that the events occurred at about the same time and *cannot* be ordered. Node m calculates a time-stamp difference of *two* ticks and concludes that *e1 has definitely occurred* before *e2*. The two nodes j and m have an *inconsistent* view about the order of event occurrence.

Agreement Protocol To arrive at a *consistent view* of the order of *non-sparse events* within a distributed computer system (which does not necessarily reflect the temporal order of event occurrence), the nodes must execute an *agreement protocol*. The first phase of an agreement protocol requires an information interchange among the nodes of the distributed system with the goal that every node acquires the differing local views about the state of the world from every other node. In the fault-free case, at the end of this first phase, every correct node possesses exactly the same information as every other node. In the second phase of the agreement protocol, each node applies a deterministic algorithm to this consistent information to reach the same conclusion about the assignment of the event to an active interval of the sparse time base—the commonly agreed value. In case faulty nodes have to be tolerated, an agreement algorithm requires additional round(s) of information exchange as well as the resources for executing the agreement algorithm.

Agreement algorithms are costly, both in terms of communication requirements and processing requirements and—worst of all—in terms of the additional delay they introduce into a control loop. It is therefore expedient to look for solutions to the consistent temporal ordering problem in distributed computer systems that do not require these additional overheads. The sparse time model, introduced below, provides for such a solution.

3.3.2 Sparse Time Base

Consider a distributed system that consists of two clusters: cluster A generates events, and cluster B observes these generated events. Each one of the clusters has its own cluster-wide synchronized time with a granularity g, but these two cluster-wide time bases are not synchronized with each other. Under what circumstances is it possible for the nodes in the observing cluster to reestablish consistently the *intended temporal order* of the generated events without the need to execute an agreement protocol?

If two nodes, nodes j and k of cluster A, generate two events at the same cluster-wide tick t_i, i.e., at tick t_i^j and at tick t_i^k, then these two events can be, at most, a distance Π apart from each other, where $g > \Pi$, the granularity of the cluster-wide time. Because there is no intended temporal order among the events that are generated at the same cluster-wide tick of cluster A, the observing cluster B should *never* establish a temporal order among the events that have occurred at about the same time. On the other hand, the observing cluster B should *always* reestablish the temporal order of the events that have occurred at different cluster-wide ticks. Is it sufficient if cluster *A* generates a *1g/3g* precedent event set, i.e., after every cluster-wide tick at which events are allowed to be generated, there will be silence for at least three granules?

If cluster *A* generates a *1g/3g* precedent event set, then it is possible that two events that are generated at the same cluster-wide granule at cluster *A* will be time-stamped by cluster *B* with time-stamps that differ by *two* ticks. The observing cluster *B* should not order these events (although it could), because they have been generated at the same cluster-wide granule. Events that are generated by cluster *A* at different cluster-wide granules (*3g* apart) and therefore should be ordered by cluster *B* could also obtain time-stamps that differ by *two* ticks. Cluster *B* cannot decide whether or not to order events with a time-stamp difference of *two* ticks. To resolve this situation, cluster *A* must generate a *1g/4g* precedent event set. Cluster *B* will not order two events if their time-stamps differ by ≤*2* ticks, but will order two events if their time-stamps differ by ≥*3* ticks, thus reestablishing the temporal order that has been intended by the sender.

3.3.3 Space-Time Lattice

The ticks of the global clock can be seen as generating a space-time lattice, as depicted in Fig. 3.8. A node is allowed to generate an event (e.g., send a message) at the filled dots and must be silent at the empty dots. This rule makes it possible for the receiver to establish a consistent temporal order of events without executing an agreement protocol. Although a sender is allowed to generate an event only at the filled dots, this is still much faster than executing an agreement protocol, provided a global time base of sufficient precision is available. Events that are generated at the filled dots of the sparse time lattice are called *sparse events*.

Events that occur outside the sphere of control of the computer system cannot be confined to a sparse time base: they happen on a dense time base and are therefore not *sparse events*. To generate a consistent view of events that occur in the controlled object and that are observed by more than one node of the distributed computer system, the execution of an agreement protocol is unavoidable at the interface between the computer system and the controlled object or other systems that do not participate in the global time. Such an agreement protocol transforms a *non-sparse event* into a *sparse event*.

Fig. 3.8 Sparse time base

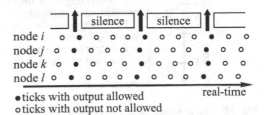

node *i* o • o o o • o o o • o o
node *j* o • o o o • o o o • o o
node *k* o • o o o • o o o • o o
node *l* o • o o o • o o o • o o

• ticks with output allowed
o ticks with output not allowed

real-time

3.3.4   Cyclic Representation of Time

Many processes in the technical and biological world are cyclic [Win01]. A cyclic process is characterized by a regular behavior, where a similar set of action patterns is repeated in every cycle.

> **Example**: In a typical control system, real time is partitioned into a sequence of control cycles (Fig. 3.9). Every control cycle starts with reading the state variables of the controlled object, proceeds with the execution of the control algorithm, and finishes with the output of new set points to the actuators at the interface between the computer system and the controlled object.

In the cyclic representation of time, the linear time is partitioned into cycles of equal duration. Every cycle is represented by a circle, where an instant within a cycle is denoted by the *phase*, i.e., the angular deviation of the instant from the beginning of the cycle. *Cycle* and *phase* thus denote an instant in a cyclic representation. In the cyclic representation of sparse time, the circumference of the circle is not a *dense* line, but a *dotted* line, where the size and the distance between dots are determined by the precision of the clock synchronization.

A sequence of consecutive processing and communication actions, such as the actions in Fig. 3.9, are *phase-aligned*, if the termination of one action is immediately followed by the start of the next consecutive action. If the actions within a *RT transaction* (see Sect. 1.7.3) are phase-aligned, then the overall duration of the RT transactions is minimized.

If we look at Fig. 3.9, we see that communication services in a typical control loop are periodically required only in the intervals B and D of a cycle. The *shorter* these intervals B and D, the *better*, since the dead time of the control loop is reduced. This requirement leads to the model of *pulsed data streams*, where, in a time-triggered system, the highest possible bandwidth is allocated periodically in the intervals B and D, while, during the rest of the cycle, the communication bandwidth can be allocated to other requests [Kop06].

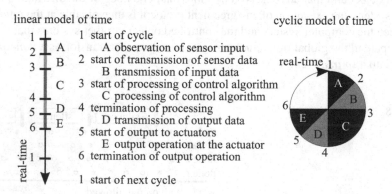

Fig. 3.9 Linear versus cyclic representation of time in a control system

An extension of the cyclic representation is the *spiral representation of time*, where a third axis is introduced to depict the linear progression of the cycles.

3.4 Internal Clock Synchronization

The purpose of internal clock synchronization is to ensure that the global ticks of all correct nodes occur within the specified *precision* Π, despite the varying drift rate of the local real-time clock of each node. Because the availability of a proper global time base is crucial for the operation of a distributed real-time system, the clock synchronization should not depend on the correctness of a single clock, i.e., it should be fault-tolerant.

Every node of a distributed system has a local oscillator that (micro)ticks with a frequency determined by the physical parameters of the oscillator. A subset of the local oscillator's microticks called the ticks (or macroticks—see Sect. 3.2.1) is interpreted as the global time ticks at the node. These global time ticks increment the local node's global time counter.

3.4.1 The Synchronization Condition

The global time ticks of each node must be periodically resynchronized within the ensemble of nodes to establish a global time base with specified precision. The period of resynchronization is called the *resynchronization interval* R_{int}. At the end of each resynchronization interval, the clocks are adjusted to bring them into better agreement with each other. The *convergence function* Φ denotes the offset of the time values immediately after the resynchronization. Then, the clocks drift again apart until they are resynchronized at the end of the next resynchronization interval R_{int} (Fig. 3.10). The *drift offset* Γ indicates the maximum accumulated divergence of any two good clocks from each other during the resynchronization interval R_{int}, where the clocks are free running. The drift offset Γ depends on the length of the resynchronization interval R_{int} and the maximum specified drift rate ρ of the clock:

$$\Gamma = 2\rho R_{int}$$

An ensemble of clocks can only be synchronized if the following *synchronization condition* holds between the *convergence function* Φ, the *drift offset* Γ, and the *precision* Π:

$$\Phi + \Gamma \leq \Pi$$

Assume that at the end of the resynchronization interval, the clocks have diverged so that they are at the edge of the precision interval Π (Fig. 3.10). The synchronization condition states that the synchronization algorithm must bring the clocks so

Fig. 3.10 Synchronization condition

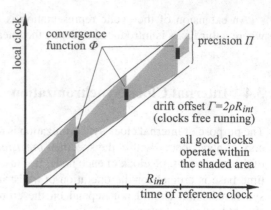

Fig. 3.11 Behavior of a malicious clock

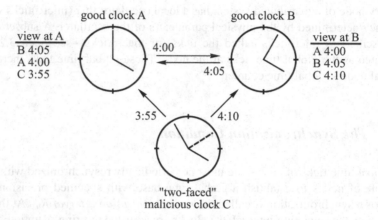

close together that the amount of divergence during the next free-running resynchronization interval will not cause a clock to leave the precision interval.

Byzantine Error The following example explains how, in an ensemble of three nodes, a malicious node can prevent the other two nodes from synchronizing their clocks since they cannot satisfy the synchronization condition [Lam82]. Assume an ensemble of three nodes and a convergence function where each of the three nodes sets its clock to the average value of the ensemble. Clocks *A* and *B* are good, while clock *C* is a malicious *two-faced* clock that disturbs the other two good clocks in such a manner that neither of them will ever correct their time value (Fig. 3.11) and will thus eventually violate the synchronization condition.

Such a malicious, *two-faced* manifestation of behavior is sometimes called a *malicious error* or a *Byzantine error* (see also Sect. 6.1.3). During the exchange of the synchronization messages, a Byzantine error can lead to inconsistent views of the state of the clocks among the ensemble of nodes. A special class of algorithms, the *interactive-consistency algorithms* [Pea80], inserts additional rounds of information exchanges to agree on a consistent view of the time values at all nodes. These additional rounds of information exchanges increase the quality of the precision at the expense of additional communication overhead. Other algorithms work with inconsistent information and establish bounds for the maximum error introduced by the inconsistency. An example of such an algorithm is the fault-tolerant average algorithm, described later in this section. It has been shown [Lam85] that clock synchronization can only be guaranteed in the presence of Byzantine errors if the total number of clocks $N \geq (3k + 1)$, where k is the number of Byzantine faulty clocks.

3.4.2 Central Master Synchronization

This is a simple non-fault-tolerant synchronization algorithm. A unique node, the central master, periodically sends the value of its time counter in synchronization messages to all other nodes, the slave nodes. As soon as a slave node receives a new synchronization message from the master, the slave records the time-stamp of message arrival. The difference between the master's time, contained in the synchronization message, and the recorded slave's time-stamp of message arrival, corrected by the known latency of the message transport, is a measure of the deviation of the clock of the master from the clock of the slave. The slave then corrects its clock by this deviation to bring it into agreement with the master's clock.

The convergence function Φ of the central master algorithm is determined by the difference between the fastest and slowest message transmission to the slave nodes of the ensemble, i.e., the *latency jitter* ε between the event of writing the synchronization time value by the master and the events of message arrival time-stamping at all slaves.

Applying the *synchronization condition*, the precision of the central master algorithm is given by

$$\Pi_{central} = \varepsilon + \Gamma$$

The central master synchronization is often used in the startup phase of a distributed system. It is simple, but not fault-tolerant, since a failure of the master ends the resynchronization, causing the free-running clocks of the slaves to leave the precision interval soon thereafter. In a variant of this algorithm, a multi-master strategy is followed: if the active master fails silently and the failure is detected by a local time-out at a *shadow master*, one of the *shadow masters* assumes the role of the master and continues the resynchronization.

3.4.3 Fault-Tolerant Synchronization Algorithms

Typically, distributed fault-tolerant clock resynchronization proceeds in three distinct phases. In the first phase, every node acquires knowledge about the state of the global time counters in all the other nodes by the exchange of messages among the nodes. In the second phase, every node analyzes the collected information to detect errors and executes the convergence function to calculate a correction value for the local global time counter. A node must deactivate itself if the correction term calculated by the convergence function is larger than the specified precision of the ensemble. Finally, in the third phase, the local time counter of the node is adjusted by the calculated correction value. Existing algorithms differ in the way in which the time values are collected from the other nodes, in the type of convergence function used, and in the way in which the correction value is applied to the local time counter.

Reading the Global Time   In a local area network, the most important term affecting the precision of the synchronization is the jitter of the time messages that carry the current time values from one node to all the other nodes. The known minimal delay for the transport of a time message between two nodes can be compensated by an a priori known delay-compensation term [Kop87] that compensates for the delay of the message in the transmission channel and in the interface circuitry. The delay jitter depends more than anything else on the system level, at which the synchronization message is assembled and interpreted. If this is done at a high level of the architecture, e.g., in the application software, all random delays caused by the scheduler, the operating system, the queues in the protocol software, the message retransmission strategy, the media-access delay, the interrupt delay at the receiver, and the scheduling delay at the receiver accumulate and degrade the quality of the time values, thus deteriorating the precision of the clock synchronization. Table 3.1 gives approximate value ranges for the jitter that can be expected at the different levels [Kop87].

Since a small jitter is important to achieve high precision in the global time, a number of special methods for jitter reduction have been proposed. Cristian [Cri89] proposed the reduction of the jitter at the application software level using a probabilistic technique: a node queries the state of the clock at another node by a query-reply transaction, the duration of which is measured by the sender. The received time value is corrected by the synchronization message delay that is assumed to be half the round-trip delay of the query-reply transaction (assuming that the delay distribution is the same in both directions). A different approach is taken in the time-triggered architecture. A special clock synchronization unit has been implemented to support the segmentation and assembly of synchronization messages at the hardware level, thereby reducing the jitter to a few microseconds. The IEEE 1588 standard for clock synchronization limits the jitter by hardware-assisted time-stamping [Eid06].

Impossibility Result   The important role of the latency jitter ε for internal synchronization is emphasized by an impossibility result by Lundelius and Lynch [Lun84]. According to this result, it is not possible to internally synchronize the clocks of an ensemble consisting of N nodes to a better precision than

$$\Pi = \varepsilon \left(1 - \frac{1}{N} \right)$$

(measured in the same units as ε) even if it is assumed that all clocks have perfect oscillators, i.e., the drift rates of all the local clocks are zero.

The Convergence Function The construction of a convergence function is demonstrated by the example of the distributed fault-tolerant average (FTA) algorithm in a system with N nodes where k Byzantine faults should be tolerated. The FTA algorithm is a one-round algorithm that works with inconsistent information and bounds the error introduced by the inconsistency. At every node, the N measured time differences between the node's clock and the clocks of all other nodes are collected (the node considers itself a member of the ensemble with time difference zero). These time differences are sorted by size. Then the k largest and the k smallest time differences are removed (assuming that an erroneous time value is either larger or smaller than the rest). The remaining $N-2k$ time differences are by definition within the precision window definition (since only k values are assumed to be erroneous and an erroneous value is larger or smaller than a good value). The average of these remaining $N-2k$ time differences is the correction term for the node's clock.

> **Example**: Figure 3.12 shows an ensemble of *seven* nodes and one tolerated Byzantine fault. The FTA takes the average of the five accepted time values shown.

The worst-case scenario occurs if all good clocks are at opposite ends of the precision window Π, and the Byzantine clock is seen at different corners by two nodes. In the example of Fig. 3.13, node j will calculate an average value of $4\Pi/5$, and node k will calculate an average value of $3\Pi/5$; the difference between these two terms, caused by the Byzantine fault, is thus $\Pi/5$.

Precision of the FTA Assume a distributed system with N nodes, each one with its own clock (all time values are measured in seconds). At most k out of the N clocks behave in a Byzantine manner.

A single Byzantine clock will cause the following difference in the calculated averages at two different nodes in an ensemble of N clocks:

$$E_{\text{byz}} = \Pi / (N - 2k)$$

In the worst case, a total of k Byzantine errors will thus cause an error term of

Table 3.1 Approximate jitter of the synchronization message

Synchronization message assembled and interpreted	Approximate range of jitter
At the application software level	500 μsec–5 msec
In the kernel of the operating system	10 μsec–100 μsec
In the hardware of the communication controller	Less than 1 μsec

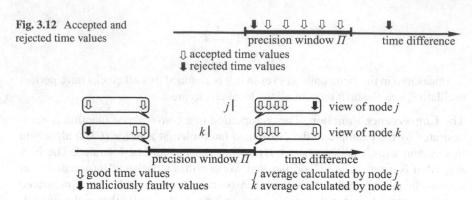

Fig. 3.12 Accepted and rejected time values

Fig. 3.13 Worst possible behavior of a malicious (Byzantine) clock

$$E_{k-byz} = k\Pi / (N - 2k)$$

Considering the jitter of the synchronization messages, the convergence function of the FTA algorithm is given by

$$\Phi(N,k,\varepsilon) = \left(\frac{k\Pi}{N-2k}\right) + \varepsilon$$

Combining the above equation with the synchronization condition (Sect. 3.4.1) and performing a simple algebraic transformation, we get the precision of the FTA algorithm:

$$\Pi(N,k,\varepsilon,\Gamma) = (\varepsilon + \Gamma)\frac{N-2k}{N-3k} = (\varepsilon + \Gamma)\mu(N,k).$$

where $\mu(N, k)$ is called the *Byzantine error term* and is tabulated in Table 3.2.

The Byzantine error term $\mu(N, k)$ indicates the loss of quality in the precision due to the inconsistency arising from the Byzantine errors. In a real environment, at most one Byzantine error is expected to occur in a synchronization round (and even this will happen very, very infrequently), and thus, the consequence of a Byzantine error in a properly designed synchronization system is not serious.

The drift offset Γ of a clock to a *perfect clock* (i.e., a clock without drift) is determined by the quality of the selected oscillator and the length of the resynchronization interval. If a standard quartz oscillator with a nominal drift rate of 10^{-4} sec/sec is used and the clocks are resynchronized every second, then Γ is about *100 μsec* to the perfect clock (and *200 μsec* between any two non-perfect clocks). Because the stochastic drift rate of a crystal is normally two orders of magnitude smaller than the nominal drift rate that is determined by the systematic error of the quartz oscillator, it is possible to reduce the drift offset Γ by up to two orders of magnitude by performing systematic error compensation.

Many other convergence functions for the internal synchronization of the clocks have been proposed and analyzed in the literature [Sch88].

3.4.4 State Correction Versus Rate Correction

The correction term calculated by the convergence function can be applied to the local-time value immediately (*state correction*), or the rate of the clock can be modified so that the clock speeds up or slows down during the next resynchronization interval to bring the clock into better agreement with the rest of the ensemble (*rate correction*).

State correction is simple to apply, but it has the disadvantage of generating a discontinuity in the time base. If clocks are set backward and the same nominal-time value is reached twice, then pernicious failures can occur within the real-time software (see the example in Sect. 3.1.4). It is therefore advisable to implement rate correction with a bound on the maximum value of the clock drift so that the error in interval measurements is limited. The resulting global time base then maintains the chronoscopy property despite the resynchronization. Rate correction can be implemented either in the digital domain by changing the number of microticks in some of the (macro)ticks or in the analog domain by adjusting the voltage of the crystal oscillator.

3.5 External Clock Synchronization

External synchronization links the global time of a cluster to an external standard of time. For this purpose it is necessary to access a *timeserver*, i.e., an external time source that periodically broadcasts the current reference time in the form of a *time message*. This time message must raise a synchronization event (such as the beep of a wristwatch) in a designated node of the cluster and must identify this synchronization event on the agreed timescale. Such a timescale must be based on a widely accepted measure of time, e.g., the physical second, and must relate the synchronization event to a defined origin of time, the *epoch*. The interface node to a timeserver is called a *time gateway*. In a fault-tolerant system, the time gateway should be a fault-tolerant unit (FTU—see Sect. 6.4.2).

3.5.1 External Time Sources

Assume that the time gateway is connected to a GPS (global positioning system). The accuracy of a GPS receiver is better than 100 nanoseconds, and it has an authoritative long-term stability—in some sense, GPS is the worldwide measurement

Table 3.2 Byzantine error term $\mu(N, k)$

Faults	Number of nodes in the ensemble (N)							
(k)	4	5	6	7	10	15	20	30
1	2	1.5	1.33	1.25	1.14	1.08	1.06	1.03
2				3	1.5	1.22	1.14	1.08
3					4	1.5	1.27	1.22

Fig. 3.14 Flow of external synchronization

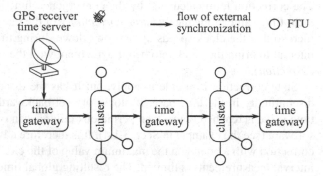

standard for measuring the progression of time. Alternatively, the external time source can be a temperature-compensated crystal oscillator (TCXO) with a drift rate of better than 1 ppm, causing a *drift offset* of better 1 μsec per second, or an atomic clock, e.g., a rubidium clock with a drift rate in the order of 10^{-12} sec/sec, causing a *drift offset* of about 1 μsec in 10 days (more expensive atomic clocks are even better). The time gateway periodically broadcasts time messages containing a synchronization event, as well as the information to place this synchronization event on the TAI scale. The time gateway must synchronize the global time of its cluster with the time received from the external time source. This synchronization is unidirectional, and therefore asymmetric, as shown in Fig. 3.14. It can be used to adjust the rate of the clocks without any concern for the occurrence of emergent instability effects.

If another cluster is connected to this *primary* cluster by a secondary time gateway, then the unidirectional synchronization functions in the same manner. The secondary time gateway considers the synchronized time of the primary cluster as its time reference and synchronizes the global time of the secondary cluster.

While internal synchronization is a cooperative activity among all the members of a cluster, external synchronization is an authoritarian process: the time gateway forces its view of external time on all its subordinates. From the point of view of fault tolerance, such an authoritarian regime introduces a problem: if the authority sends an incorrect message, then all its *obedient* subordinates will behave incorrectly. However, for external clock synchronization, the situation is under control because of the *inertia* of time. Once a cluster has been synchronized, the fault-tolerant global time base within a cluster acts as a *monitor* of the time gateway. An external synchronization message will only be accepted if its content is sufficiently close to the cluster's view of the external time. The time gateway has only a limited authority to correct the drift rate of a cluster. The enforcement of a *maximum common-mode correction rate* is required to keep the error in relative time measurements small. The software in each node of the cluster checks the maximum correction rate.

The implementation must guarantee that it is impossible for a faulty external synchronization to interfere with the proper operation of the internal synchronization, i.e., with the generation of global time within a cluster. The worst possible failure scenario occurs if the external timeserver fails maliciously—a very low probability failure mode if the external timeserver is GPS. This leads to a common-mode deviation of the global time from the external time base with the *maximum permitted deviation rate*. In a properly designed synchronization system, this drift from the external time base will not affect the internal synchronization within a cluster.

3.5.2 Time Gateway

The time gateway must control the timing system of its cluster in the following ways:

(i) It must initialize the cluster with the current external time.
(ii) It must periodically adjust the rate of the global time in the cluster to bring it into agreement with the external time and the standard of time measurement, the second.
(iii) It must periodically send the current external time in a time message to the nodes in the cluster so that a reintegrating node can reinitialize its external time value.

The time gateway achieves this task by periodically sending a time message with a rate-correction byte. This rate-correction byte is calculated in the time gateway's software. First, the difference between the occurrence of a significant event, e.g., the exact start of the full second in the timeserver, and the occurrence of the related significant event in the global time of the cluster is measured by using the local time base (microticks) of the gateway node. Then, the necessary rate adjustment is calculated, bearing in mind the fact that the rate adjustment is bounded by the agreed maximum rate correction. This bound on the rate correction is necessary to keep the maximum deviation of relative time measurements in the cluster below an agreed threshold and to protect the cluster from faults of the server.

3.5.3 Time Formats

Over the last few years, a number of external time formats have been proposed for external clock synchronization. The most important one is the standard for the time format proposed in the network time protocol (NTP) of the Internet [Mil91]. This time format (Fig. 3.15) with a length of 8 bytes contains two fields: a 4-byte full

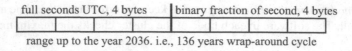

full seconds UTC, 4 bytes	binary fraction of second, 4 bytes

range up to the year 2036. i.e., 136 years wrap-around cycle

Fig. 3.15 Time format in the network time protocol (NTP)

second field, where the seconds are represented according to UTC, and a fraction of
a second field, where the fraction of a second is represented as a binary fraction with
a resolution of about *232* picoseconds. On January 1, 1972, at midnight the NTP
clock was set to *2,272,060,800.0* s, i.e., the number of seconds since January 1,
1900, at 00:00 h.

The NTP time is not chronoscopic because it is based on UTC which has to
accommodate for the switching second. The occasional insertion of a leap second
into UTC can disrupt the continuous operation of a time-triggered real-time system.

Another time format is the IEEE 1588 standard time format [Eid06]. In this time
format, the epoch starts on January 1, 1970, at 00:00 hours or is user defined. The
full seconds are counted according to TAI, while the unit of the fraction of a second
is the nanosecond. This leads to abrupt change in the representation whenever a full
second is reached.

The time-triggered architecture (TTA) uses a time format that is a combination
of IEEE 1588 and NTP. The full seconds are counted as in TAI (such as IEEE 1588),
but parts of a second are represented in a binary fraction of the full second (such as
NTP). It is thus chronoscopic and conforms fully to the dual system.

Points to Remember

- An event happens at an *instant*, i.e., at a point of the timeline. A *duration* is a
  section of the timeline delimited by two instants.
- A consistent delivery order of a set of events in a distributed system does not
  necessarily reflect the temporal or causal order of the events.
- A *physical clock* is a device for time measurement that contains a counter and a
  physical oscillation mechanism that periodically generates an event to increase
  the counter.
- Typical maximum drift rates ρ of physical clocks are in the range from 10^{-2} to
  10^{-7} sec/sec, or lower, depending on the quality (and price) of the resonator.
- The *precision* denotes the maximum offset of respective ticks of any two clocks
  of an ensemble during the time interval of interest.
- The *accuracy* of a clock denotes the maximum offset of a given clock from the
  external time reference during the time interval of interest.
- TAI is a *chronoscopic* timescale, i.e., a timescale without any discontinuities,
  that is derived from the frequency of the radiation of a specified transition of the
  cesium atom 133.

- UTC is a non-chronoscopic timescale that is derived from astronomical observations of the rotation of the earth in relation to the sun.
- A *global time* is an abstract notion that is approximated by properly selected microticks from the synchronized local physical clocks of an ensemble.
- The *reasonableness condition* ensures that the synchronization error is always less than one granule of the global time.
- If the difference between the time-stamps of two events is *equal to or larger than two ticks*, then that temporal order of events can be recovered, provided the global time is reasonable.
- The temporal order of events can *always* be recovered from their time-stamps if the event set is at least 0/3g precedent.
- If events happen only at properly selected points of a *sparse time base*, then it is possible to recover the temporal order of the events without the execution of an agreement protocol.
- The *convergence function* Φ denotes the offset of the time values immediately after the resynchronization.
- The *drift offset* Γ indicates the maximum divergence of any two good clocks from each other during the resynchronization interval R_{int}, in which the clocks are free running.
- The synchronization condition states that the synchronization algorithm must bring the clocks so close together that the amount of divergence during the next free-running resynchronization interval will not cause a clock to leave the precision interval.
- Clock synchronization is only possible if the total number of clocks N is larger or equal to $(3k + 1)$, if k is the number of clocks behaving *maliciously* faulty.
- The most important term affecting the precision of the synchronization is the *latency jitter* of the synchronization messages that carry the current time values from one node to all other nodes of an ensemble.
- When applying the fault-tolerant average algorithm, the *Byzantine error factor* $\mu(N, k)$ denotes the loss of quality in the precision caused by the Byzantine errors.
- *State correction* of a clock has the disadvantage of generating a discontinuity in the time base.
- While internal synchronization is a cooperative activity among all members of a cluster, external synchronization is an authoritarian process: the timeserver forces its view of external time on all its subordinates.
- The *NTP time*, based on UTC, is not chronoscopic. The occasional insertion of a leap second can disrupt the continuous operation of a time-triggered real-time system.
- The *time gateway* maintains the external synchronization by periodically sending a time message with a rate-correction byte to all the nodes of a cluster.

Bibliographic Notes

The problem of generating a global time base in a distributed system has first been analyzed in the context of the SIFT [Wen78] and FTMP [Hop78] projects. A VLSI chip for clock synchronization in distributed systems was developed by Kopetz and Ochsenreiter [Kop87]. The network time protocol of the Internet was published in 1991 by Mills [Mil91]. Kopetz presented the concept of a sparse time model first in [Kop92]. Steiner and Rushby formally verified the fundamental limits of time measurement in [Ste11b] and corrected minor inaccuracies. The excellent book by Eidson [Eid06] covers the IEEE 1588 protocol for clock synchronization in detail. The integration of internal and external clock synchronization is discussed in [Kop04]. For a more philosophical treatment of the problem of time, the reader is advised to study the excellent book by Withrow in [Wit90] entitled *The Natural Philosophy of Time*.

Review Questions and Problems

3.1. What is the difference between an *instant* and an *event*?

3.2. What is the difference between *temporal* order, *causal* order, and a consistent *delivery* order of messages? Which of the orders implies another?

3.3. How can clock synchronization assist in finding the *primary* event of an alarm shower?

3.4. What is the difference between UTC and TAI? Why is TAI better suited as a time base for distributed real-time systems than UTC?

3.5. Define the notions of *offset*, *drift*, *drift rate*, *precision*, and *accuracy*.

3.6. What is the difference between *internal* synchronization and *external* synchronization?

3.7. What are the fundamental limits of time measurement?

3.8. When is an event set ε/Δ-*precedent*?

3.9. What is an *agreement protocol*? Why should we try to avoid agreement protocols in real-time systems? When is it impossible to avoid agreement protocols?

3.10. What is a *sparse time base*? How can a sparse time base help to avoid agreement protocols?

3.11. Give an example that shows that, in an ensemble of three clocks, a Byzantine clock can disturb the two good clocks such that the synchronization condition is violated.

3.12. Given a clock synchronization system that achieves a precision of *90* microseconds, what is a reasonable granularity for the global time? What are the limits for the observed values for a time interval of *1.1* msec?

3.13. What is the role of the *convergence function* in internal clock synchronization?

3.14. Given a latency jitter of 20 μsec, a clock drift rate of 10^{-5} sec/sec, and a resynchronization period of 1 s, what precision can be achieved by the central master algorithm?

3.15. What is the effect of a Byzantine error on the quality of synchronization by the FTA algorithm?

3.16. Given a latency jitter of 20 μsec, a clock drift rate of 10^{-5} sec/sec, and a resynchronization period of 1 s, what precision can be achieved by the FTA algorithm in a system with *ten* clocks where *one* clock could be malicious?

3.17. Discuss the consequences of an error in the external clock synchronization. What effect can such an error have on the internal clock synchronization in the worst possible scenario?

Chapter 4
Real-Time (RT) Model

Overview

The objective of this chapter is to introduce the reader to a cross-domain architecture model of the behavior of a real-time system. This model will be used throughout the rest of the book. The model is based on three basic concepts, the concept of a *computational component*, the concept of *state*, and the concept of a *message*. Large systems can be built by the recursive composition of components that communicate by the exchange of messages. Components can be reused on the basis of their interface specification without having to understand the component internals. Concerns about the understandability have been of utmost importance in the development of this model.

The chapter is structured as follows. In Sect. 4.1 we give a broad outline of the model, describing the essential characteristics of a *component* and a *message*. Related components that work toward a joint objective are grouped into *clusters*. The differences between *temporal control* and *logical control* are explained. The following Sect. 4.2 elaborates on the close relationship between *real time* and the *state of a component*. The importance of a well-defined *ground state* for the dynamic reintegration of a component is highlighted. Section 4.3 refines the message concept and introduces the notions of event-triggered messages, time-triggered messages, and data streams. Section 4.4 presents the four interfaces of a component, two operational interfaces and two control interfaces. Section 4.5 deals with the concept of a *gateway component* that links two clusters that adhere to different *architectural styles*. Section 4.6 deals with the specification of the linking interface of a component. The linking interface is the most important interface of a component. It is relevant for the integration of a component within a cluster and contains all the information that is needed for the use of a component. The linking interface specifications consist of three parts: (i) the transport specification that contains the information for the transport of the messages, (ii) the operational specification that is concerned with interoperability of components and the establishment of the message variables, and (iii) the meta-level specification that assigns meaning to the

H. Kopetz, W. Steiner, *Real-Time Systems*,
https://doi.org/10.1007/978-3-031-11992-7_4

message variables. Points to consider when composing a set of components to build *systems of subsystems* or *system of systems* are discussed in Sect. 4.7. In this section the four principles of composability are introduced and the notion of a multilevel system is explained.

4.1 Model Outline

Viewed from the perspective of an outside observer, a real-time system can be decomposed into three communicating subsystems: a *controlled object* (the *physical* subsystem, the behavior of which is governed by the laws of physics), a *"distributed" computer subsystem* (the *cyber system*, the behavior of which is governed by the programs that are executed on digital computers), and a *human user or operator.* The distributed computer system consists of computational nodes that interact by the exchange of messages. A *computational node* can host one or more *computational components.*

4.1.1 Components and Messages

We call the process of executing an algorithm by a processing unit a *computation.* Computations are performed by *components.* In our model, a component is a self-contained hardware/software unit that interacts with its environment exclusively by the exchange of messages. We call the *timed sequence of output messages* that a component produces at an interface with its environment the *behavior* of the component at that interface. The *intended behavior* of a component is called its *service.* An unintended behavior is called a *failure.* The internal structure of a component, whether complex or simple, is neither visible nor of concern to a user of a component.

A component consists of a *design* (e.g., the software) and an *embodiment* (e.g., the hardware, including a processing unit, memory, and an I/O interface). A *real-time component* contains a real-time clock and is thus aware of the progression of real time. After *power-up*, a component enters a *ready-for-start* state to wait for a triggering signal that indicates the start of execution of the component's computations. Whenever the triggering signal occurs, the component starts its predefined computations at the *start instant.* It then reads input messages and its internal state, produces output messages and an updated internal state, and so on until it terminates its computation—if ever—at a *termination instant.* It then enters the *ready-for-start* state again to wait for the next triggering signal. In a cyclic system, the real-time clock produces a triggering signal at the start of the next cycle.

An important principle of our model is the *consequent separation of the computational components* from the *communication infrastructure* in a distributed computer system. The communication infrastructure provides for the transport of unidirectional messages from a sending component to one or more receiving components (*multicasting*) within a given interval of real time. *Unidirectionality* of messages supports the unidirectional reasoning structure of *causal chains* and eliminates any dependency of the sender on the receiver(s). This property of *sender independence* is of utmost importance in the design of fault-tolerant systems, because it avoids error backpropagation from a faulty receiving component to a correct sending component by design.

Multicasting is required for the following reasons:

(i) Multicasting supports the *nonintrusive observation* of component interactions by an independent observer component, thus making the interactions of components perceptually accessible and removing the barrier to understanding that has its origin in *hidden interactions* (see Sect. 2.1.3 on cognitive complexity).

(ii) Multicasting is required for the implementation of fault tolerance by active redundancy, where each single message has to be sent to a set of replicated components.

A message is sent at a *send instant* and arrives at the receiver(s) at some later instant, the *receive instant*. The message paradigm combines the temporal control and the value aspect of an interaction into a single concept. The temporal properties of a message include information about the send instants, the temporal order, the inter-arrival time of messages (e.g., periodic, sporadic, aperiodic recurrence), and the latency of the message transport. Messages can be used to synchronize a sender and a receiver. A message contains a data field that holds a data structure that is transported from the sender to the receiver. The communication infrastructure is *agnostic* about the contents of the data field. The message concept supports *data atomicity* (i.e., atomic delivery of the complete data structure contained in a message). A single well-designed message-passing service provides a simple interface of a component to other components inside and outside a node and to the environment of a component. It facilitates encapsulation, reconfiguration, and the recovery of component services.

4.1.2   Cluster of Components

A cluster is a *set of related components* that have been grouped together in order to achieve a common objective (Fig. 4.1). In addition to the set of components, a cluster must contain an intra-cluster communication system that provides for the transport of messages among the components of the cluster. The components that form a computational cluster agree on a common *architectural style* (see the last paragraphs of Sect. 2.2.4).

> **Example:** Figure 4.1 depicts an example of a computational cluster within a car. This cluster consists of a computational component, the *assistant system*, and gateway components to the man-machine interface (the driver), the physical subsystems of the car, and a gateway to other cars via a wireless vehicle-to-vehicle communication link.

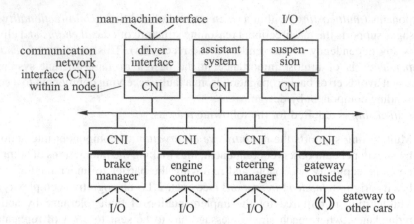

Fig. 4.1 Example of an *in-car cluster*

4.1.3   Temporal Control Versus Logical Control

Let us revisit the rolling mill example of Fig. 1.9 of Sect. 1.7.3 and specify a relation between measured variables that must be monitored by an alarm-monitoring task in the MMI component. Assume that the pressures p_1, p_2, and p_3, between the roles of the three drives, are measured by the three controller components of Fig. 1.9. The measurements are sent to the man-machine interface (MMI) component for checking the following alarm condition:

when $((p_1 < p_2) \wedge (p_2 < p_3))$
then *everything ok*
else *raise pressure alarm*;

This looks like a reasonable specification at the user level. Whenever the pressure between the rolls does not satisfy the specified condition, a pressure alarm must be raised.

During the refinement of this specification by a system architect, four different tasks (three *measurement* tasks in the three control nodes and one *alarm-monitoring* task in the MMI node of Fig. 1.9) must be designed. The following questions concerning the temporal activation of these tasks arise:

(i) What is the maximum tolerable time interval between the occurrence of the alarm condition in the controlled object and the raising of the alarm at the MMI? Because the communication among the components takes a finite amount of time, some time intervals are unavoidable!

(ii) What are the maximum tolerable time intervals between the three pressure measurements at the three different control nodes? If these time intervals are not properly controlled, false alarms will be generated or important alarms will be missed.

(iii) When and how often do we have to activate the pressure measurement tasks at the three control nodes?

(iv) When do we have to activate the alarm-monitoring task at the alarm-monitoring component (the MMI component in Fig. 1.9)?

Because these questions are not answered by the given specification, it is evident that this specification lacks precise information concerning the architectural requirements in the temporal domain. The temporal dimension is buried in the ill-specified semantics of the **when** statement. In this example, the **when** statement is intended to serve two purposes. It is specifying:

(i) The point in time when the alarm condition must be raised.

(ii) The conditions in the value domain that must be monitored.

It thus intermingles two separate issues, the behavior in the time domain and the behavior in the value domain. A clean distinction between these two issues requires a careful definition of the concepts of *temporal control* and *logical control*.

Temporal control is concerned with determining the instants in the domain of real time when computations must be performed, i.e., when tasks must be activated. These instants are derived from the dynamics of the application. In the above example, the decision regarding the instants at which the pressure measuring tasks and the alarm-monitoring task must be activated is an issue of temporal control. Temporal control is related to the progression of *real time*.

Logical control is concerned with the control flow *within* a task that is determined by the given task structure and the particular input data, in order to realize the desired computation. In the above example, the evaluation of the branch condition and the selection of one of the two alternatives is an example of *logical control*. The time interval needed for the execution of a task that performs the logical control is determined by the frequency of the oscillator that drives the processing unit—we call this time interval the *execution time*. The execution time is determined by the given implementation and will change if we replace the given processor by a faster one.

Since *temporal control* is related to *real time*, while *logical control* is related to *execution time*, a careful distinction must be made between these two types of control. A good design will separate these two control issues in order to decouple the reasoning about temporal constraints dictated by the application, from the reasoning about logical issues inside the algorithmic part of a program. Synchronous real-time languages, such as LUSTRE [Hal92], ESTEREL [Ber85], and SL [Bou96], distinguish cleanly between logical control and temporal control. In these languages, the progression of real time is partitioned into an (infinite) sequence of intervals of specified real-time duration, which we call steps. Each step begins with a tick of a real-time clock that starts a computational task (logical control). The computational model assumes that a task, once activated by the tick of a real-time clock (temporal control), finishes its computation *quasi-immediately*. Practically this means that a task must terminate its executions before the next triggering signal (the next tick of the real-time clock) initiates the next execution of the task.

The periodic finite state machine (PFSM) model [Kop07] extends the classic FSM, which is concerned with *logical control*, by introducing a new dimension for the progression of a global sparse time to cover *temporal control* issues.

If the issues of temporal control and logical control are intermingled in a program segment, then it is not possible to determine the worst-case execution time (WCET—see Sect. 10.2) of this program segment without analyzing the behavior of the environment of this program segment. This is a violation of the design principle *separation of concerns* (see Sect. 2.5).

> **Example:** A *semaphore wait statement* is a temporal control statement. If a *semaphore wait statement* is contained in a program segment that also includes logical control (algorithmic) statements, then the temporal behavior of this program segment depends on both the progress of *execution time* and the progress of *real time* (see also Sects. 9.2 and 10.2).

4.1.4   Event-Triggered Control Versus Time-Triggered Control

In Sect. 4.1.1, we introduced the notion of a *triggering signal*, i.e., a control signal that indicates the instant when an activity should start in the temporal domain. What are the possible origins of such a triggering signal? The triggering signal can be associated either with the occurrence of a *significant event*—we call this *event-triggered control*—or with the arrival of a *specified instant* on the timeline—we call this *time-triggered control*.

The significant events that form the basis of event-triggered control can be the arrival of a particular message, the completion of an activity inside a component, the occurrence of an external interrupt, or the execution of a *send message statement* by the application software. Although the occurrences of significant events are normally sporadic, there should be a minimal real-time interval between two successive events so that an overload of the communication system and the receiver of the events can be avoided. We call such an event stream, for which a minimum inter-arrival time between events is maintained, a *rate-controlled* event stream.

Time-triggered control signals are derived from the progression of the global time that is available in every component. Time-triggered control signals are normally cyclic. A cycle can be characterized by its *period*, i.e., the real-time interval between two successive *cycle starts*, and by its *phase*, that is, the interval between the start of the period, expressed in the global time, and the *cycle start* (see also Sect. 3.3.4). We assume that a cycle is associated with every time-triggered activity.

4.2   Component State

The concept of *state* of a component is introduced in order to separate *past behavior* from *future behavior* of a real-time component. The concept of state requires a clear distinction between past events and future events, i.e., there must be a consistent temporal order among the events of significance (refer to Sect. 3.3.2).

4.2.1   Definition of State

The notion of *state* is widely used in the computer science literature, albeit some-times with meanings that are different from the meaning of *state* that is useful in a real-time system context. In order to clarify the situation, we follow the precise defi-nition of Mesarovic [Mes89, p. 45], which is the basis for our elaborations:

> The state enables the determination of a future output solely on the basis of the future input
> and the state the system is in. In other words, the state enables a "decoupling" of the past
> from the present and future. The state embodies all past history of a system. Knowing the
> state "supplants" knowledge of the past. . . . Apparently, for this role to be meaningful, the
> notion of past and future must be relevant for the system considered.

The sparse time model introduced in Sect. 3.3.2 makes it possible to establish the consistent system-wide separation of the past from the future that is necessary to define a *consistent system state* in a distributed real-time computer system.

4.2.2   The Pocket Calculator Example

Let us look at the familiar example of a pocket calculator to investigate the concept of *state* in more detail. An operand, i.e., a number of keyboard digits, must be entered into the calculator before the selected operator, e.g., a key for the trigono-metric function *sine*, can be pressed to initiate the computation of the selected func-tion. After the computation terminates, the result is displayed at the calculator display. If we consider the computation to be an atomic operation and observe the system immediately before or after the execution of this atomic operation, the inter-nal state of this simple calculator device is empty at the points of observation.

Let us now observe the pocket calculator (Fig. 4.2) during the interval between the *start* of the computation and the *end* of the computation. If the internals of the device can be observed, then a number of intermediate results that are stored in the local memory of the pocket calculator can be traced during the series expansion of the *sine* function. If the computation is interrupted at an instant between the instants *start* and *end*, the contents of the program counter and all memory cells that hold the intermediate results form the *state at this instant* of interruption. After the end instant of the computation, the contents of these intermediate memory cells are no longer relevant, and the state is empty again. Figure 4.3 depicts a typical *expansion* and *contraction* of the state during a computation.

Let us now analyze the state of a pocket calculator used to sum up a set of num-bers. When entering a new number, the sum of the previously entered numbers must be stored in the device. If we interrupt the work after having added a subset of num-bers and continue the addition with a new calculator, we first have to input the intermediate result of the previously added numbers. At the user level, the state consists of the intermediate result of the previous additions. At the end of the

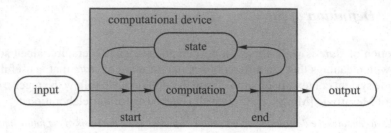

Fig. 4.2 Model of a pocket calculator

Fig. 4.3 Expansion and
contraction of the *state*
during a computation

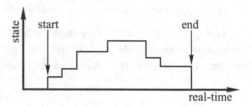

operation, we receive the final result and clear the memory of the calculator. The
state is empty again.

From this simple example, we can conclude that the size of the state of a system
depends on the *instant of observation* of the system. If the granularity of observations is increased and if the observation points are selected immediately before or
after an atomic operation at the chosen level of abstraction, then the size of the state
can be reduced.

The state at any instant of interruption is contained in the contents of the program
counter and all *state variables* that must be loaded into a *virgin* hardware device to
resume the operation from the instant of interruption onward. If an interruption is
caused by a failure of a component and we must reintegrate the component into a
running system, then the size of the state that must be reloaded into the repaired
component is of concern.

If our hardware device is a programmable computer, we must first load the software, i.e., operating system, the set of application programs, and the initial values
for all state variables, into a *virgin* hardware device before we can start a computation. We call the *totality of software* that has to be loaded into a virgin hardware
device the *core image* or the *job*. Normally, the job is a data structure that is static,
i.e., it is not changed during the execution of the software. In some embedded hardware devices, the job is stored in a ROM (read-only memory), and thus the *software
becomes literally a part of the hardware*.

4.2.3   Ground State

In order to facilitate the dynamic reintegration of a component into a running system, it is necessary to design periodic *reintegration instants* into the behavior, where the component's state at the reintegration instant contains a small set of well-defined application-specific *state variables*. We call the state at the reintegration instant the *ground state* (*g-state*) of a component and the temporal interval between two reintegration points the *ground cycle*.

The ground state at the reintegration point is stored in a declared *g-state data structure*. Designing a minimal ground state data structure is the result of an explicit design effort that involves a semantic analysis of the given application. The designer has to find periodic instants where there is a maximum decoupling of future behavior from past behavior. This is relatively easy in *cyclic applications*, such as in control applications and multimedia applications. In these applications, a natural reintegration instant is immediately after the termination of one cycle and before the beginning of the next cycle. Design techniques for the minimization of the ground state are discussed in Sect. 6.6.

At the reintegration instant, no task may be active, and all communication channels must be flushed, i.e., there are no messages in transit [Ahu90]. Consider a node that contains a number of concurrently executing tasks that exchange messages among each other and with the environment of the node. Let us choose a level of abstraction that considers the execution of a task as an *atomic action*. If the execution of the tasks is asynchronous, then the situation depicted in the upper part of Fig. 4.4 can arise; at every instant, there is at least one active task, thus implying that there is no instant when the ground state of the node can be defined.

In the lower part of Fig. 4.4, there is an instant when no task is active and when all channels are empty, i.e., when the system is in the *g-state*. If a node is in the *g-state*, then the entire state that is essential for the future operation of the node is contained in the declared ground state data structure.

Example: Consider the relation between the size of the g-state and the duration of the ground (g) cycle in the design of a clock. If the g-cycle is 24 h and the start of a new day is the reintegration instant, then the g-state is empty. If every complete hour is a reintegration instant, then the g-state contains 5 bits (to identify one out of 24 h per day). If every complete minute is a reintegration instant, then the g-state is 11 bits (to identify one of 1440 min per day). If every complete second is a reintegration instant, then the g-state is 17 bits (to

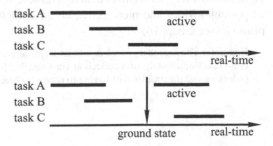

Fig. 4.4 Task executions: without (above) and with (below) ground state

Table 4.1 Comparison of g-state recovery and checkpoint recovery

	G-state recovery	Checkpoint recovery
Data selection	Application-specific small data set that is essential for the future operation of the system	All data elements that have been modified since the start of the computation
Data modification	G-state data is modified to establish consistency between the g-state and the state of the environment at the future reintegration instant. Rollback of the environment is not possible in a real-time system	No modification of checkpoint data. Consistency is established by rolling the (data) environment back to the instant when the checkpoint data was captured

identify one out of 86400 sec per day). It depends on the application characteristics to determine which one of the above alternatives is optimal for a given application. If the clock is an alarm clock that can store up to four alarms and the accuracy of an alarm is 5 minutes, then the g-state for every alarm is 10 bits (9 bits for the alarm and 1 bit to denote whether the alarm is *on* or *off*). If we assume that the reintegration cycle is 1 second and four alarms must be supported, then the g-state message in this simple example is 57 bits in lengths. This g-state can be stored in an 8-byte data structure. In the restart message, the time field must be corrected such that it contains the precise time value at the restart instant.

Table 4.1 shows that g-*state recovery* is substantially different from *checkpoint recovery* that is used to establish a consistent state after a failure in a non-real-time data-intensive system.

4.2.4   Database Components

We call a component where the number of *dynamic data elements*, i.e., data elements that are modified by the computations, is too large for storing them in a single ground state message a *database component*. The dynamic data elements that are contained in a database component can be either part of the *state* or *archival data*.

The term *archival data* refers to data that has been collected for archival purposes and does and not have a direct influence on the future behavior of the component. Archival data is needed to record the history of production process variables in order to be able to analyze a production process at some later time off-line. It is a good practice to send archival data to a remote storage site for archival data as soon as possible. More and more often, cloud computing and storage is used for this purpose (see Chap. 14).

Example: The legendary *black box* in an airplane contains archival data. One could send these data immediately after collection via a satellite link to a storage site on the ground in order to avoid the problems of having to recover a black box after an accident.

4.3   The Message Concept

The concept of a *message* is the third basic concept of our model. A *message* is an atomic data structure that is formed for the purpose of communication, i.e., data transmission and synchronization among components.

4.3.1   Message Structure

The concept of a *message* is related to the concept of a *letter* in the postal system. A message consists of a header, a data field, and a trailer. The header, corresponding to the envelope of a letter, contains the *port address* of the receiver (the mailbox) where the message must be delivered, information about how the message must be handled (e.g., a registered letter), and may contain the address of the sender. The data field contains the application-specific data of the message, corresponding to the content of a letter. The trailer, corresponding to the signature in a letter, contains information that allows the receiver to detect whether the data contained in the message is uncorrupted and authentic. There are different types of trailers in use: the most common trailer is a CRC field that allows the receiver to determine whether the data field has been corrupted during transport. A message may also contain an *electronic signature* in the trailer that makes it possible to determine whether the *authenticated contents* of the message have not been *altered* (see Sect. 6.2). The notion of *atomicity* implies that a message is delivered either in its entirety or not at all. If a message is corrupted or only parts of the message arrive at the receiver's site, the whole message is discarded. It is the purpose of said trailer to ensure *atomicity* with sufficiently high probability.

The temporal dimension of the message concept relates to the instants when a message is sent by the sender and received by the receiver and consequently how long the message has been in transit. We call the interval between the *send instant* and the *receive instant* the *transport delay*. A second aspect of the temporal dimension relates to the rate of message production by the sender and message consumption by the receiver. If the sending rate is constrained, then we speak about *a rate-constrained* message system. In case the sender's rate is not constrained, the sender may overload the transport capacity of the communication system (we call this *congestion*) or the processing capacity of the receiver. In case the receiver cannot keep up with the message production rate of the sender, the receiver can send a control message to the sender telling the sender to slow down (*back pressure flow control*). Alternatively, the receiver or the communication system may simply discard messages that exceed its processing capacity.

4.3.2   Event Information Versus State Information

The state of a dynamic system changes as real time progresses. Let us assume that we periodically observe the state variables of a system with a duration d between two successive observation instants. If we observe that the value of all state variables is the same in two successive observations, then we infer that no *event*, i.e., *change of state*, has occurred in the last observation interval d. This conclusion is only valid, if the dynamics of the system is slow compared to our observation interval d (refer to *Shannon's theorem* [Jer77]). If two successive observations of the values of some state variables differ, then we conclude that at least one event has occurred in the last observation interval d. We can report about the occurrence of an event, i.e., a change of state, in two different ways: either by sending a single message containing *event information* or by sending a sequence of messages containing *state information*.

We talk about *event information* if the information conveys the *difference in values* of the previous state observation and the current state observation. The instant of the current (later) observation is postulated to be the instant of event occurrence. This assumption is not fully accurate, since the event may have occurred at any instant during the last interval of duration d. We can reduce this temporal observation error of an event by making the interval d smaller, but we cannot fully eliminate the temporal uncertainty about the observation of events. This holds true even if we use the interrupt system of a processor to report about an event. The input signal that relays the occurrence of an interrupt is not sensed continuously by the processor, but only after the termination of the execution of an instruction. This delay is introduced in order to reduce the amount of *processor state* that has to be saved and restored in the processor in order to be able to continue the interrupted task after the interrupt has been served. As outlined in Sect. 4.2.3, the state is minimal immediately before or after the execution of an atomic operation—in this case, the execution of a complete instruction by a processor.

If the precise timing of an event is critical, we can provide a separate dedicated hardware device to time-stamp the observed state change of the interrupt line immediately and thus reduce the temporal observation error to a value that is in the order of magnitude of the cycle time of the hardware. Such hardware devices are introduced to achieve precise clock synchronization in distributed systems, where the precision of the distributed clocking system must be in the nanosecond range.

> **Example:** The IEEE 1588 standard for clock synchronization suggests the implementation of a separate hardware device to precisely capture the arrival instant of a clock synchronization message.

We talk about *state information* if the information conveys the *values* of the current state variables. If the data field of a message contains *state information*, it is up to the receiver to compare two successive state observations and determine whether an event has occurred or not. The temporal uncertainty about the event occurrence is the same as above.

4.3.3   Event-Triggered (ET) Message

A message is called *event-triggered (ET)* if the triggering signal for sending the message is derived from the occurrence of a significant event, such as the execution of a *send message* command by the application software.

ET messages are well suited to transport *event information*. Since an event refers to a *unique* change of state, the receiver must consume every single event message, and it is not allowed to duplicate an event message. We say that an event message must adhere to the *exactly once* semantics. The event message model is the standard model of message transmission that is followed in most non-real-time systems.

> **Example:** The event message *valve must be closed by 5 degrees* means that the new intended position of the valve equals the current position plus 5 degrees. If this event message is lost or duplicated, then the image of the state of the valve position in the computer will differ from the actual state of the valve position in the environment by 5 degrees. This error can be corrected by *state alignment*, i.e., the (full) state of the intended valve position is sent to the valve.

In an event-triggered system, error detection is in the responsibility of the sender who must receive an explicit acknowledgment message from the receiver telling the sender that the message has arrived correctly. The receiver cannot perform error detection, because the receiver cannot distinguish between *no activity by the sender* and *loss of message*. Thus the control flow must be *bidirectional* control flow, even if the data flow is only *unidirectional*. The sender must be *time-aware*, because it must decide within a finite interval of real time that the communication has failed. This is one reason why we cannot build fault-tolerant system that is unaware of the progression of *real time*.

4.3.4   Time-Triggered (TT) Message

A message is called *time-triggered (TT)* if the triggering signal for sending the message is derived from the progression of real time. There is a *cycle*, characterized by its *period* and *phase*, assigned to every time-triggered message before the system starts operating. At the instant of cycle start, the transmission of the message is initiated automatically by the operating system. There is no *send message command* necessary in TT message transmission.

TT messages are well suited to transport *state information*. A TT message that contains state information is called a *state message*. Since a new version of a state observation normally replaces the existing older version, it is reasonable that a new state message *updates-in-place* the older message. On reading, a state message is *not consumed*; it remains in the memory until it is updated by a new version. The semantics of state messages is similar to the semantics of a *program variable* that can be read many times without consuming it. Since queues are not necessary in state message transmissions, queue overflow is no issue. Based on the a priori known cycle of state messages, the *receiver can perform error detection autonomously* to detect the loss of a state message. State messages support the *principle of independence* (refer to Sect. 2.5) since sender and receiver can operate at different (independent) rates and there is no means for a receiver to influence the sender.

> **Example:** A temperature sensor observes the state of a temperature sensor in the environment every second. A state message is well suited to transport this observation to a user and store it in a program variable named *temperature*. The user program can read this variable *temperature* whenever it needs to refer to the current temperature of the environment, knowing that the value stored in this variable is *up to date* to within about 2 s. If a single state message is lost, then for one cycle the value stored in this variable is *up to date* to only within about 3 s. Since, in a time-triggered system, the communication system knows a priori when a new state message must arrive, it can associate a *flag* with the variable temperature to inform the user if the variable temperature has been properly updated in the last cycle.

4.4   Component Interfaces

Let us assume that the design of a large component-based system is partitioned into two distinct design phases, *architecture design* and *component design* (see also Sect. 11.2 on *system design*). At the end of the architecture design phase, a *platform-independent model* (PIM) of a system is available. The PIM is an executable model that partitions the system into clusters and components and contains the precise interface specification (in the domains of value and time) of the linking interfaces of the components. The linking interface specification of the PIM is *agnostic* about the component implementation and can be expressed in a high-level executable system language, e.g., in *System C*. A PIM component that is transformed to a form that can be executed on the final execution platform is called a *platform-specific model* (PSM) of the component. The PSM has the same interface characteristics as the PIM. In many cases, an appropriate compiler can transform the PIM to the PSM automatically.

An interface should serve a *single well-defined purpose* (principle of *separation of concerns*; see Sect. 2.5). Based on *purpose*, we distinguish between the following four message interfaces of a component (Fig. 4.5):

- The *linking interface* (LIF) provides the specified service of the component at the considered level of abstraction. This interface is agnostic about the component implementation. It is the same for the PIM and the PSM.
- The *technology-independent interface* (TII) is used to configure and control the execution of the component. This interface is agnostic about the component implementation. It is the same for the PIM and the PSM.
- The *technology-dependent interface* (TDI) is used to provide access to the internals of a component for the purpose of maintenance and debugging. This interface is implementation specific.

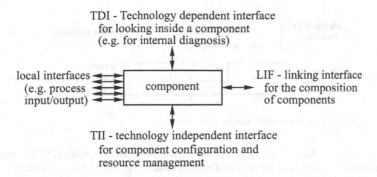

TDI - Technology dependent interface
for looking inside a component
(e.g. for internal diagnosis)

local interfaces   LIF - linking interface
(e.g. process   component   for the composition
input/output)   of components

TII - technology independent interface
for component configuration and
resource management

Fig. 4.5 The four interfaces of a component

- The *local interface* links a component to the external world that is the external environment of a cluster. This interface is syntactically specified at the PSM level only, although the *semantic content* of this interface is contained in the LIF.

The LIF and the *local interface* are *operational interfaces*, while the TII and TDI are *control interfaces*. The control interfaces are used to control, monitor, or debug a component, while the operational interfaces are in use during the normal operation of a component. Before discussing these four interfaces in detail, we elaborate on some general properties of message interfaces.

4.4.1 Interface Characterization

Push Versus *Pull Interface:* There are two options to handle the arrival of a new message at the interface of a receiving component:

- *Information push*: The communication system raises an interrupt and forces the component to immediately act on the message. Control over the temporal behavior of the component is delegated to the environment outside of the component.
- *Information pull*: The communication system puts the message in an intermediate storage location. The component looks periodically if a new message has arrived. Temporal control remains inside the component.

In real-time systems, the information pull strategy should be followed whenever possible. Only in situations when an immediate action is required and the delay of one cycle that is introduced by the information pull strategy is not acceptable, one should resort to the information push strategy. In the latter case, mechanisms must be put into place to protect the component from erroneous interrupts caused by failures external to the component (see also Sect. 9.5.3). The information push strategy violates the principles of independence (see Sect. 2.5).

Example: An engine control component for an automotive engine worked fine as long as it was not integrated with the theft-avoidance system. The message interface between the

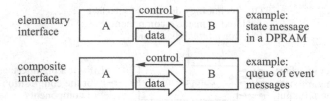

Fig. 4.6 Elementary vs. composite interface

engine controller and the theft-avoidance system was designed as a push interface, causing the sporadic interruption of a time-critical engine control task when a message arrived from the theft-avoidance system at an ill-timed instant. As a consequence the engine controller sporadically missed a deadline and failed. Changing the interface to a *pull interface* solved the problem.

Elementary Versus Composite Interface In a distributed real-time system, there are many situations where a simple unidirectional data flow between a sending and a receiving component must be implemented. We call such an interface *elementary* if both the data flow and the control flow are unidirectional. If, in a unidirectional data flow scenario, the control flow is bidirectional, we call the interface *composite* (Fig. 4.6) [Kop99].

Elementary interfaces are *inherently simpler* than composite interfaces, because there is *no dependency* of the behavior of the sender on the behavior of the receiver. We can reason about the correctness of the sender without having to consider the behavior of the receiver. This is of particular importance in safety-critical systems.

4.4.2 Linking Interface (LIF)

The *services* of a component are accessible at its *cluster LIF*. The cluster LIF of a component is an *operational message-based* interface that interconnects a component with the other components of the cluster and is thus the interface for the integration of components into the cluster. The LIF of a component abstracts from the internal structure and the local interfaces of the component. The specification of the LIF must be self-contained and cover not only the functionality and timing of the component itself but also the semantics of its local interfaces. The LIF is *technology agnostic* in the sense that the LIF does not expose implementation details of the internals of the component or of its local interfaces. A technology agnostic LIF ensures that different implementations of computational components (e.g., general-purpose CPU, FPGA, ASIC) and different local input/output subsystems can be connected to a component without any modification to the other components that interact with this component across its message-based LIF.

> **Example:** In an input/output component, the external input and output signals are connected by a local point-to-point wiring interface. The introduction of a bus system, e.g., a

CAN bus, will not change the cluster LIF of the input/output component, as long as the temporal properties of the data appearing at the LIF are the same.

4.4.3 Technology-Independent Interface (TII)

The technology-independent interface is a *control interface* that is used to configure a component, e.g., assign the proper names to a component and its input/output ports; to *reset, start,* and *restart* a component; and to monitor and control the resource requirements (e.g., power) of a component during runtime, if required. Furthermore, the TII is used to configure and reconfigure a component, i.e., to assign a specific *job* (i.e., core image) to the programmable component hardware.

The messages that arrive at the TII communicate either directly with the component hardware (e.g., *reset*), with the component's operating system (e.g., *start a task*), or with the *middleware* of the component, but not with the application software. The TII is thus orthogonal to the LIF. This strict separation of the application-specific message interfaces (LIF) from the system control interface of a component (TII) simplifies the application software and reduces the overall complexity of a component (see also the *principle of separation of concerns* in Sect. 2.5).

4.4.4 Technology-Dependent Interface (TDI)

The TDI is a *special control interface* that provides a means to look inside a component and to observe the internal variables of a component. It is related to the *boundary scan interface* that is widely used for testing and debugging large VLSI chips and has been standardized in the IEEE standard 1149.1 (also known as the JTAG standard). The TDI is intended for the person who has a deep understanding of the component internals. The TDI is of no relevance for the user of the LIF services of the component or the system engineer who configures a component. The precise specification of the TDI depends on the technology of the component implementation and will be different if the same functionality of a component is realized by software running on a CPU, by an FPGA or by an ASIC.

4.4.5 Local Interfaces

The local interfaces establish a connection between a component and its outside environment, e.g., the sensors and actuators in the physical plant, the human operator, or another computer system. A component that contains a local interface is called a *gateway component* or an *open component*, in contrast to a component that does not contain a local interface, which is called a *closed component*. The distinction between open and closed components is important from the point of view of the specification of the semantics of the LIF of the component. Only closed components can be fully specified without knowing the *context of use* of the component.

From the point of view of the *cluster LIF*, only the *timing* and the *semantic content*, i.e., the meaning of the information exchanged across a local interface, are of relevance, while the detailed structure, naming, and access mechanisms of the local interface are *intentionally left unspecified* at the cluster level. A modification of the local access mechanisms, e.g., the exchange of a CAN bus by Ethernet, will not have any effect on the LIF specification, and consequently on the users of the LIF specification, as long as the *semantic content* and the *timing* of the relevant data items are the same (see also the *principle of abstraction* in Sect. 2.5).

> **Example:** A component that calculates a trigonometric function is a *closed component*. Its functionality can be formally specified. A component that reads a temperature sensor is an *open component*. The meaning of *temperature* is application specific, since it depends on the position where the sensor is placed in the physical plant.

4.5   Gateway Component

Viewed from the perspective of a cluster, a gateway component is an *open component* that links two worlds, the *internal world* of the cluster and the *external world* of the cluster environment. A gateway component acts as a *mediator* between these two worlds. It has two operational interfaces, the LIF message interface to the cluster and the local interface to the external world, which can be the physical plant, a man-machine interface, or another computer system (Fig. 4.7). Viewed from the outside of a cluster, the *role of the interfaces is reversed*. The previous local interface becomes the new LIF and the previous LIF becomes the new local interface.

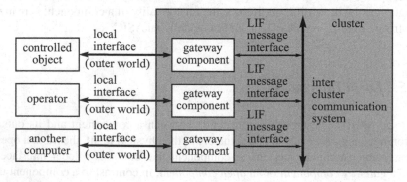

Fig. 4.7 Gateway components between the LIF and the local external-world interface

4.5.1 Property Mismatches

Every system is developed according to an *architectural style*, i.e., a set of adopted rules and conventions for the concept formation, representation of data, naming, programming, interaction of components, semantics of the data, and many more. The architectural style characterizes the *properties* of the entities that represent the design. The details of the architectural style are sometimes explicitly documented, but more often only implicitly followed by the development team, reflecting the unstated conventions that govern the development community (see also the last paragraphs of Sect. 2.2.4).

Whenever a communication channel links two systems, developed by two different organizations, it is highly probable that some of the properties of the messages that are exchanged across this channel are in disagreement. We call any disagreement in a property of the data or the protocol between the sender and the receiver of a message a *property mismatch*. It is up to a gateway component to resolve property mismatches.

> **Example:** Assume that there is a difference in *endianness*, i.e., the byte ordering of data, between the sender and the receiver of a message. If the sender assumes *big endian*, i.e., the most significant byte is first, and the receiver assumes *little endian*, i.e., the least significant byte is first, then this *property mismatch* has to be resolved either by the sender, by the receiver, or by an intermediate connector system, i.e., the gateway component.

Property mismatches occur at the borders where systems interact, not inside a well-designed system. Inside a system, where all partners respect the rules and constraints of the architectural style, property mismatches are normally no issue. Property mismatches should be resolved in the gateway components that link two systems in order to maintain the integrity of the architectural styles within each of the interacting subsystems.

4.5.2 LIF Versus Local Interface of a Gateway Component

As noted before, the set of related components that form a cluster share a common *architectural style* at their cluster LIFs. This means that *property mismatches* among LIF-cluster messages are rare.

This is in contrast to the messages that cross a gateway component. These messages come from two different worlds that are characterized by two different architectural styles. Property mismatches, syntactic incompatibility, incoherent naming, and differences in representation between these two sets of messages are the rule rather than the exception. It is the main task of a gateway component to translate the architectural style of one world to the architectural style of the other world without changing the *semantic content* of the message variables.

There are situations when the architectural styles of the two interfacing worlds are based on a different *conceptualization* of reality, i.e., there are not only differences in the *names* and the *representations* of the same concepts (as shown in the example in Sect. 2.2.4), but the concepts themselves, i.e., the semantic contents are dissimilar.

Viewed from a given cluster, the gateway component hides the details of the external world (the local interface) from the standardized message formats within the computational cluster and filters the incoming information: only the information that is relevant for the operation of the cluster under consideration is exposed in the form of cluster-standard messages at the cluster LIF of the gateway component.

An important special case of an external-world interface is a process I/O interface, which establishes a link between the cyber world and the physical world. Table 4.2 depicts some of the differences between the LIF messages and a process-control I/O interface to the physical plant.

Example: In the external world of a process plant, the value of the process variable *temperature* may be coded in an analog 4–20 mA sensor signal, where 4 mA means 0% and 20 mA means 100% of the selected measurement range, i.e., between 0° Celsius and 100° Celsius. This analog representation must be converted to the standard representation of temperature that has been defined in the architectural style of the given cluster, which might be degrees *Kelvin*.

Example: Let us look at an important interface, the man-machine interface (MMI), in order to distinguish between the *local interface* and the *LIF message interface* of a gateway MMI component (Fig. 4.8). At the level of architectural modeling, we are not interested in the representational details of the local external-world interface, but only in the *semantic content* and temporal properties of the message variables at the *LIF message interface*. An important message is sent to the MMI component. It is somehow relayed to the operator's mind. A response message from the operator is expected within a given time interval at the *LIF message interface*. All intricate issues concerning the representation of the information at the graphic user interface (GUI) of the operator terminal are irrelevant from the point of view of architecture modeling of the interaction patterns between the operator and the cluster. If the purpose of our model were the study of human factors governing the specific man-machine interaction, then the form and attributes of the information representation at the GUI (e.g., shape and placement of symbols, color, and sound) would be relevant and could not be disregarded.

A gateway component can link two different clusters that have been designed by two different organizations using two different architectural styles. Which one of

Table 4.2 Characteristics of a LIF versus a local process I/O interface

Characteristic	Local process I/O interface	LIF message interface
Information representation	Unique, determined by the given physical interface device	Uniform within the whole cluster
Coding	Analog or digital, unique	Digital, uniform codes
Time base	Dense	Sparse
Interconnection pattern	One-to-one	One-to-many
Coupling	Tight, determined by the specific hardware requirements and the I/O protocol of the connected device	Weaker, determined by the LIF message communication protocol

the two interfaces of such a gateway component is considered the LIF and which one is the local interface depends on the view taken. As already mentioned, if we change the view, the LIF becomes the local interface and the local interface becomes the LIF.

4.5.3 Standardized Message Interface

To improve the compatibility between components designed by different manufacturers, to enhance the interoperability of devices, and to avoid property mismatches, some international standard organizations have attempted to standardize message interfaces. An example of such a standardization effort is the SAE J 1587 Message Specification. The Society of Automotive Engineers (SAE) has standardized the message formats for heavy-duty vehicle applications in the J 1587 standard. This standard defines message names and parameter names for many data elements that occur in the application domain of heavy vehicles. Besides data formats, the range of the variables and the update frequencies are also covered by the standard.

4.6 Linking Interface Specification

As noted in Sect. 4.1.1, the *timed sequence of messages* that a component exchanges across an interface with its environment defines the *behavior* of the component at that interface [Kop03]. The *interface behavior* is thus determined by the properties of all messages that cross an interface. We distinguish three parts of an interface specification: (i) the *transport specification* of the messages, (ii) the *operational specification* of the messages, and (iii) the *meta-level specification* of the messages.

The *transport specification* describes all properties of a message that are needed to transport the message from the sender to the receiver(s). The transport specification covers the addressing and temporal properties of a message. If two components are linked by a communication system, the transport specification suffices to

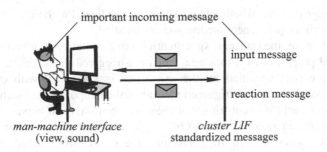

man-machine interface
(view, sound)

cluster LIF
standardized messages

Fig. 4.8 Standardized LIF versus concrete MMI

describe the requested services from the communication system. The communication system is *agnostic* about the contents of the data field of a message. For the communication system, it does not matter whether the data field contains multimedia data, such as voice or video, numerical data, or any other data type.

Example: The Internet provides a defined message-transport service between two end systems, not knowing what types of digital data are transported.

In order to be able to interpret the data field of a message at the endpoints of the communication, we need the *operational* and the *meta-level specification*. The *operational specification* informs about the syntactic structure of the message that is exchanged across the LIF and establishes the *message variables*. Both the transport and the operational specification must be *precise* and *formal* to ensure the *syntactic interoperability* of components. The *meta-level specification* of a LIF assigns meaning to the *message variable names* introduced by the operational specification. It is based on an interface model of the user environment. Since it is impossible to formalize all aspects of a real-world user environment, the meta-level specification will often contain natural language elements, which lack the precision of a formal system. Central concepts of the application domains and applications can be specified using *domain-specific ontologies*.

4.6.1   Transport Specification

The *transport specification* contains the information needed by the communication system to transport a message from a sender to the receiver(s). An interface contains a number of ports where the messages destined for this interface arrive or where messages are placed for sending. The following attributes must be contained in the transport specification:

- Port address and direction.
- Length of the data field of the message.
- Type of message (e.g., *time-triggered* or *event-triggered* or *data stream*).
- Cycle of a time-triggered message.
- Queue depth for an event-triggered message or a data stream.

It is a design decision whether these attributes are linked with the message or are associated with the port where the message is handled.

As noted above, the transport specification must contain the information about the temporal properties of a message. For time-triggered messages, the temporal domain is precisely specified by the cycle that is associated with every time-triggered message. For event-triggered messages, the temporal specification is more difficult, particularly if event-triggered messages can arrive in bursts. In the long run, the message arrival rate must not be larger than the message consumption rate by the receiver, since an event-triggered message must conform to the *exactly once semantics* (see Sect. 4.3.3). In the short run, an arriving message burst can be buffered in the receiver queue until the queue is full. The specification of the proper queue length for bursty event-triggered messages is very important.

4.6.2   Operational Specification

From the point of view of communication, the data field of an arriving message can be considered as an *unstructured bit vector.* At the endpoints of the communication, the operational specification determines how this bit vector must be structured into *message variables.* A *message variable* is a syntactic unit that consists of a fixed part and a variable part (see Sect. 2.2.4). The information about how the data field of a message is structured in syntactic units is contained in a *message-structure declaration (MSD).* The MSD contains the *message variable names* (i.e., the fixed part of the message variable) that point to the relevant concepts on one side and, on the other side, specifies which part of the unstructured bit vector denotes *the value (the variable part) of a message variable.* In addition to the structure information, the MSD may contain *input assertions* for checking the validity of incoming data (e.g., to test if the data is within a permitted data domain that the receiving component is capable to handle) and *output assertions* for checking outgoing data. An incoming data element that passes the input assertion is called a *permitted data element.* An outgoing data element that passes the output assertion is called a *checked data element.* The formalism used for specifying the data structures and the assertions in the MSD depends on the available programming environment.

In many real-time systems, the MSD is static, i.e., it does not change over the lifetime of the system. In these systems, for performance reasons, the MSD is not transmitted in the message but stored in the memories of the communicating partners where the data fields must be handled.

The link between the *unstructured bit vector* arriving at a port and the associated MSD can be established by different means:

- The *MSD name* is assigned to the input port name. In this case only a single message type can be accepted at a port.
- The *MSD name* is contained in the data field of the message. In this case different message types can be received at the same port. This approach is followed in CAN (see Sect. 7.3.1).
- The *MSD name* is assigned to the cyclic arrival instant of a time-triggered message. In this case different message types can be received at the same port, without the need to store the MSD name in the data field of the message. This approach is followed in TTP (see Sect. 7.5.1).
- The *MSD name* is stored in a server that can be accessed by the receiver of a message. This approach is followed in CORBA [Sie00].
- The MSD itself is part of the message. This is the most flexible arrangement, at the cost of having to send the full MSD in every message. This approach is followed in service-oriented architectures (SOA) [Ray10].

4.6.3   Meta-Level Specification

The meta-level LIF specification assigns a meaning to the message variables exchanged between two communicating LIFs at the operational level and thus establishes *semantic interoperability*. It thus bridges the gap between the syntactic units and the user's mental model of the service provided at the interface. Central to this meta-level specification is the LIF service model. The LIF service model defines the concepts that are associated with the *message variable names* contained in the operational specification. These concepts will be qualitatively different for *closed components* and *open components* (see Sect. 4.4.5).

The LIF service model for a *closed* component can be formalized, since a closed component does not interact with the external environment. The relationship between the LIF inputs and LIF outputs depends on the discrete algorithms implemented within the closed component. There is no input from the external environment that can bring unpredictability into the component behavior. The sparse time base within a cluster is discrete and supports a consistent temporal order of all events.

The LIF service model for an *open* component is fundamentally different since it must encompass the inputs from the external environment, the local interfaces of the component in its interface specification. *Without knowing the context of use of an open component, only the operational specification of an open component can be provided.* Since the external physical environment is not rigorously definable, the interpretation of the external inputs depends on human understanding of the natural environment. The concepts used in the description of the LIF service model must thus fit well with the accustomed concepts within a user's *internal conceptual landscape* (see Sect. 2.2); otherwise the description will not be understood.

The discussion that follows focuses on LIFs of open components, since the systems we are interested in must interact with the external environment. The LIF service model of an open component must meet the following requirements:

- *User orientation*: Concepts that are familiar to a prototypical user must be the basic elements of the LIF service model. For example, if a user is expected to have an engineering background, terms and notations that are *common knowledge* in the chosen engineering discipline should be used in presenting the model.
- *Goal orientation*: A user of a component employs the component with the intent to achieve a goal, i.e., to contribute to the solution of her/his problem. The relationship between user intent and the services provided at the LIF must be exposed in the LIF service model.
- *System view*: A LIF service user (the system architect) needs to consider the system-wide effects of an interaction of the component with the external physical environment, i.e., effects that go beyond the component. The LIF service model is different from the model describing the algorithms implemented within a component, since these algorithms are within the component's boundaries.

Example: Let us analyze the simple case of a variable that contains a temperature. As any variable, it consists of the two parts, the *static variable name* and the *dynamic variable value*. The MSD contains the *static variable name* (let us assume the variable is named

Temperature-11) and the position where the *dynamic variable value* is placed in an arriving bit stream of the message. The meta-level specification explains the meaning of *Temperature-11* (see also the examples in Sect. 2.2.4).

4.7 Component Integration

A component is a self-contained validated unit that can be used as a building block in the construction of larger systems. In order to enable a straightforward composition of a component into a cluster of components, the following four *principles of composability* should be observed.

4.7.1 Principles of Composability

(i) *Independent Development of Components:* The architecture must support the precise specification of the linking interface (LIF) of a component in the domains of value and time. This is a necessary prerequisite for the independent development of components on one side and the reuse of existing components that is based solely on their LIF specification on the other side. While the operational specification of the value domain of interacting messages is *state-of-the-art* in embedded system design, the temporal properties of these messages are often ill defined. Many of the existing architectures and specification technologies do not deal with the temporal domain with the appropriate care. Note that the transport specification and the operational LIF specification are independent of the context of use of an open component, while the meta-level LIF specification of an open component depends on the context of use. *Interoperability* of open components is thus not the same as *interworking* of open component, since the latter assumes the compatibility of the meta-level specifications.

(ii) *Stability of Prior Services:* The *stability of prior services principle* states that the services of a component that have been validated in isolation (i.e., prior to the integration of the component into the larger system) remain intact after the integration (see the example in Sect. 4.4.1).

(iii) *Non-interfering Interactions:* If there exist two disjoint subgroups of cooperating components that share a common communication infrastructure, then the communication activities within one subgroup must not interfere with the communication activities within the other subgroup. If this principle is not satisfied, then the integration within one component-subgroup will depend on the proper behavior of the other (functionally unrelated) component-subgroups. These global interferences compromise the composability of the architecture.

Example: In a communication system where a single communication channel is shared by all components on a *first-come first-serve basis*, a *critical instant* is defined as an instant, when all senders start sending messages simultaneously. Let us assume that in such a system, ten components are to be integrated into a cluster. A given communication system is capable to handle the critical instant if eight components are active. As soon as the ninth and tenth components are integrated, sporadic timing failures are observed.

(iv) *Preservation of the Component Abstraction in the Case of Failures:* In a composable architecture, the introduced abstraction of a component must remain intact, even if a component becomes faulty. It must be possible to diagnose and replace a faulty component without any knowledge about the component internals. This requires a certain amount of redundancy for error detection within the architecture. This principle constrains the implementation of a component, because it restricts the implicit sharing of resources among components. If a shared resource fails, more than one component can be affected by the failure.

Example: In order to detect a faulty component that acts like a *babbling idiot*, the communication system must contain information about the permitted temporal behavior of every component. If a component fails in the temporal domain, the communication system cuts off the component that violates its temporal specification, thus maintaining the timely communication service among the correct components.

4.7.2   Integration Viewpoints

In order to bring an understandable structure into a large system, it makes sense to view— as seen from an integration viewpoint—a (original) cluster of components as a single gateway component. The integration viewpoint establishes a new cluster that consists of the respective gateway components of the original clusters. Viewed from an original cluster, the external (local) interface of the gateway becomes the LIFs of the new cluster, while, viewed from the new cluster, the LIF of the gateway to the original cluster is the local interface of the new cluster (see Sect. 4.5.2). The gateways to the new cluster make only those information items available to the new cluster that are of relevance for the operation of the new cluster.

Example: Figure 4.1 depicts a cluster of components that form the control system within an automobile. The vehicle-to-vehicle gateway component (the right lower component in Fig. 4.1) establishes a wireless link to other vehicles. In this example, we distinguish the following two levels of integration: (i) the integration of components into the cluster depicted in Fig. 4.1 and (ii) the integration of a car into a *dynamic system of cars* that is achieved via the car-to-car (C2C) gateway component. If we look at the integration of components within the cluster of Fig. 4.1, then the communication network interface (CNI) of the C2C gateway component is the *cluster LIF*. From the C2C communication viewpoint, the cluster LIF is the (unspecified) local interface of the C2C gateway component (see also the last paragraph of Sect. 4.5.2).

The hierarchical composition of components and clusters that leads to *distinct integration levels* is an important special case of the integration of components. Multilevelness is an important organizing principle in large systems. At the lowest integration level, primitive components (i.e., components that are considered to be atomic units and are not composed any further) are integrated to form a cluster. One distinct component of this cluster is a gateway component that forms, together with distinct gateway components of other clusters, a *higher-level cluster*. This process of integration can be continued recursively to construct a hierarchical system with distinct levels of integration (see also Sect. 11.7.2 on the recursive integration of components in the time-triggered architecture).

> **Example:** In the GENESYS [Obe09, p. 44] architecture, three integration levels are introduced. At the lowest level, the chip level, the components are IP cores of an MPSoC that interact by a network on chip. At the next higher level, chips are integrated to form a device. At the third integration level, devices are integrated to form *closed* or *open* systems. A *closed* system is a system in which the subsystems that form the systems are known a priori. In an open system, subsystems (or devices) join and leave the system dynamically, leading to a system of systems.

4.7.3   System of Systems

There are two reasons for the rising interest in *systems of systems*: (i) the realization of new functionality and (ii) the control of the complexity growth of large systems caused by their continuous evolution. The available technology (e.g., the Internet) makes it possible to interconnect *independently developed systems (legacy systems)* to form new *system of systems (SoS)*. The integration of different *legacy systems* into an SoS promises more efficient economic processes and improved services.

The continuous adaptations and modifications that are necessary to keep the services of a large system relevant in a dynamic business environment bring about a growing complexity that is hard to manage in a monolithic context [Leh85]. One promising technique to attack this complexity problem is to break a single large monolithic system up into a set of nearly autonomous constituent systems that are connected by well-defined message interfaces. As long as the *relied-upon properties* at these message interfaces meet the user intentions, the internal structure of the constituent systems can be modified without any adverse effect on the global system-level services. Any modification of the *relied-upon properties* at the message interfaces is carefully coordinated by a meta-entity that monitors and coordinates the system evolution. The SoS technology is thus introduced to get a handle on the complexity growth, caused by the necessary evolution of large systems, by introducing structure and applying the simplification principles of *abstraction, separation of concerns*, and *observability of subsystem interactions* (see Sect. 2.5).

The distinction between a *system of subsystems* and a *system of systems* is thus based on the degree of autonomy of the constituent systems [Mai98]. We call a monolithic system made out of subsystems that are designed according to a master plan and are in the *sphere of control* of a single development organization a *system of subsystem*. If the *systems* that cooperate are in the *sphere of control* of different development organizations, we speak of a *system of (constituent) systems*. Table 4.3 compares the different characteristics of a *monolithic system* versus a *system of systems*. The interactions of the autonomous constituent systems can evoke planned or unanticipated emergent behavior, e.g., *cascade effect* [Fis06], that must be detected and controlled (see also Sect. 2.4).

In many distributed real-time applications, it is not possible to bring *temporally accurate real-time information* to a central point of control within the available time interval between the *observation of the local environment* and the need to *control the local environment*. In these applications central control by a monolithic control system is not possible. Instead, the autonomous distributed controllers must cooperate to achieve the desired effects.

> **Example:** It is not possible to control the movement of the cars in an open road system, where cyclists and pedestrians interfere with the traffic flow of the cars, by a monolithic central control system because the amount and timeliness of the real-time information that must be transported to and processed by the central control system is not manageable within the required response times. Instead, each car performs autonomous control functions and cooperates with other cars in order to maintain an efficient flow of traffic. In such a system, a *cascade effect* of a traffic jam can occur due to emergent behavior if the traffic density increases beyond a tipping point.

Any ensemble of constituent systems that form an SoS must agree on a shared purpose, a chain of trust, and a shared ontology on the semantic level. These global properties must be established at the meta-level and are subject of a carefully managed continuous evolution. A new entity must be established at the meta-level that monitors and coordinates the activities of the constituent systems in order that the shared purpose can be achieved.

Table 4.3 Comparison of a *monolithic system* and a *system of systems*

Monolithic system	System of systems (SoS)
Sphere of control and system responsibility within a single development organization. Subsystems are obedient to a central authority	Constituent systems are in the sphere of control of different development organizations. Subsystems are autonomous and can only be influenced, but not controlled, by other subsystems
The architectural styles of the subsystems are aligned. Property mismatches are the exception	The architectural styles of the constituent systems are different. Property mismatches are the rule, rather than the exception
The LIFs that effectuate the integration are controlled by the responsible system organization	The LIFs that effectuate the integration are established by international standard organizations and outside the control of a single system supplier
Normally hierarchical composition that leads to levels of integration	Normally the interactions among the constituent systems follow a mesh network structure without clear integration levels
Subsystems are designed to interact in order to achieve the system goal: *Integration*	Constituent systems have their own goals that are not necessarily compatible with the SoS goal. Voluntary cooperation of systems to achieve a common purpose: *Interoperation*
Evolution of the components that form the subsystems is coordinated	Evolution of the constituent systems that form the SoS is uncoordinated
Emergent behavior controlled	Emergent behavior often planned, but sometimes unanticipated

An important characteristic of an SoS is the independent development and uncoordinated evolution of the constituent systems (Table 4.3). The focus in SoS design is on the linking interface behavior of the monolithic systems. The monolithic system themselves can be heterogeneous. They are developed according to different architectural styles by different organizations. If the monolithic systems are interconnected via open communication channels, then the topic of security is of utmost concern, since an outside attacker can interfere with the system operation, e.g., by executing a *denial-of-service* attack (see Sect. 6.2.2).

[Sel08, p. 3] discusses two important properties of an *evolutionary architecture*: (i) the complexity of the overall framework does not grow as constituent systems are added or removed, and (ii) a given constituent system does not have to be reengineered if other constituent systems are added, changed, or removed. This implies a precise specification and continuous revalidation of the interface properties (in the domains of value and time) of the constituent systems. The evolution of a constituent system will have no adverse effect on the overall behavior if the *relied-upon interface properties* of this constituent system are not modified. Since the precise specification of the temporal dimension of the relied-upon interface properties requires a time reference, the availability of a synchronized global time in all constituent systems of a large SoS is helpful, leading to a *time-aware architecture* (TAA; see Sect. 11.7). Such a global time can be established by reference to the global GPS signals (see Sect. 3.5). We call an SoS where all constituent systems have access to a synchronized global time a *time-aware SoS*.

The preferred interconnection medium for the construction of *systems of systems* is the Internet, leading to the Internet of Things (IoT). Chap. 13 is devoted to the topic of the *Internet of Things*.

Points to Remember

- A real-time component consists of a *design* (e.g., the software), an *embodiment* (e.g., the hardware, including a processing unit, memory, and an I/O interface), and a real-time clock that makes the component aware of the progression of real time.
- The *timed sequence of output messages* that a component produces at an interface with its environment is the *behavior* of the component at that interface. The *intended behavior* is called the *service*. An unintended behavior is called a *failure*.
- *Temporal control* is concerned with determining the instants in the domain of real time when tasks must be activated, while *logical control* is concerned with the control flow *within* a task.
- *Synchronous programming languages* distinguish cleanly between temporal control, which is related to the progression of real time, and logical control, which is related to execution time.

- A cycle, characterized by its *period* and *phase*, is associated with every time-triggered activity.
- At a given instant, the *state of a component* is defined as a data structure that contains the information of the past that is relevant for the future operation of the component.
- In order to enable the dynamic reintegration of a component into a running system, it is necessary to design periodic *reintegration instants* into the behavior, where the state at the reintegration instant is called the *ground state* of the component.
- A *message* is an atomic data structure that is formed for the purpose of communication, i.e., data transmission and synchronization, among components.
- *Event information* conveys the *difference* of the previous state observation and the current state observation. Messages that contain event information must adhere to the *exactly once* semantic.
- *State messages* support the *principle of independence* because sender and receiver can operate at different (independent) rates and there is no danger of buffer overflow.
- In real-time systems, the *information pull strategy* should be followed whenever possible.
- *Elementary interfaces* are *inherently simpler* than composite interfaces, because there is *no dependency* of the behavior of the sender on the behavior of the receiver.
- The *services* of a component are offered at its *cluster LIF* to the other components of the cluster. The cluster LIF is an *operational message-based* interface that is relevant for the integration of components into the cluster. The detailed structure, naming, and access mechanisms of the local interface of a component are *intentionally left unspecified* at its cluster LIF.
- Every system is developed according to an *architectural style*, i.e., a set of adopted rules and conventions for the conceptualization, representation of data, naming, programming, interaction of components, semantics of the data, and many more.
- Whenever a communication channel links two systems developed by two different organizations, it is highly probable that some of the properties of the messages that are exchanged across this channel are in disagreement because of the differences in architectural styles.
- A *gateway component* resolves property mismatches and exposes the external-world information in the form of cluster-standard messages at the cluster LIF of the gateway components.
- We distinguish between three parts of a LIF specification: (i) the *transport specification* of the messages, (ii) the *operational specification* of the messages, and (iii) the *meta-level specification* of the messages.
- Only the *operational specification* of an open component can be provided without knowing the context of use of the open component.
- The information on how the data field of a message is structured into syntactic units is contained in a *message-structure declaration (MSD)*. The MSD estab-

lishes the *message variable names* (i.e., the fixed part of the message variable) that point to the respective concepts and specify which part of the unstructured bit vector denotes the *variable part of a message variable*.

- The four principles of composability are (i) independent development of components, (ii) stability of prior services, (iii) non-interfering interactions, and (iv) preservation of the component abstraction in case of failures.
- Multilevelness is an important organizing principle in large systems.
- The distinction between a *system of subsystems* and a *system of systems* is based more on organizational than on technical grounds.

Bibliographic Notes

The presented real-time model of computation has been developed over the past 25 years and is documented in a number of publications, starting with *The Architecture of Mars* [Kop85] and further in the following publications: *Real-time Object Model* [Kim94], the *Time-Triggered Model of Computation* in [Kop98], *Elementary* versus *Composite Interfaces in Distributed Real-Time Systems* [Kop99], and *Periodic Finite State Machines* [Kop07].

Review Questions and Problems

4.1. How is a *real-time system component* defined? What are elements of a component? How is the behavior of a component specified?

4.2. What are the advantages of separating computational components from the communication infrastructure? List some of the consequences of this separation.

4.3. What is the difference between *temporal control* and *logical control?*

4.4. What is the definition of the *state* of a real-time system? What is the relationship between *time* and *state*? What is the *ground state*? What is a database component?

4.5. What is the difference between *event information* and *state information*? What is the difference in the handling of an *event message* from the handling a *state message*?

4.6. List and describe the properties of the four interfaces of a component? Why are the local interfaces of a component intentionally left *unspecified* at the architectural level?

4.7. What are the differences between an *information push interface* and an *information pull interface*? What are the differences between an *elementary interface* and a *composite interface*?

4.8. What do we mean by the term *architectural style*? What is a *property mismatch*?

4.9. What are the characteristics of a local process I/O interface and the LIF message interface?

4.10. What is the role of a gateway component?

4.11. What are the three parts of a linking interface specification?

4.12. What is the *message-structure declaration (MSD)?* How do we associate the MSD with the bit vector contained in a message?

4.13. List the *four principles of composability?*

4.14. What is an integration level? How many integration levels are introduced in the GENESYS architecture?

4.15. Assume that the pressures p_1 and p_2 between the first two pairs of rolls in Fig. 1.9 are measured by the two controller nodes and sent to the man-machine interface (MMI) node for verifying the following alarm condition:

> **when** $(p_1 < p_2)$
>   **then** *everything ok*
>   **else** *raise pressure alarm*;

The rolling mill is characterized by the following parameters: maximum pressure between the rolls of a stand = 1000 kp cm^{-2} [kp is kilopond], absolute pressure measurement error in the value domain = 5 kp cm^{-2}, and maximum rate of change of the pressure = 200 kp cm^{-2} sec^{-1}. It is required that the error due to the imprecision of the points in time when the pressures are measured at the different rolls should be of the same order of magnitude as the measurement error in the value domain, i.e., 0.5% of the full range. The pressures must be *continuously* monitored, and the first alarm must be raised by the alarm monitor within 200 msec (at the latest) after a process has possibly left the normal operating range. A second alarm must be raised within 200 msec after the process has definitely entered the alarm zone:

(a) Assume an event-triggered architecture. Each node contains a local real-time clock, but no global time is available. The minimum time d_{min} for the transport of a single message by the communication system is 1 msec. Derive the temporal control signals for the three tasks.

(b) Assume a time-triggered architecture. The clocks are synchronized with a precision of 10 µsec. The time-triggered communication system is characterized by a TDMA round of 10 msec. The time for the transport of a single message by the communication system is 1 msec. Derive the temporal control signals for the three time-triggered tasks.

(c) Compare the solutions of 4.15. (a) and 4.15. (b) with respect to the generated computational load and the load on the communication system. How sensitive are the solutions if the parameters, e.g., the jitter of the communication system or the duration of the TDMA round, are changed?

Chapter 5
Temporal Relations

Overview

The behavior of a real-time cluster must be based on timely information about the state of its physical environment and the state of other cooperating clusters. Real-time data is *temporally accurate* for a limited real-time interval only. If real-time data is used outside this application-specific time interval, the system will fail. It is the objective of this chapter to investigate the temporal relations among state variables in the different parts of a cyber-physical system.

In this chapter, the notions of a *real-time (RT) entity*, a *real-time (RT) image*, and a *real-time (RT) object* are introduced, and the concept of *temporal validity* of an RT image is established. The temporal validity of the RT image can be extended by state estimation. A real-time clock is associated with every RT object. The object's clock provides periodic temporal control signals for the execution of the object procedures, particularly for state estimation purposes. The granularity of the RT object's clock is aligned with the dynamics of the RT entity in the controlled object that is associated with the RT object. The notions of parametric and phase-sensitive observations of RT entities are introduced, and the concept of *permanence* of an observation is discussed. The duration of the action delay, which is the time interval between the transmission of a message and the instant when this message becomes *permanent*, is estimated.

The final section of this chapter is devoted to an elaboration of the concept of determinism. Determinism is a desired property of a computation that is needed if fault tolerance is to be achieved by the replication of components. Determinism is also helpful for testing and for understanding the operation of a system. A set of replicated RT objects is *replica deterministic* if the objects visit the same state at approximately the same future point in time. The main causes for a loss of determinism are failures that the fault-tolerance mechanisms are intended to mask and *nondeterministic design constructs* which must be avoided in the design of deterministic systems.

© The Author(s), under exclusive license to Springer Nature Switzerland AG 2022
H. Kopetz, W. Steiner, *Real-Time Systems*,
https://doi.org/10.1007/978-3-031-11992-7_5

5.1   Real-Time Entities

A *real-time (RT) entity* is a *state variable* of relevance for the given purpose. It is
located either in the environment of the computer system or in the computer system
itself. Examples of RT entities are the flow of a liquid in a pipe, the set point of a
control loop that is selected by the operator, or the intended position of a control
valve. An RT entity has static attributes that do not change during the lifetime of the
RT entity and dynamic attributes that change with the progression of real time.
Examples of static attributes are the name, the type, the value domain, and the maximum rate of change. The value set at a particular instant is the most important
dynamic attribute. Another example of a dynamic attribute is the rate of change at a
chosen instant.

5.1.1   Sphere of Control

Every RT entity is in the sphere of control (SOC) of a subsystem that has the authority to set the value of the RT entity [Dav79]. Outside its SOC, the RT entity can only
be observed, but the *semantic content* of the RT entity cannot be modified. At the
chosen level of abstraction, syntactic transformations of the representation of the
value of an RT entity that do not change its *semantic content* (see Sect. 2.2.4) are
disregarded.

> **Example**: Figure 5.1 shows another view of Fig. 1.8 and represents the small control sys-
> tem that controls the flow of a liquid in a pipe according to a set point selected by the opera-
> tor. In this example, there are three RT entities involved: the *flow in the pipe* is in the SOC
> of the controlled object, the *set point* for the flow is in the SOC of the operator, and the
> *intended position* of the control valve is in the SOC of the computer system.

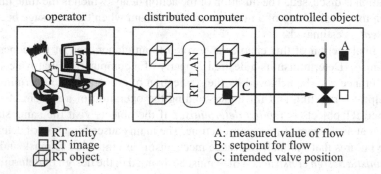

Fig. 5.1 RT entities, RT images, and RT objects

Fig. 5.2 Discrete RT entity

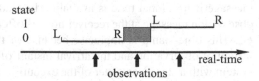

5.1.2   Discrete and Continuous Real-Time Entities

An RT entity can have a discrete value set (*discrete RT entity*) or a continuous value set (*continuous RT entity*). If the timeline proceeds from left to right, then the value set of a discrete RT entity is constant during an interval that starts with a left event (*L_event*) and ends with a right event (*R_event*)—see Fig. 5.2.

In the interval between an *R_event* and the next *L_event*, the set of values of a discrete RT entity is undefined. In contrast, the set of values of a continuous RT entity is always defined.

> **Example**: Consider a garage door. Between the defined states specified by *door closed* and *door open*, there are many intermediate states that can be classified neither as *door open* nor as *door closed*.

5.2   Observations

The information about the state of an RT entity at a particular instant is captured by the notion of an *observation*. An observation is an *atomic data structure*

$$\text{Observation} = \langle \text{Name}, t_{obs}, \text{Value} \rangle$$

consisting of the name of the RT entity, the instant when the observation was made (t_{obs}), and the observed value of the RT entity. A continuous RT entity can be observed at any instant, while the observation of a discrete RT entity gives a meaningful value only in the interval between a *L_event* and an *R_event* (see Fig. 5.2).

We assume that an *intelligent sensor node* is associated with a *physical sensor* to capture the physical signal, to generate the time-stamp, and to transform the physical signal to meaningful digital technical units. An observation should be transported in a single message from this sensor node to the rest of the system because the message concept provides for the needed atomicity of an observation message.

5.2.1   Untimed Observation

In a distributed system without global time, a time-stamp can only be interpreted within the scope of the node that created the time-stamp. The time-stamp of a sender that made an observation is thus meaningless at the receiver of the observation message if no global time is available. Instead, the time of arrival of an untimed observation message at the receiver node is often taken to be the time of observation t_{obs}. This time-stamp is imprecise because of the delay and the jitter between the instant of observation and the arrival instant of the message at its destination. In a system with a significant jitter of the execution time of the communication protocol (in comparison to the median execution time) and without access to a global time base, it is not possible to determine the instant of observation of an RT entity precisely. This imprecision of time measurement can reduce the quality of the observation (see Fig. 1.6).

5.2.2   *Indirect Observation*

In some situations, it is not possible to observe the value of an RT entity directly. Consider, for example, the measurement of the internal temperature within a slab of steel. This internal temperature (the value of the RT entity) must be measured indirectly.

The three temperature sensors T_1, T_2, and T_3 measure the change of temperature of the surface (Fig. 5.3) over a period of time. The value of the temperature T within the slab and the instant of its relevance must be inferred from these surface measurements by using a mathematical model of heat transfer.

5.2.3   *State Observation*

An observation is a *state observation* if the value of the observation contains the state of the RT entity. The time of the state observation refers to the point in real time when the RT entity was sampled (observed). Every reading of a state observation is self-contained because it carries an *absolute value*. Many control algorithms require a sequence of equidistant state observations, a service provided by periodic time-triggered readings.

The semantics of state observations matches well with the semantics of the state messages introduced in Sect. 4.3.4. A new reading of a state observation replaces the previous readings because clients are normally interested in the most recent value of a state variable.

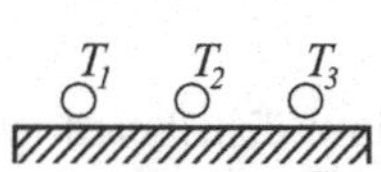

Fig. 5.3 Indirect measurement of an RT entity

temperature T

5.2.4 Event Observation

An *event* is an occurrence (a state change) that happens at an instant. Because an observation is also an event, it is not possible to observe an event in the controlled object directly. It is only possible to observe the *consequences of the controlled object's event* (Fig. 5.4), i.e., the subsequent state. An observation is an *event observation* if it contains the *change in value* between the *old* and the *new* states. The instant of the event observation denotes the best estimate of the instant of the occurrence of this event. Normally, this is the time of the L-event of the new state.

There are a number of problems with event observations:

(i) Where do we get the precise time of the event occurrence? If the event observation is event-triggered, then the time of event occurrence is assumed to be the rising edge of the interrupt signal. Any delayed response to this interrupt signal will cause an error in the time-stamp of the event observation. If the event observation is time-triggered, then the time of *event occurrence* can be at any point within the sampling interval.

(ii) Since the value of an event observation contains the *difference* between the old state and the new state (and not the absolute state), the loss or duplication of a single event observation causes the loss of state synchronization between the state of the observer and the state of the receiver. From the point of view of reliability, event observations are more fragile than state observations.

(iii) An event observation is only sent if the RT entity changes its value. The latency for the detection of a failure (e.g., a *crash*) of the observer node cannot be bounded because the receiver assumes that the RT entity has not changed its value if no new event message arrives.

On the other hand, event observations are more *data-efficient* than state observations in the case where the RT entity does not change frequently.

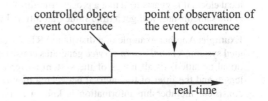

Fig. 5.4 Observation of an event

controlled object event occurence

point of observation of the event occurence

real-time

5.3   Real-Time Images and Real-Time Objects

5.3.1   Real-Time Images

A real-time (RT) image is a *current* picture of an RT entity. An RT image is *valid* at a given instant if it is an accurate representation of the corresponding RT entity, both in the value and the time domains. The notion of *temporal accuracy* of an RT image will be discussed in detail in the next section. While an observation records a fact that remains valid forever (a statement about an RT entity that has been observed at a particular instant), the validity of an RT image is *time dependent* and thus invalidated by the progression of real time. RT images can be constructed from up-to-date state observations or from up-to-date event observations. They can also be estimated by a technique called *state estimation* that will be discussed in Sect. 5.4.3. RT images are stored either inside the computer system or in the environment (e.g., in an actuator).

5.3.2   Real-Time Objects

A real-time (RT) object is a *container* within a node of the distributed computer system that holds an RT image or an RT entity [Kop90]. A real-time clock with a specified granularity is associated with every RT object. Whenever this object clock ticks, a temporal control signal is relayed to the object to activate an object procedure [Kim94].

Distributed RT Objects   In a distributed system, an RT object can be replicated in such a manner that every local site has its own version of the RT object to provide the specified service to the local site. The quality of service of a distributed RT object must conform to some specified consistency constraints.

> **Example**: A good example of a distributed RT object is *global time*; every node has a local clock object that provides a synchronized time service with a specified precision Π (quality of service attribute of the internal clock synchronization). Whenever a process reads its local clock, it is guaranteed that a process running on another node that reads its local clock at the same instant will get a time value that differs by at most one tick.

> **Example**: Another example of a distributed RT object is a *membership service* in a distributed system. A *membership service* generates consistent information about the state (operational or failed) of all nodes of the system at agreed instants (*membership points*). The length and the jitter of the interval between a membership point and the instant when the consistent membership information is known at the other nodes are *quality of service parameters* of the membership service. A responsive membership service has a small maximum delay between the instant of a relevant state change of a node (failure or join) and the instant at which all other nodes have been informed in a consistent manner of this state change.

5.4 Temporal Accuracy

Temporal accuracy denotes the temporal relationship between an *RT entity* and its associated *RT image*. Because an RT image is stored in an RT object, the temporal accuracy can also be viewed as a relation between an RT entity and an RT object.

5.4.1 Definition

The temporal accuracy of an RT image is defined by referring to the *recent history* of observations of the related RT entity. A recent history RH_i at time t_i is an ordered set of instants $\{t_i, t_{i-1}, t_{i-2}, \ldots t_{i-k}\}$, where the length of the recent history, $d_{acc} = z(t_i) - z(t_{i-k})$, is called the *temporal accuracy interval* or the *temporal accuracy* d_{acc} ($z(e)$ is the time-stamp of event e generated by the reference clock z; see Sect. 3.1.2). Assume that the RT entity has been observed at every instant of the recent history. An RT image is temporally accurate at the present time t_i if

$$\exists t_j \in RH_i : \text{Value}\left(RT\,\text{image at}\,t_i\right) = \text{Value}\left(RT\,\text{entity at}\,t_j\right)$$

The current value of a temporally accurate RT image is a member of the set of observations in the recent history of the corresponding RT entity. Because the transmission of an observation message from the observing node to the receiving node takes some amount of time, the RT image lags behind the RT entity (see Fig. 5.5).

> **Example**: Let us assume that the temporal accuracy interval of a temperature measurement is 1 min. An RT image is temporally accurate if the value contained in the RT image has been observed at most a minute ago, i.e., it is still in the recent history of the corresponding RT entity.

Temporal Accuracy Interval The size of the admissible temporal accuracy interval d_{acc} is determined by the dynamics of the RT entity in the controlled object. The delay between the observation of the RT entity and the use of the RT image causes an error(t) of the RT image that can be approximated by the product of the gradient of the value v of the RT entity multiplied by the length of the interval between the instant of observation and the instant of its use (see also Fig. 1.6):

Fig. 5.5 Time lag between RT entity and RT image

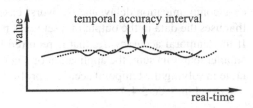

$$\text{error}(t) = \frac{\mathrm{d}v(t)}{\mathrm{d}t}\left(z(t_{use}) - z(t_{obs})\right)$$

If a temporally valid RT image is used, the worst-case error,

$$\text{error} = \left(\max_{\forall t} \frac{\mathrm{d}v(t)}{\mathrm{d}t} d_{acc}\right),$$

is given by the product of the maximum gradient and the temporal accuracy d_{acc}. In a balanced design, this worst-case error caused by the temporal delay is in the same order of magnitude as the worst-case measurement error in the value domain and is typically a fraction of a percentage point of the full range of the measured variable.

If the RT entity changes its value quickly, a short accuracy interval must be maintained. Let us call t_{use} the instant when the result of a computation using an RT image is applied to the environment. For the result to be accurate, it must be based on a temporally accurate RT image, i.e.:

$$z(t_{obs}) \leq z(t_{use}) \leq \left(z(t_{obs}) + d_{acc}\right)$$

where d_{acc} is the accuracy interval of the RT image. If this important condition is transformed, it follows that

$$\left(z(t_{use}) - z(t_{obs})\right) \leq d_{acc}.$$

Phase-Aligned Transaction Consider the case of an RT transaction consisting of the following *phase synchronized tasks*: the computational task at the sender (observing node) with a worst-case execution time WCET_{send}, the message transmission with a worst-case communication delay WCCOM, and the computational task at receiver (actuator node) with a worst-case execution time WCET_{rec} (Fig. 5.6). Such a transaction is called a *phase-aligned transaction*.

In such a transaction, the worst-case difference between the point of observation and the point of use,

$$\left(t_{use} - t_{obs}\right) = \text{WCET}_{send} + \text{WCCOM} + \text{WCET}_{rec},$$

is given by the sum of the worst-case execution time of the sending task, the worst-case communication delay, and the worst-case execution time of the receiving task that uses the data in the output of a set point to the actuator in the controlled object. If the temporal accuracy d_{acc} that is required by the dynamics of the application is smaller than this sum, the application of a new technique, *state estimation*, is inevitable in solving the temporal accuracy problem. The technique of state estimation is discussed in Sect. 5.4.3.

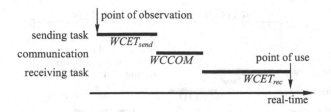

Fig. 5.6 Synchronized actions

Table 5.1 Temporal accuracy intervals in engine control

RT image within computer	Max. change	Accuracy	d_{acc}
Position of piston within cylinder	6000 rpm	0.1°	3 μsec
Position of accelerator pedal	100%/sec	1%	10 msec
Engine load	50%/sec	1%	20 msec
Temperature of the oil and the coolant	10%/minute	1%	6 s

Example: Let us analyze the required temporal accuracy intervals of the RT images that are used in a controller of an automobile engine (Table 5.1) with a maximum rotational speed of 6000 revolutions per minute (rpm). There is a difference of more than six orders of magnitude in the temporal accuracy intervals of these RT images. It is evident that the d_{acc} of the first data element, namely, the position of the piston within the cylinder, requires the use of state estimation.

5.4.2   Classification of Real-Time Images

Parametric RT Image   Assume that an RT image is updated periodically by a state observation message from the related RT entity with an update period d_{update} (Fig. 5.7), and assume that the transaction is phase-aligned at the sender. If the temporal accuracy interval d_{acc} satisfies the condition

$$d_{acc} > (d_{update} + WCET_{send} + WCCOM + WCET_{rec}),$$

then we call the RT image *parametric* or *phase-insensitive*.

A parametric RT image can be accessed at the receiver at any time without having to consider the phase relationship between the incoming observation message and the point of use of the data.

Example: The RT transaction that handles the position of the accelerator pedal (observation and preprocessing at sender, communication to the receiver, processing at the receiver, and output to the actuator) takes an amount of time

$WCET_{send} + WCCOM + WCET_{rec} = 4$ msec.

Because the accuracy interval of this observation is 10 msec (Table 5.1), messages sent with periods less than 6 msec will make this RT image *parametric*.

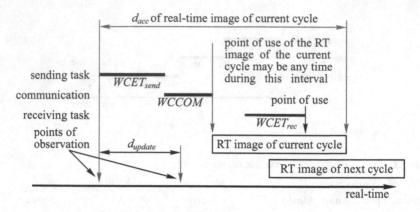

Fig. 5.7 Parametric real-time image

If components are replicated, then care must be taken that all replicas access the same version of a parametric RT image; otherwise the *replica determinism* (see Sect. 5.6) will be lost.

Phase-Sensitive RT Image   Assume an RT transaction that is phase-aligned at the sender. The RT image at the receiver is called *phase-sensitive* if

$$d_{acc} \leq \left(d_{update} + \text{WCET}_{send} + \text{WCCOM} + \text{WCET}_{rec} \right) \text{and}$$
$$d_{acc} > \left(\text{WCET}_{send} + \text{WCCOM} + \text{WCET}_{rec} \right)$$

In this case, the phase relationship between the moment at which the RT image is updated and the moment at which the information is used must be considered. In the above example, an update period of more than 6 msec, e.g., 8 msec, would make the RT image phase-sensitive.

Every phase-sensitive RT image imposes an additional constraint on the scheduling of the real-time task that uses this RT image. The scheduling of a task that accesses phase-sensitive RT images is thus significantly more complicated than the scheduling of tasks using parametric RT images. It is good practice to minimize the number of RT images that are phase-sensitive. This can be done, within the limits imposed by d_{update}, by either increasing the update frequency of the RT image or by deploying a state estimation model to extend the temporal accuracy of the RT image. While an increase in the update frequency puts more load on the communication system, the implementation of a state estimation model puts more load on the processor. A designer has the choice to find a tradeoff between utilizing communication resources and processing resources.

5.4.3 State Estimation

State estimation involves the building of a model of an RT entity inside an RT object to compute the probable state of an RT entity at a selected future instant and to update the corresponding RT image accordingly. The state estimation model is executed periodically within the RT object that stores the RT image. The control signal for the execution of the model is derived from the tick of the real-time clock that is associated with the RT object (see Sect. 5.3.2). The most important future instant where the RT image must be in close agreement with the RT entity is t_{use}, the instant where the value of the RT image is used to deliver an output to the environment. State estimation is a powerful technique to extend the temporal accuracy interval of an RT image, i.e., to bring the RT image into better agreement with the RT entity.

> **Example**: Assume that the crankshaft in an engine rotates with a rotational speed of 3000 revolutions per minute, i.e., 18 degrees per millisecond. If the time interval between the instant of observation, t_{obs}, of the position of the crankshaft and the instant of use, t_{use}, of the corresponding RT image is 500 microseconds, we can update the RT image by 9 degrees to arrive at an estimate of the position of the crankshaft at t_{use}. We could improve our estimate if we also consider the angular acceleration or deceleration of the engine during the interval $[t_{obs}, t_{use}]$.

An adequate state estimation model of an RT entity can only be built if the behavior of the RT entity is governed by a known and regular process, i.e., a well-specified physical or chemical process. Most technical processes, such as the above-mentioned control of an engine, fall into this category. However, if the behavior of the RT entity is determined by chance events, then the technique of state estimation is not applicable.

Input to the State Estimation Model The most important dynamic input to the state estimation model is the precise length of the time interval $[t_{obs}, t_{use}]$. Because t_{obs} and t_{use} are normally recorded at different nodes of a distributed system, a communication protocol with minimal jitter or a global time base with a good precision is a prerequisite for state estimation. This prerequisite is an important requirement in the design of a field bus.

If the behavior of an RT entity can be described by a continuous and differentiable function $v(t)$, the first derivative dv/dt is sometimes sufficient in order to obtain a reasonable estimate of the state of the RT entity at the instant t_{use} in the neighborhood of the instant of observation:

$$v(t_{use}) \approx v(t_{obs}) + (t_{use} - t_{obs})\frac{dv}{dt}$$

If the precision of such a simple approximation is not adequate, a more elaborate series expansion around t_{obs} can be carried out. In other cases a more detailed mathematical model of the process in the controlled object may be required. The execution of such a mathematical model can demand considerable processing resources.

Fig. 5.8 Latency at sender and receiver

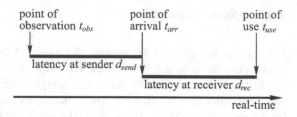

5.4.4 Composability Considerations

Assume a time-triggered distributed system where an RT entity is observed by the sensor node, and the observation message is then sent to one or more nodes that interact with the environment. The length of the relevant time interval $[t_{obs}, t_{use}]$ is thus the sum of the delay at the sender, given by the length $[t_{obs}, t_{arr}]$, and the delay at the receiver, given by the length $[t_{arr}, t_{use}]$ (the communication delay is subsumed in the sender delay). In a time-triggered architecture, all these intervals are static and known a priori (Fig. 5.8).

If the state estimation is performed in the RT object at the receiver, then any modification in the delay at the sender will cause a modification of the time interval that must be compensated by the state estimation of the receiver. The receiver software must be changed if a latency change takes place inside the sender node. To decrease this coupling between the sender and the receiver, the state estimation can be performed in two steps: the sender performs a state estimation for the interval $[t_{obs}, t_{arr}]$ and the receiver performs a state estimation for the interval $[t_{arr}, t_{use}]$. This gives the receiver the illusion that the RT entity has been observed at the point of arrival of the observation message at the receiver. The point of arrival is then the implicit time-stamp of the observation, and the receiver is not affected by a schedule change at the sender. Such an approach helps to unify the treatment of sensor data that are collected via a field bus as well as directly by the receiving node.

5.5 Permanence and Idempotency

5.5.1 Permanence

Permanence is a relation between a particular message arriving at a node and the set of all messages that have been sent to this node before this particular message. A particular message becomes *permanent* at a given node at that point in time when the node knows that all the messages that have been sent to it prior to the send time of this message have arrived (or will never arrive) [Ver94].

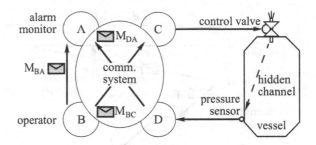

Fig. 5.9 Hidden channel in the controlled object

Example: Consider the example of Fig. 5.9, where the pressure in a vessel is monitored by a distributed system. The alarm-monitoring node (node A) receives a periodic message M_{DA} from the pressure-sensor node (node D). If the pressure changes abruptly for no apparent reason, the alarm-monitoring node A should raise an alarm. Suppose that the operator node B sends a message M_{BC} to node C to open the control valve in order to release the pressure. At the same time, the operator node B sends a message M_{BA} to node A, to inform node A about the opening of the valve, so that node A will not raise an alarm due to the anticipated drop in pressure.

Assume that the communication system has a minimum protocol execution time d_{min} and a maximum protocol execution time d_{max}, i.e., a jitter $d_{jit} = d_{max} - d_{min}$. Then the situation depicted in Fig. 5.10 could occur. In this figure, the message M_{DA} from the pressure sensor node arrives at the alarm-monitoring node A before the arrival of the message M_{BA} from the operator (that informs the alarm-monitoring node A of the anticipated drop in pressure). The transmission delay of the *hidden channel* in the controlled object between the opening of the valve and the changing of the pressure sensor is shorter than the maximum protocol execution time. Thus, to avoid raising any false alarms, the alarm-monitoring node should delay any action until the alarm message M_{DA} has become *permanent*.

Action Delay The time interval between the start of transmission of a given message and the point in time when this message becomes permanent at the receiver is called the *action delay*. The receiver must delay any action on the message until *after* the action delay has passed to avoid an incorrect behavior.

Irrevocable Action An *irrevocable action* is an action that cannot be undone. An irrevocable action causes a lasting effect in the environment. An example of an irrevocable action is the activation of the firing mechanism on a firearm. It is particularly important that an irrevocable action is triggered only after the action delay has passed.

Example: The pilot of a fighter aircraft is instructed to eject from the airplane (irrevocable action) immediately after a critical alarm is raised. Consider the case where the alarm has been raised by a message that has not become permanent yet (e.g., event *4* in Fig. 5.10). In this example, the hidden channel, which was not considered in the design, is the cause for the loss of the aircraft.

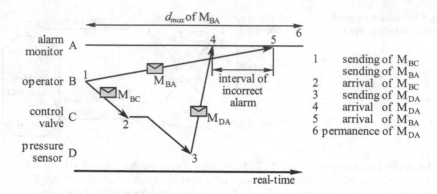

Fig. 5.10 Permanence of messages

5.5.2 *Duration of the Action Delay*

The duration of the action delay depends on the jitter of the communication system and the temporal awareness of the receiver. Let us assume the position of the omniscient outside observer who can see all significant events.

Systems with a Global Time In a system with global time, the send time t_{send} of the message, measured by the clock of the sender, can be part of the message and can be interpreted by the receiver. If the receiver knows that the maximum delay of the communication system is d_{max}, then the receiver can infer that the message will become permanent at $t_{permanent} = t_{send} + d_{max} + 2g$, where g is the granularity of the global time base (see Sect. 3.2.4 to find out where the $2g$ comes from).

Systems Without a Global Time In a system without global time, the receiver does not know when the message has been sent. To be on the safe side, the receiver must wait $d_{max} - d_{min}$ time units after the arrival of the message, even if the message has already been d_{max} units in transit. In the worst case, as seen by the outside observer, the receiver thus has to wait for an amount of time

$$t_{permanent} = t_{send} + 2d_{max} - d_{min} + g_l$$

before the message can be used safely (where g_l is the granularity of the local time base). Since in an *event-triggered communication system* ($d_{max} - d_{min} + g_l$) is normally much larger than $2g$, where g is the granularity of the global time, a system without a global time base is significantly *slower* than a system with a global time base. (In this case, the implementation of a time-triggered communication system is not possible, since we operate under the assumption that no global time base is available.)

5.5.3   Accuracy Interval Versus Action Delay

An RT image may only be used if the message that transported the image is permanent and the image is temporally accurate. In a system without state estimation, both conditions can only be satisfied in the time window $(t_{permanent}, t_{obs} + d_{acc})$. The temporal accuracy d_{acc} depends on the dynamics of the control application, while $(t_{permanent} - t_{obs})$ is an implementation-specific duration. If an implementation cannot meet the temporal requirements of the application, then state estimation may be the only alternative left in order to design a correct real-time system.

5.5.4   Idempotency

Idempotency is the relationship among the members of a set of replicated messages arriving at the same receiver. A set of replicated messages is *idempotent* if the effect of receiving more than one copy of a message is the same as receiving only a single copy. If messages are idempotent, the implementation of fault tolerance by means of replicating messages is simplified. No matter whether the receiver receives one or more of the replicated messages, the result is always the same.

> **Example**: Let us assume that we have a distributed system without synchronized clocks. In such a system, only untimed observations can be exchanged among nodes, and the time of arrival of an observation message is taken as the time of observation. Assume a node observe an RT entity, e.g., a valve, and report this observation to other nodes in the system. The receivers use this information to construct an updated version of the local RT image of the RT entity in their RT objects. A state message might contain the absolute value *position of valve at 45°* and will replace the old version of the image. An event message might contain the relative value *valve has moved by 5°*. The contents of this event message are added to the previous contents of the state variable in the RT object to arrive at an updated version of the RT image. While the state message is idempotent, the event message is not. A loss or duplication of the event message results in a permanent error of the RT image.

5.6   Determinism

Determinism is a *property* of a computation that makes it possible to predict the future result of a computation, given that the *initial state* and all *timed inputs* are known. A given computation is either *deterministic* or *not deterministic*.

> **Example**: Consider the case of a fault-tolerant brake-by-wire system in a car. After agreement on the sensor inputs (e.g., brake pedal position, speed of the car, etc.), three independent but synchronized channels process *identical* inputs. The three outputs are presented to four smart voting actuators (Fig. 9.8), one at the brake cylinder of each one of the four wheels of the car. After the arrival of the first output message at a voting actuator, an *acceptance window* is opened. The duration of the acceptance window is determined by the differences in the execution speeds and the jitter of the communication system of the three

channels, provided they operate correctly. Every correct deterministic channel will deliver the same result before the end of the acceptance window. If one channel fails, one of the three arriving result messages will contain a value that is different from the other two (*value failure*), or only two (identical) result messages will arrive during the *acceptance window* (*timing failure*). By selecting the majority result at the end of the *acceptance window*, the voter will mask a failure of any one of the three channels. The *endpoint* of the acceptance window is the *significant event* when the voting actions can be performed and the result can be transmitted to the environment. If the computations and the communication system have a large jitter, then this endpoint of the acceptance window is far in the future, and the responsiveness of the computer system is reduced.

5.6.1   Definition of Determinism

In Sect. 2.5 under the topic of *How can we achieve simplicity?*, the *principle of causality* has been introduced. Causality refers to the unidirectional relationship that connects an *effect* to a *cause* [Bun08]. If this relationship is one of *logical and temporal entailment*, we speak of *determinism*, which we define as follows: *A physical system behaves deterministically if, given an initial state at instant t and a set of future timed inputs, then the future states and the values and times of future outputs are entailed.* The words *time* and *instants* refer to the progression of dense (physical) time. Many natural laws of physical systems conform to this definition of determinism. In a digital computer model of a physical system, there is no dense time. In a deterministic distributed computer system, we must assume that all events, e.g., the observation of the initial state at instant *t* and the timed inputs, are *sparse events* on a *sparse* global time base (see Sect. 3.3) in order that the temporal properties of and the relations (such as *simultaneity*) among the events that occur in the different nodes of the distributed system can be precisely specified despite the finite precision of the clock synchronization and the discrete time base. This transformation of dense events in the physical world to sparse events in the cyber world (the distributed computer system), performed by an agreement protocol (see Sect. 3.3.1), reduces the *faithfulness* of the computer model, since events that are closer than the granularity of the time base cannot be ordered consistently.

In a real-time context, the concept of determinism requires that the behavior of a system is predictable in the domains of *values* and *time*. Neglecting the temporal dimension leads to a reduced notion of determinism—we call it *logical (L) determinism*. L-determinism can be defined as follows: *A system behaves L-deterministically if, given an initial state and a set of ordered inputs, then the subsequent states and the values of subsequent outputs are entailed.*

The use of the word *determinism* in everyday language relates the future behavior of a system as a consequence of its present state. Since in a timeless system the concept of *future* does not exist, *L-determinism* does not capture the everyday meaning of the word *determinism*.

Example: In the above example of a braking system, it is not sufficient for the establishment of correctness to demand that the braking action will *eventually* take place. The maintenance of an upper real-time bound for the start of the braking action (the endpoint of the acceptance window), e.g., *that the braking action will start 2 milliseconds after the brake pedal has been pressed*, is an integral part of correct behavior.

Deterministic behavior of a component is desired for the following reasons:

*   An *entailment relation* between initial state, input, output, and time simplifies the *understanding* of the real-time behavior of the component (see also Sect. 2.1.1).
*   Two replicated components that start from the same initial state and receive the same timed inputs will produce the same results at about the same time. This property is important if the results of a faulty channel are to be masked (outvoted) by the correct results of two correct channels (see Sect. 6.4.2) as exemplified in the above example on the braking system of a car.
*   The testability of the component is simplified, since every test case can be reproduced, eliminating the appearance of spurious *Heisenbugs* (see Sect. 6.1.2).

Determinism is a *desired property* of behavior. The implementation of a computation will achieve this desired property with an *estimated probability*.

An implementation can fail to meet this *desired property of determinism* for the following reasons:

  (i)   The initial states of the computations are not precisely defined.
 (ii)   The hardware fails due to a random physical fault.
(iii)   The notion of *time* is unclear.
(iv)   The system (software) contains design errors or *nondeterministic design constructs* (NDDCs) that lead to unpredictable behavior in the value domain or in the temporal domain.

From the point of view of fault tolerance, every loss of determinism of a replicated channel is tantamount to a failure of that channel that eliminates the further fault-masking capability of the fault-tolerant system.

In order to realize replica-deterministic behavior in an implementation of a fault-tolerant distributed real-time computer system, we must ensure that:

*   The initial state of all involved computations is defined consistently. It is *impossible* to build a replica-deterministic distributed real-time system without the establishment of some sort of a sparse global time base for the consistent time-stamping of the events in order to be able to determine whether an event is included in the initial state or not. Without a sparse global time base and sparse events, simultaneity cannot be resolved consistently in a distributed system, possibly resulting in an inconsistent temporal order of the replicated messages that report about these simultaneous events. Inconsistent ordering results in the loss of *replica determinism*.
*   The assignment of events to a sparse global time base can be established at the system level by the generation of *sparse events* or at the application level by the execution of agreement protocols which assign consistently dense events to sparse intervals.

- The message-transport system among the components is *predictable*, i.e., the instants of message delivery can be foreseen, and the temporal order of the received messages is the same as the temporal order of the sent messages *across all* channels.
- The computer system and the observer (user) agree on a precise notion of real time.
- All involved computations are *certain*, i.e., there are no program constructs in the implementation that produce arbitrary results or contain NDDCs and the final result of a computation will be available during the anticipated acceptance window.

If any one of the above conditions is not satisfied, then the fault-masking capability of a fault-tolerant system may be reduced or lost.

5.6.2   Consistent Initial States

Correct replicated channels that are introduced to mask a failure will only produce identical results if they start from the same initial state and receive the same inputs at the same instants.

According to Sect. 4.2.1, the *state* of a component can only be defined if there is a consistent separation of *past events* from *future events*. The sparse time model, introduced in Sect. 3.3, provides for such a consistent separation of past events from future events and makes it possible to define the instants where the initial state of a distributed system is consistently defined. Without a sparse global time, the establishment of a consistent initial state of replicated components of a distributed system is difficult.

A sensor is a physical device that will eventually fail. In order to mask the failure of a sensor, multiple sensors must be provided in a fault-tolerant system that measure, either directly or indirectly, the same physical quantity. There are two reasons why redundant observations of a physical quantity by replicated sensors will deviate:

(i) It is impossible to build *perfect* sensors. Every *real sensor* has a finite measurement error that limits the accuracy of the observed value.
(ii) The quantities in the physical world are normally *analog values*, but their representations in cyberspace are *discrete values*, leading to a *discretization error*.

It is therefore necessary to execute agreement protocols at the boundary between the physical world and cyberspace in order that all replicated channels receive the *consistent (exactly the same) agreed input data* (see Sect. 9.6). These agreement protocols will present the same set of values at the same sparse time interval to all replicated channels.

5.6.3 Nondeterministic Design Constructs (NDDCs)

A distributed computation that starts from a well-defined initial state can fail to reach the envisioned goal state for the following reasons:

(i) A hardware fault or design error causes the computation to crash, to deliver an incorrect result, or to delay the computation beyond the end of the agreed temporal acceptance window. It is the goal of a fault-tolerant design to mask these kinds of failures.

(ii) The communication system or the clocking system fails.

(iii) A nondeterministic design construct (NDDC) destroys the determinism. A loss of determinism, caused by an NDDC, eliminates the fault-masking capability of a fault-tolerant system.

The undesired effect of an NDDC can be in the *value domain* or in the *temporal domain*. A basic assumption in the design of a fault-tolerant system that masks failures by comparing the results of replica-deterministic channels is the statistical independence of failures in different channels. This assumption is violated if an NDDC is the cause of the loss of determinism, because the same NDDC may appear in all replicated channels. This leads to dangerous correlated failures of the replicated channels.

The following list is indicative of constructs that can lead to a loss of determinism in the value domain (i.e., *L-determinism*):

(i) *Random number generator:* If the result of a computation depends on a random number that is different for each channel, then the determinism is lost. Communication protocols that resolve a media-access conflict by reference to a random number generator, such as the bus-based CSMA/CD Ethernet protocol, exhibit *nondeterminism*.

(ii) *Nondeterministic language features:* The use of a programming language with nondeterministic language constructs, such as the SELECT statement in an ADA program, can lead to the loss of replica determinism. Since the programming language does not define which alternative is to be taken at a decision point, it is left up to the implementation to decide the course of action to be taken. Two replicas may take different decisions.

(iii) *Major decision point:* A *major decision point* is a decision point in an algorithm that provides a choice between a set of significantly different courses of action. If the replicated components select different computational trajectories at a major decision point, then the states of the replicas will start to diverge.

Example: Consider the case where the *result of a timeout check* determines whether a process continues or backtracks. This is an example for a major decision point.

(iv) *Preemptive scheduling:* If dynamic preemptive scheduling is used, then the points in the computations where an external event (interrupt) is recognized may differ at the different replicas. Consequently, the interrupting processes see different states at the two replicas at the point of interruption. They may reach different results at the next major decision point.

(v) *Inconsistent message order:* If the message order in the replicated communication channels is not identical, then the replicated channels may produce different results.

Most of the above constructs can also cause a loss of determinism in the temporal domain. Additionally, the following mechanisms and inadequacies must be considered that can cause a loss of the *temporal dimension of determinism*, even if the system is *L-deterministic*:

(i) *Task preemption and blocking:* Task preemption and blocking extend the execution time of tasks and may delay a result until the acceptance window has been closed.

(ii) *Retry mechanisms:* Any retry mechanism in hardware or software leads to an extension of the execution time and can cause an unacceptable delay of a value-correct result.

(iii) *Race conditions*: A *semaphore wait operation* can give rise to nondeterminism, because of the uncertain outcome regarding the process that will win the race for the semaphore. The same argument applies to communication protocols that resolve the access conflict by relying on the outcome of nondeterministic temporal decisions, such as CAN or, to a lesser degree, ARINC 629.

In some designs where NDDCs are present, an attempt is made to reestablish replica determinism by explicitly coordinating the decisions that could lead to a loss of determinism among the replicas, e.g., by distinguishing between a *leader* process and a *follower* process [Pow91]. If at all possible, inter-replica coordination should be avoided because it compromises the independence of the replicas and requires additional time and additional communication bandwidth.

5.6.4 Recovery of Determinism

A loss of determinism in an L-deterministic system can be avoided if the *acceptance window* is extended such that the probability of a *deadline miss* (i.e., that the result is available at the end of the acceptance window) is reduced to an acceptably low value. This technique is often used to reestablish determinism at the macro-level, even if the precise temporal behavior at the micro-level cannot be predicted. The main disadvantage of this technique is the increased delay until a result is delivered, which causes an increase in the *dead time* of control loops and the reaction time of reactive systems.

Example: Many *natural laws* at the level of Newtonian physics are considered to be *deterministic*, although the underlying quantum-mechanical processes at the micro-level are *nondeterministic*. The abstraction of *deterministic behavior at the macro-level* is possible because the large number of involved particles and the large time spans at the macro level, relative to the duration of the processes at the micro-level, make it highly improbable that nondeterministic behavior can be observed at the macro-level.

Example: In a server farm of a cloud, where more than 100,000 *L-deterministic* virtual machines (VMs) can be active at any instant, a failed VM can be reconfigured and restarted such that the intended result is still made available within the *specified acceptance window*. Such a system will have a deterministic behavior at the *external level* (see the four-universe model in Sect. 2.3.1), although the implementation at the lower *informational level* behaves nondeterministically.

The recovery of determinism at the *external level* (see the *four-universe model* in Sect. 2.3.1) of systems that behave nondeterministically at the level of the implementation is an important strategy when developing an understandable model of the behavior of a system of systems at the user level.

Points to Remember

* An observation of an RT entity is an atomic triple <Name, t_{obs}, Value> consisting of the name of the RT entity, the point in real time when the observation was made (t_{obs}), and the observed value of the RT entity. A continuous RT entity has always a defined value set and can be observed at any instant, whereas a discrete RT entity can only be observed between the *L_event* and the *R_event*.
* An observation is a *state observation* if the value of the observation contains the absolute state of the RT entity. The time of the state observation refers to the point in real time when the RT entity was sampled.
* An observation is an *event observation* if it contains information about the *change of value* between the *old state* and the *new state*. The time of the event observation denotes the best estimate of the instant of this event.
* A *real-time (RT) image* is a *current* picture of an RT entity. An RT image is *valid* at a given instant if it is an accurate representation of the corresponding RT entity, both in the value domain and time domain.
* A *real-time (RT) object* is a container within a node of the distributed computer system holding an RT image or an RT entity. A real-time clock with a specified granularity is associated with every RT object.
* The present value of a temporally accurate RT image is a member of the set of values that the RT entity had in its recent history.
* The delay between the observation of the RT entity and the use of the RT image can cause, in the worst case, a maximum error error(t) of the RT image that can be approximated by the product of the maximum gradient of the value v of the RT entity multiplied by the length of the accuracy interval.
* Every phase-sensitive RT image imposes an additional constraint on the scheduling of the real-time task that uses this RT image.
* *State estimation* involves the building of a model of an RT entity inside an RT object to compute the probable state of an RT entity at a selected future instant and to update the corresponding RT image accordingly.
* If the behavior of an RT entity can be described by a continuous and differentiable variable $v(t)$, the first derivative dv/dt is sometimes sufficient to get a rea-

sonable estimate of the state of the RT entity at the point t_{use} in the neighborhood of the point of observation.

- To decrease the coupling between sender and receiver, the state estimation can be performed in two steps: the sender performs a state estimation for the interval $[t_{obs}, t_{arr}]$, and the receiver performs a state estimation for the interval $[t_{arr}, t_{use}]$.
- A particular message becomes *permanent* at a given node at that instant when the node knows that all the messages that were sent to it, prior to the send time of this message, have arrived (or will never arrive).
- The time interval between the start of transmission of a message and the instant when this message becomes permanent at the receiver is called the *action delay*. To avoid incorrect behavior, the receiver must delay any action on the message until *after* the action delay has passed.
- An RT image may only be used if the message that transported the image has become permanent and the image is temporally accurate. In a system without state estimation, both conditions can be satisfied only in the time window $[t_{permanent}, t_{obs} + d_{acc}]$.
- No matter whether the receiver receives one or more out of set of replicated *idempotent* messages, the result will always be the same.
- *Determinism* is a *desired property of a computation* that enables the prediction of the output at a future instant on the basis of a given initial state and timed inputs.
- The basic causes for replica nondeterminism are inconsistent inputs, a difference between the computational progress and the progress of the physical time in the replicas (caused by differing oscillator drifts), and NDDCs.
- If at all possible, inter-replica coordination should be avoided because it compromises the independence of the replicas and requires additional time and additional communication bandwidth.

Bibliographic Notes

The concept of *temporal accuracy* of a real-time object has been introduced in the real-time object model presented in [Kop90]. Kim has extended this model and analyzed the temporal properties of real-time applications using this model [Kim94]. The problem of replica determinism has been extensively studied in [Pol95]. An interesting philosophical treatment of the topics of causality and determinism can be found in [Bun08] and [Hoe10].

Review Questions and Problems

5.1. Give examples of RT entities that are needed to control an automotive engine. Specify the static and dynamic attributes of these RT entities, and discuss the temporal accuracy of the RT images associated with these RT entities.

5.2. What is the difference between a *state observation* and an *event observation*? Discuss their advantages and disadvantages.

5.3. What are the problems with event observations?

5.4. Give an informal and a precise definition of the concept of *temporal accuracy*. What is the *recent history*?

5.5. What is the difference between a *parametric* RT image and a *phase-sensitive* RT image? How can we create parametric RT images?

5.6. What are the inputs to a state estimation model? Discuss state estimation in a system with and without a global time base.

5.7. Discuss the interrelationship between state estimation and composability.

5.8. What is a *hidden* channel? Define the notion of *permanence*.

5.9. Calculate the action delay in a distributed system with the following parameters: $dmax = 20$ msec, $dmin = 1$ msec:

(a) No global time available, and the granularity of the local time is 10 μsec

(b) Granularity of the global time 20 μsec

5.10. What is the relationship between *action delay* and *temporal accuracy*?

5.11. Define the notion of *determinism*! What is *L-determinism*?

5.12. Give an example that shows that a local time-out can lead to replica nondeterminism.

5.13. What mechanisms may lead to replica nondeterminism?

5.14. How can we build a replica-deterministic system?

5.15. Why should explicit inter-replica coordination be avoided?

Chapter 6
Dependability

Overview

It is said that Nobel Laureate *Hannes Alfven* once remarked that *in Technology Paradise no acts of God can be permitted and everything happens according to the blueprints.* The real world is no technology paradise—*components* can fail and *blueprints* (software) can contain design errors. This is the subject of this chapter. The chapter introduces the notions of *fault, error,* and *failure* and discusses the important concept of a *fault-containment unit.* It then proceeds to investigate the topic of *security* and argues that a security breach can compromise the safety of a safety-critical embedded system. The direct connection of many embedded systems to the Internet—the Internet of Things (IoT)—makes it possible for a distant attacker to search for *vulnerabilities* and, if the intrusion is successful, to exercise remote control over the physical environment. Security is thus becoming a prime concern in the design of embedded systems that are connected to the Internet. The following section deals with the topic of anomaly detection. An anomaly is an out-of-norm behavior that indicates that some exceptional scenario is evolving. Anomaly detection can help to detect the early effects of a random failure or the activities of an intruder that tries to exploit system vulnerabilities. Whereas an anomaly lies in the gray zone between correct behavior and failure, an error is an incorrect state that requires immediate action to mitigate the consequences of the error. Error detection is based on knowledge about the intended state or behavior of a system. This knowledge can stem either from a priori established regularity constraints and known properties of the correct behavior of a computation or from the comparison of the results that have been computed by two redundant channels. Different techniques for the detection of temporal failures and value errors are discussed. The following two sections deal with the design of fault-tolerant systems that are capable of masking faults that are contained in the given fault hypothesis. The most important fault-tolerance strategy is *triple modular redundancy* (TMR), which requires a deterministic behavior of replicated components and a deterministic communication infrastructure. We then discuss the robustness and resilience concepts and conclude this chapter with the topic of component reintegration.

6.1 Basic Concepts

The seminal paper by Avizienis et al. [Avi04] establishes the fundamental concepts in the field of dependable computing. The core concepts of this paper are *fault*, *error*, and *failure* (Fig. 6.1).

Computer systems are provided to offer a dependable service to system users. A user can be a human user or another computer system. Whenever the *behavior* of a system (see Sect. 4.1.1), as seen by the user of the system, deviates from the *intended service*, the system is said to have *failed*. A failure can be *pinned down* to an *unintended state* within the system, which is called an *error*. An error is *caused* by an adverse phenomenon, which is called a *fault*.

We use the term *intended* to state the *correct* state or behavior of a system. Ideally, this correct state or behavior is documented in a precise and complete specification. However, sometimes the specification itself is wrong or incomplete. In order to include specification errors in our model, we introduce the word *intended* to establish an abstract reference for correctness.

If we relate the terms *fault*, *error*, and *failure* to the levels of the four-universe model (Sect. 2.3.1), then the term *fault* refers to an adverse phenomenon at any level of the model, while the terms *error* and *failure* are reserved for adverse phenomena at the digital logic level, the informational level, or the external level. If we assume that a sparse global time base is available, then any adverse phenomenon at the digital logic level and above can be identified by a specific bit pattern in the value domain and by an instant of occurrence on the sparse global time base. This cannot be done for phenomena occurring at the physical level.

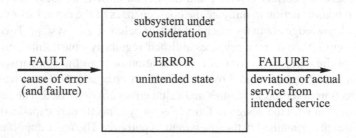

Fig. 6.1 Faults, errors, and failures

6.1.1 Faults

We assume that a system is built out of components. A component is a *fault-containment unit (FCU)*, if the direct effect of a single fault influences only the operation of a single component. Multiple FCUs should fail independently.

Figure 6.2 depicts a classification of faults.

Fault Space It is important to distinguish faults that are related to a deficiency *internal to the FCU* or to some adverse phenomena occurring *external to the FCU*. An *internal fault* of a component, i.e., a fault within the FCU, can be a *physical fault*, such as the random break of a wire, or a *design fault* either in the software (a *program error*) or in the hardware (an *erratum*). An *external fault* can be a physical disturbance, e.g., a lightning stroke causing spikes in the power supply or the impact of a cosmic particle. The provision of incorrect input data is another class of an external fault. *Fault containment* refers to design and engineering efforts that ensure that the immediate consequences of a fault are limited to a single FCU. Many reliability models make the tacit assumption that FCUs fail independently, i.e., there is no single fault that can affect more than one FCU. This *FCU independence assumption* must be justified by the design of the system.

> **Example:** The physical separation of the FCUs of a fault-tolerant system reduces the probability for *spatial proximity faults*, such that fault at a single location (e.g., impact in case of an accident) cannot destroy more than a single FCU.

Fault Time In the temporal domain, a fault can be *transient* or *permanent*. Whereas physical faults can be *transient* or *permanent*, design faults (e.g., software errors) are always *permanent*.

A *transient fault* appears for a short interval at the end of which it disappears without requiring any explicit repair action. A *transient* fault can lead to an *error*, i.e., the corruption of the *state* of an FCU, but leaves the physical hardware undamaged (*by definition*). We call a transient external physical fault a *transitory fault*. An example for a transitory fault is the impact of a cosmic particle that corrupts the state of an FCU. We call a transient internal physical fault an *intermittent fault*.

Fig. 6.2 Classification of faults

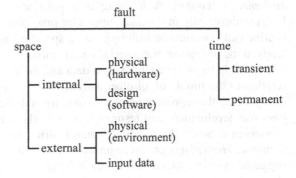

Examples for intermittent faults are oxide defects, corrosion, or other fault mechanisms that have not yet developed to a stage where the hardware fails permanently. According to [Con02], a substantial number of the transient faults observed in the field are intermittent faults. Whereas the failure rate of transitory faults is constant, the failure rate for intermittent faults increases as a function of time. An increasing intermittent failure rate of an electronic hardware component is an indication for the *wear-out* of the component. It suggests that preventive maintenance—the replacement of the faulty component—should be performed in order to avoid a permanent fault of the component.

A *permanent fault* is a fault that remains in the system until an explicit repair action has taken place that removes the fault. An example for a *permanent external fault* is a lasting breakdown of the power supply. A *permanent internal fault* can be in the physical embodiment of the hardware (e.g., a break of an internal wire) or in the design of the software or hardware. The mean time it takes to repair a system after the occurrence of a permanent fault is called *MTTR* (*mean time to repair*).

6.1.2   Errors

The immediate consequence of a fault is an *incorrect state* in a component. We call such an incorrect state, i.e., a wrong data element in the memory, a register or, in a flip-flop circuit of a CPU, an *error*. As time progresses, an error is *activated* by a computation, *detected* by some error-detection mechanism, or *wiped out*.

An error is *activated* if a computation accesses the *error*. From this instant onward, the computation itself becomes incorrect. If a fault impacts the contents of a memory cell or a register, the consequent error will be activated when this memory cell is accessed by a computation. There can be a long time interval between error occurrence and error activation (the *dormancy* of an error) if a memory cell is involved. If a fault impacts the circuitry of the CPU, an immediate activation of the fault may occur, and the current computation will be corrupted. As soon as an incorrect computation writes data into the memory, this part of memory becomes *erroneous* as well.

We distinguish between two types of software errors, called *Bohrbugs* and *Heisenbugs* [Gra86]. A *Bohrbug* is a software error that can be reproduced L-deterministically in the data domain by providing a specific input pattern to the routine that contains the Bohrbug, i.e., a specific pattern of input data that always leads to the activation of the underlying Bohrbug. A *Heisenbug* is a software error that can only be observed if the input data and the exact timing of the input data—in relation to the timing of all other activities in the computer—are reproduced precisely. Since the reproduction of a *Heisenbug* is difficult, many software errors that pass the development and testing phase and show up in operational systems are *Heisenbugs*. Since the temporal control structure in event-triggered systems is dynamic, *Heisenbugs* are more probable in event-triggered systems than in time-triggered systems, which have a data-independent static control structure.

Example: A typical example for a *Heisenbug* is an error in the synchronization of the data accesses in a concurrent system. Such an error can only be observed if the temporal relationships between the tasks that access the mutually exclusive data are reproduced precisely.

An error is *detected* when a computation accesses the *error* and finds out that the results of the computation deviate from the *expectations* or the *intentions of the user*, either in the domain of value or the domain of time. For example, a simple parity check detects an error if it can be assumed that the fault has corrupted only a single bit of a data word. The time interval between the instant of *error (fault) occurrence* and the instant of *error detection* is called the *error-detection latency*. The probability that an error is detected is called *error-detection coverage*. Testing is a technique to detect *design faults* (software errors and hardware errata) in a system.

An error is *wiped out* if a computation overwrites the error with a new value before the error has been activated or detected. An error that has neither been activated, detected, nor wiped out is called a *latent error*. A latent error in the *state* of a component results in a *silent data corruption* (SDC), which can lead to serious consequences.

Example: Let us assume that a *bitflip* occurs in a memory cell that is not protected by a parity bit and that this memory cell contains sensory input data about the intended acceleration of an automotive engine. The consequent *silent data corruption* can result in an unintended acceleration of the car.

6.1.3   Failures

A *failure* is an event that denotes a *deviation* between the actual behavior and the intended behavior (the *service*) of a component, occurring at a particular instant. Since, in our model, the behavior of a component denotes the sequence of messages produced by the component, a failure manifests itself by the production of an *unintended* (or *no intended*) message. Figure 6.3 classifies the *failure* of a component.

Domain   A failure can occur in the *value domain* or in the *temporal domain*. A value failure means that an *incorrect value* is presented at the component-user interface. (Remember, the user can be another system.) A *temporal failure* means that a value is presented outside the *intended interval of real time*. Temporal failures only exist if the system specification contains information about the intended temporal behavior of the system. Temporal failures can be subdivided into *early temporal failures* and *late temporal failures*. A component that contains internal error-detection mechanisms in order to detect any *error* and suppresses a result that contains a *value error* or an *early temporal failure* will only exhibit a *late temporal failure*, i.e., an *omission*, at the interface to its users. We call such a failure an *omission failure*. A component that only makes omission failures is called a *fail-silent component*. If a component stops working after the first omission failure, it is called

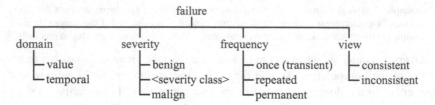

Fig. 6.3 Classification of failures

a *fail-stop component*. The corresponding failure is sometimes called a *clean failure* or a *crash failure*.

> **Example:** A self-checking component is a component that contains internal failure-detection mechanisms such that it will only exhibit *omission failures* (or *clean failures*) at the *component-user interface*. A self-checking component can be built out of two deterministic FCUs that produce two results at about the same time and where the two results are checked by a *self-checking* checker.

> **Example:** Fault-injection experiments of the MARS architecture have shown that between 1.9% and 11.6% of the observed failures were temporal failures, meaning that a message was produced at an unintended instant. An independent time-slice controller, a guardian, has detected all of these temporal failures [Kar95, p. 326].

Severity Depending on the effect a failure has on its environment, we distinguish between two extreme cases, *benign* and *malign* failures (see also Sect. 1.5). The cost of a benign failure is of the same order of magnitude as the loss of the normal utility of the system, whereas a malign failure can result in failure costs that are orders of magnitude higher than the normal utility of a system, e.g., a malign failure can cause a catastrophe such as an accident. We call applications where malign failures can occur *safety-critical* applications. The characteristics of the application determine whether a failure is benign or malign. In between these two extreme cases of benign and malign, we can assign a *severity class* to a failure, e.g., based on the *monetary impact* of a failure or the impact of the failures on the *user experience*.

> **Example:** In a multimedia system, e.g., a digital television set, the failure of a single pixel that is overwritten in the next cycle is masked by the human perception system. Such a failure is thus of negligible severity.

Frequency Within a given time interval, a failure can occur only *once* or *repeatedly*. If it occurs only once, it is called a *single* failure. If a system continues to operate after the failure, we call the failure a *transient failure*. A frequently occurring transient failure is called a *repeated failure*. A special case of a single failure is a *permanent* one, i.e., a failure after which the system ceases to provide a service until an explicit repair action eliminates the cause of the failure.

View If more than one user looks at a failing component, two cases can be distinguished: all users see the *same failing behavior*—we call this a *consistent failure*— or different users see *different behaviors* (we call this an *inconsistent failure*). In the literature, different names are used for an *inconsistent failure*: *two-faced failure*, *Byzantine failure*, or *malicious failure*. Inconsistent failures are most difficult to handle, since they have the potential to *confuse* the correct components (see Sect. 3.4.1). In high-integrity systems, the occurrence of Byzantine failures must be considered [Dri03].

> **Example:** Let us assume that a system contains three components. If one of them fails in an *inconsistent failure mode*, the other two correct components will have different views of the behavior of the failing component. In an extreme case, one correct component classifies the failing component as correct, while the other correct component classifies the failing component as erroneous, leading to an inconsistent view of the failed component among the correct components.

> **Example:** A *slightly-out-of-specification (SOS) failure* is a special case of a Byzantine. SOS failures can occur at the interface between the analog level and the logical level of the four-universe model (see Sect. 2.3.1). If, in a bus system, the voltage of the high-level output of a sender is slightly below the level specified for the high-level state, then some receivers might still accept the signal, assuming the value of the signal is *high*, while others might not accept the signal, assuming the value is *not high*. SOS failures are of serious concern if signals are marginal w.r.t. voltage or timing.

Propagation   If an error inside a component is activated and propagates outside the confines of the component that has been affected by the fault, then we speak of *error propagation*. Let us make the simplifying assumption that a component communicates with its environment solely by the exchange of messages and there is no other means of interaction of components (such as a common memory). In such a system, an error can propagate outside the affected component solely by the transmission of an *incorrect message*.

In order to avoid that a propagated error infects other—up to that time healthy—components and thus invalidates the *component independence assumption, error propagation boundaries* must be erected around each component. A message can be incorrect either in the *value domain* (the data field of the message contains a corrupted value) or in the *time domain*, i.e., the message is sent at an unintended instant or not at all (omission failure). Temporal message failures can be detected by the communication system, provided the communication system has a priori knowledge about the correct temporal behavior of a component. Since a communication system is *agnostic* about the contents of the value field of a message (see Sect. 4.6.2), it is the responsibility of the receiver of the message to detect corrupted values, i.e., *errors*, in the data field of a message.

In a cyclic system, the corruption of the *g-state* (see Sect. 4.2.3) is of particular concern, since the g-state contains the information of the current cycle that influences the behavior of the next cycle. Since a *latent error* in the g-state can become an incorrect input to a computation in the next cycle, a gradual increase in the number of errors in the g-state can occur (called *state erosion*). If the g-state is empty, then there is no possibility of error propagation of an error from the current cycle to the next cycle. In order to avoid error propagation from one cycle to the next cycle, the integrity of the g-state should be monitored by a special error-detection task of an independent diagnostic component.

6.2 Information Security

Information security deals with the *authenticity*, *integrity*, *confidentiality*, *privacy*, and *availability* of information and services that are provided by a computer system. In the following section, we always mean *information security* when we use the term *security*. We call a *deficiency* in the design or operation of a computer system that can lead to a security incident a *vulnerability* and the successful exploitation of a vulnerability an *intrusion*. The following reasons make clear why information security has become *a prime concern* in the design and operation of embedded systems [Car08]:

(i) *Controllers are computers*: Over the past few years, hard-wired electronic controllers have been replaced by programmable computers with non-perfect operating systems, making it possible for an outsider to exploit the vulnerabilities of the software system.

(ii) *Embedded systems are distributed*: Most embedded systems are distributed, with wire-bound or wireless channels connecting the nodes. An outside intruder can use these communication channels to gain access to the system.

(iii) *Embedded systems are connected to the Internet*: The connection of an embedded system to the Internet makes it possible for an intruder anywhere in the world to attack a remote system and to systematically exploit any detected vulnerability.

As of today, there is normally a human mediator between the *cyberspace (e.g., the Internet)* and *actions in the physical world*. Humans are supposed to have *common sense* and *responsibility*. They are able to recognize an evidently wrong computer output and will not set any actions in the physical world based on such a wrong output. The situation is different in embedded systems connected directly to the Internet—the *Internet of Things (IoT)*, where the *smart object* at the edge of the Internet (e.g., a robot) can immediately interact with the physical world. An adversary can compromise the integrity of the embedded system by breaching the security walls, thus becoming a safety hazard. Alternatively, an adversary can carry out a *denial-of-service* attack and thus bring down the availability of an important service. *Security and safety are thus interrelated and of utmost concern in embedded systems that are connected to the Internet.*

> **Example:** Let us assume that an owner of a vacation home can set the temperature of the thermostat of his electric furnace in the vacation home remotely via the Internet. If an adversary gets control of the thermostat, he can elevate the temperature to a high level and increase the energy consumption significantly. If the adversary executes this attack on all vacation homes in a neighborhood, then the total power consumption in the neighborhood

might increase beyond the critical level that leads to a blackout (example taken from [Koo04]).

Standard security techniques are based on a sound security architecture that controls the information flow among subsystems of different *criticality* and *confidentiality*. The architectural decisions are implemented by the deployment of cryptographic methods, such as *encryption, random number generation*, and *hashing*. The execution of cryptographic methods requires extra energy and silicon real estate, which are not always available in a small (portable) embedded system.

6.2.1   Secure Information Flow

The main security concerns in embedded systems are the *authenticity and integrity of the real-time data* and of the *system configuration* and, to a lesser extent, the control of access to data. The security policy must specify which processes are authorized to modify data (*data integrity*) and which processes are allowed to see the data (*confidentiality of data*). A security policy for *data integrity* can be established on the basis of the Biba model, while a security policy for the confidentiality of data can be derived from the Bell-LaPadula model [Lan81]. Both models classify the *processes* and the *data files* according to an ordered sequence of levels, from highest to lowest. A process may read and modify data that is at the same level as the process. The respective security models govern the access and modification of data at a level that is different from the level of the reading or writing process.

The concern of the Biba model is the *integrity of the data*, a concern that is highly relevant in *multi-criticality embedded systems*. The classification of the data files and the processes is determined by the *criticality* from the point of view of the safety analysis (see Sect. 11.4.2). In order to ensure the integrity of a (high-critical) process, the (high-critical) process must not read data that is classified at a lower level than the classification of the (high-critical) process. In order to ensure that a (low-criticality) process will not corrupt data of a higher-criticality level, the Biba model states that no (low-criticality) process may modify data that is at a higher-criticality level than that of the (low-criticality) process.

The concern of the *Bell-LaPadula model* is the *confidentiality of the data*. The classification of the data files and the processes is determined by the *confidentiality* of the data from *top secret* to *unclassified*. In order to ensure the confidentiality of top-secret data, it must be made certain that no (unclassified) process may read data that is classified at a higher level than the classification of the (unclassified) process. In order to ensure that a (top-secret) process will not publish confidential data to a (unclassified) lower level, the *Bell-LaPadula* states no (top-secret) process may write data to a data file that is at a lower confidentiality level than that of the (top-secret) process.

The classification of processes and data from the point of view of integrity will normally be different from classification according to the point of view of confidentiality. These differences can lead to a conflict of interest. In case of such a conflict, the integrity concern is the more important concern in embedded systems.

The selected security policy must be enforced by *mechanisms* that establish the authenticity of processes and the integrity of the data that is exchanged. These mechanisms make wide use of the well-understood cryptographic methods discussed in Sect. 6.2.3.

6.2.2   Security Threats

A systematic security analysis starts with the specification of an *attack model*. The attack model puts forward an *attack hypothesis*, i.e., it lists the *threats* and makes assumptions about the *attack strategy* of an adversary. It then outlines the conjectured steps taken by an adversary to break into a system. In the next phase, a *defense strategy* is developed in order to counter the attack. There is always the possibility that the attack hypothesis is incomplete and a clever adversary finds a way to attack the system by a method that is not covered by the attack hypothesis.

The typical attacker proceeds according to the following three phases: *access to the selected subsystem, search for and discovery of a vulnerability*, and, finally, *intrusion and control of the selected subsystem*. The control can be *passive* or *active*. In *passive* control, the attacker observes the system and collects confidential information. In *active* control, the attacker modifies the behavior of the system such that the system will contribute to the attacker's mean purpose. A security architecture must contain observation mechanisms, i.e., *intrusion detection* mechanisms, to detect malicious activities that relate to any of these three phases of an attack. It also must provide firewalls and procedures that mitigate the consequences of an attack such that the system can survive.

Access to the system must be prevented by requiring strict adherence to a mandatory access control procedure, where every person or process must authenticate itself and this authentication is verified by *callback* procedures. *Security firewalls* play an important role to limit the access to sensitive subsystems to authorized users.

The attacker's *search for vulnerabilities* can be detected by *intrusion detection mechanisms*, which can be part of an anomaly-detection subsystem (see Sect. 6.3.1). Anomaly detection is needed in order to detect the consequences of random physical faults as well as the activities of a malicious intruder.

The *capture of control* over a subsystem can be prevented by a structured security architecture, where different criticality levels are assigned to different processes and a formal security policy, based on a formal model, controls the interactions among these criticality levels.

The attainment of topmost security is not only a technical challenge. It requires high-level management commitment in order to ensure that the users strictly follow the organizational rules of the given security policy. Many security violations are not caused by technical weaknesses in a computer system, but by a failure of the users to comply with the existing security policies of the organization.

Example: Beautement et al. [Bea08] state: *It is widely acknowledged in security research and practice that many security incidents are caused by human, rather than technical failures. Researchers approaching the issue from a Human-Computer Interaction (HCI) perspective demonstrated that many human failures are caused by security mechanisms that are too difficult for a non-expert to use.*

The following list of security attacks is only an indication of what has been observed. It is by no means complete:

Malicious Code Attack: A malicious code attack is an attack where an adversary inserts *malicious code*, e.g., a *virus*, a *worm*, or a *Trojan horse*, into the software in order that the attacker gets partial or full control over the system. This malicious code can be inserted statically, e.g., by a malicious maintenance action (*insider attack*), by the process of downloading a new software version, or dynamically during the operation of a system by accessing an infected Internet site or opening an infected data structure.

Spoofing Attack: In a spoofing attack, an adversary masquerades as a legitimate user in order to gain unauthorized access to a system. There are many versions of spoofing attacks in the Internet: replacement of a legitimate web page (e.g., of a bank) by a seemingly identical copy that is under the control of the adversary (also called *phishing*), the replacement of the correct address in an email by a fake address, and a *man-in-the-middle attack* where an intruder intercepts a session between two communicating partners and gains access to all exchanged messages.

Password Attack: In a password attack, an intruder tries to guess the password that protects the access to a system. There are two versions of password attacks, *dictionary attacks* and *brute force attacks*. In a dictionary attack, the intruder guesses commonly used password strings. In a brute force attack, the intruder searches systematically through the full code space of the password until he is successful.

Ciphertext Attack: In this attack model, the attacker assumes to have access to the ciphertext and tries to deduce the plaintext and possibly the encryption key from the ciphertext. Modern standardized encryption technologies, such as the AES (Advanced Encryption Standard), have been designed to make the success of ciphertext attacks highly improbable.

Denial-of-Service Attack: A *denial-of-service attack* tries to make a computer system *unavailable* to its users. In any wireless communication scenario, such as a sensor network, an adversary can jam the ether with high-power signals of the appropriate frequency in order to interfere with the communication of the targeted devices. In the Internet, an adversary can send a coordinated burst of service requests to a site to overload the site such that legitimate service requests cannot be handled any more.

Botnet Attack: A *botnet* (the word *bot* is an abbreviation of *robot*) is a set of infected networked nodes (e.g., thousands of PC or set top boxes) that are under the control of an attacker and cooperate (unknowingly to the owner of the node) to achieve a malicious mission. In a first phase, an attacker gets control over the botnet nodes and infects them with malicious code. In the second phase, he launches a *distributed denial-of-service attack* to a chosen target website to make the target website *unavailable* to legitimate users. Botnet attacks are among the most serious attack modes in the Internet.

Example: A study in Japan [Tel09, p. 213] showed that *it takes about four minutes, on average, for an unprotected PC to be infected when connected to the Internet* and that an estimated 500,000 PCs are infected. A total of around 10 Gbps of traffic from Japanese IP addresses are wasted by botnets (SPAM mail traffic via botnets is not included).

Malicious Training Data Attack: With the advent of machine learning in real-time systems, e.g., for object recognition in self-driving cars, an attacker may take advantage of malicious training data to cause a failure during system operation [Bar10], e.g., a faulty object classification.

6.2.3   Cryptographic Methods

The provision of an adequate level of integrity and confidentiality in embedded systems that are connected to the Internet, the IoT, can only be achieved by the judicious application of cryptographic methods. Compared to general computing systems, the security architecture of embedded systems must cope with the following two additional constraints:

- *Timing constraints*: the encryption and decryption of data must not extend the response time of time-critical tasks; otherwise the encryption will have a negative impact on the quality of control.
- *Resource constraints*: many embedded systems are resource constrained, concerning memory, computational power, and energy.

Basic Cryptographic Concepts   The basic cryptographic primitives that must be supported in any security architecture are *symmetric key encryption*, *public key encryption*, *hash functions*, and *random number generation*. The proper application of these primitives, supported by a *secure key management system*, can ensure the authenticity, integrity, and confidentiality of data.

In the following paragraphs, we use the term *hard* to mean: it is beyond the capabilities of the envisioned adversary to break the system within the time period during which the security must be provided. The term *strong cryptography* is used if the system design and the cryptographic algorithm and key selection justify the assumption that a successful attack by an adversary is highly improbable.

In cryptography, an algorithm for *encryption* or *decryption* is called a *cipher*. During encryption, a *cipher* transforms a *plaintext* into a *ciphertext*. The ciphertext holds all the information of the plaintext but cannot be understood without knowledge of the algorithm and the keys to decrypt it.

A *symmetric key encryption* algorithm encrypts and decrypts a *plaintext* with the same (or trivially related) keys. Therefore both the encryption and decryption key must be kept secret. In contrast, an *asymmetric key algorithm* uses different keys, a *public key* and a *private key*, for encryption and decryption. Although the two keys are mathematically related, it is *hard* to derive the private key from the knowledge of the public key. Asymmetric key algorithms form the basis for the widely used *public key encryption technology* [Riv78].

The procedure for key distribution is called *key management*. In *public key encryption systems*, the security of the system depends on the *secrecy of the private keys* and the establishment of a *trusted relationship* between the *public key* and the *identity of the owner of the respective private key*. Such a trusted relationship can be established by executing a secure network authentication protocol to an *a priori known security server*. An example of such a network authentication protocol is the KERBEROS protocol that provides mutual authentication [Neu94] and establishes a secure channel between two nodes in an open (insecure) network by using a *trusted security server*.

Random numbers are required in both symmetric and asymmetric cryptography for key generation and for the generation of *unpredictable* numbers that are used only once (called a *nonce*) in order to ensure the uniqueness of a key of a session. In *public key encryption*, the node that needs a *private key* must generate the asymmetric pair of keys out of a *nonce*. The private key is kept secret to the node, while the public key is disseminated over open channels to the public. A signed copy of the public key must be sent to a *security server* in order that other nodes can check the trusted relationship between the public key and the identity of the node that generated the public key.

In order to ensure the secrecy, a private key should not be stored in plain text but must be sealed in a *cryptographic envelope*. To operate on such an envelope, a non-encrypted key is required, which is usually called the *root key*. The root key serves as the starting point for the *chain of trust*.

The computational effort required to support *public key encryption* is substantially higher than the computation effort needed for *symmetric key encryption*. For this reason, public key encryption is sometimes only used for the secure distribution of keys, while the encryption of the data is done with symmetric keys.

A *cryptographic hash function* is an *L-deterministic* (see Sect. 5.6) mathematical function that converts a large variable-sized *input string* into a fixed-size bit string, called the *cryptographic hash value* (or for short a *hash*) under the following constraints:

- An accidental or intentional change of data in the input string will change the hash value.
- It should be *hard* to find an input string that has a given hash value.
- It should be *hard* to find two different input strings with the same hash value.

Cryptographic hash functions are required to establish the authenticity and integrity of a plaintext message by an electronic signature.

Authentication
Anyone who knows the sender's public key can decrypt a message that is encrypted with
the sender's private key. If a *trusted relationship* between the sender's public key and the
identity of the sender has been ascertained, then the receiver knows that the identified
sender has produced the message.

Digital signature If both the authenticity and integrity of a plaintext message must
be established, the plaintext is taken as the input to a cryptographic hash function.
The hash value is then encrypted with the author's private key to generate the *digital
signature* that is added to the plaintext. A receiver who is in the possession of the
author's public key must check whether the decrypted signature is the same bit
string as the recalculated hash value of the received text.

Privacy Anyone who uses a receiver's public key for the encryption of a message
can be sure that only the receiver, whose public key has been used, can decipher the
message.

Resource Requirements The computational effort for cryptographic operations
measured in terms of *required energy, time,* and *gate count of an implementation*
depends on the selected algorithm and its implementation. In 2001 the US National
Institute of Standards has selected the AES (Advanced Encryption Standard) algorithm as the Federal Information Processing Standard for symmetric encryption.
AES supports key sizes of 128, 192, and 256 bits. Table 6.1 gives an estimate of the
resource requirements of different hardware implementations for the AES. From
this table it is evident that there is an important design tradeoff between required
time and required silicon area.

The different implementations of the AES algorithm depicted in Table 6.1 show
the tradeoff between *silicon area* (gate count) and *speed* (clock cycles) that applies
to many algorithms. The resource requirements for public key encryption are higher
than listed in Table 6.1.

Security measures are readily available in today's silicon solutions for embedded
systems. For example, automotive microcontrollers commonly implement *hardware security modules (HSM)*. [Pot21] gives an overview of the HSM functionality
on the Infineon AURIX, discusses its limitations, and compares the HSM with a
software-based implementation.

Table 6.1 Comparison of requirements of different hardware AES implementations

AES 128 encryption	Gate equivalent	Clock cycles
Feldhofer	3 628	992
Mangard	10 799	64
Verbauwhede	173 000	10

Adapted from [Fel04a]

6.2.4   Network Authentication

In the following section, we outline a sample of a network authentication protocol that uses public key cryptography to establish the *trusted relationship* between a *new node* and its *public key*. For this purpose we need the *trusted security server*. Let us assume all nodes know the public cryptographic key of the security server and the security server knows the public cryptographic keys of all nodes a priori.

If a node, say node A, wants to send an encrypted message to a yet unknown node, say node B, then node A takes the following steps:

(1) Node A forms a signed message with the following content: *current time, node A wants to know what is the public key of node B?*, and *signature of node A*. It then encrypts this message with the public key of the security server and sends the ciphertext message to the security server over an open channel.
(2) The security server decrypts the message with its private key and checks whether the message has been sent recently. It then examines the signature of the message with the a priori known public key of the signature of node A to find out whether the contents of the message from node A are authentic.
(3) The security server forms a response message with the contents *current time, address of node B, public key of node B*, and *signature*, encrypts this message with the public key of node A, and sends this ciphertext message to node A over an open channel.
(4) Node A decrypts the message with its private key and checks whether the message has been sent recently and whether the signature of the security server is authentic. Since node A trusts the information of the security server, it now knows that the public key of node B is authentic. It uses this key to encrypt the messages it sends to B.

A network authentication protocol that establishes a secure channel between two nodes by using symmetric cryptography is the aforementioned KERBEROS protocol [Neu94].

6.2.5   Protection of Real-Time Control Data

Let us assume the following *attack model* for a real-time process-control system in an industrial plant. A number of sensor nodes distributed throughout the plant periodically send real-time *sensor values* by open wireless channels to a controller node which calculates the set points for the control valves. An *adversary* wants to sabotage the operation of the plant by sending counterfeit *sensor values* to the controller node.

In order to establish the authenticity and integrity of a sensor value, a standard security solution would be to append an electronic signature to the sensor value by the *genuine* sensor node and to check this signature by the controller node that receives the message. However, this approach would extend the duration of the control loop by the time it takes for generating and checking the electronic signature. Such an extension of the length of the control-loop period has a negative effect on the quality of control and must be avoided.

In a real-time control system, the design challenge is to find a solution that detects an *adversary* without any extension of the duration of the control-loop period. The above example shows that the two requirements, *real-time performance* and *security*, cannot be dealt with separately in a real-time control system.

There are characteristics of real-time control systems that must be considered when designing a security protocol:

- In many control systems, a single corrupted set-point value is not of serious concern. Only a sequence of corrupted values must be avoided.
- Sensor values have a short *temporal accuracy* (see Sect. 5.4)—often in the range of a few milliseconds.
- The resources of many mobile embedded system, both computational and energy, are constrained.

Some of these characteristics are helpful; others make it more difficult to find a solution.

Example: It is possible to take the signature generation and the signature check of real-time data out of the control loop and perform it in parallel. As a consequence, the detection of an *intrusion* will be delayed by one or more control cycles (which is acceptable considering the characteristics of control system).

Further research is needed to find effective protection techniques for real-time data under the listed constraints.

6.3   Anomaly Detection

6.3.1   What Is an Anomaly?

If we look at the state space of *real-life embedded systems*, we find many examples that show a *gray zone* between the *intended (correct) state* and an *error*. We call states in this intermediate gray zone between intended and erroneous states or behavioral patterns an *anomaly* or an *out-of-norm state* (see Fig. 6.4). Anomaly detection is concerned with the detection of states or behavioral patterns outside the expected, i.e., the normal behavioral patterns, but do not fall into the category of errors or failures. There are many reasons for the occurrence of *anomalies*: activities by an *intruder* to find a vulnerability, exceptional circumstances in the environment, user mistakes, degradation of sensors resulting in imprecise sensor readings, external perturbations, specification changes, or imminent failures caused by an error in the design or the hardware. The detection of anomalies is important, since the occurrence

Fig. 6.4 Gray zone
between intended and
erroneous states

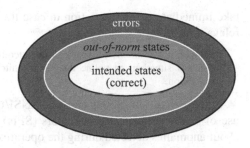

errors

out-of-norm states

intended states
(correct)

of an anomaly is an indication that some atypical scenario that may require immediate corrective action (e.g., the imminent *intrusion* by an adversary) is developing.

Application-specific a priori knowledge about the restricted ranges and the known interrelationships of the values of RT entities can be used to detect anomalies that are undetectable by syntactic methods. Sometimes these application-specific mechanisms are called *plausibility checks*.

> **Example:** The constraints imposed on the speed of change of the RT entities by the inertia of a technical process (e.g., change of temperature) form a basis for very effective plausibility checks.

Plausibility checks can be expressed in the form of *assertions* to check for the plausibility of an intermediate result or at the end of a program by applying an *acceptance test* [Ran75]. Acceptance tests are effective to detect anomalies that occur in the value domain.

Advanced dynamic anomaly-detection techniques keep track of the *operational context* of a system and autonomously *learn* about the normal behavior in specific contexts in order to be able to detect anomalies more effectively. In a real-time control system that exhibits periodic behavior, the analysis of the time series of real-time data is a very effective technique for anomaly detection. An excellent survey of anomaly-detection techniques is contained in [Cha09].

The anomaly-detection subsystem should be separated from the subsystem that performs the operational functions for the following reasons:

- The anomaly-detection subsystem should be implemented as an independent fault-containment unit, such that a failure in the anomaly-detection subsystem will have no direct effect on the operational subsystem and vice versa.
- Anomaly detection is a well-defined task that must be performed independently from the operational subsystem. Two different engineering groups should work on the operational subsystem and the anomaly-detection subsystem in order to avoid common mode effects.

The multi-cast message primitive, introduced in Sect. 4.1.1, provides a means to make the g-state of a component accessible to an independent anomaly-detection subsystem without inducing a probe effect. The anomaly-detection subsystem classifies the observed anomalies on a severity scale and reports them either to an off-line *diagnostic system* or to an online *integrity monitor*. The integrity monitor can take immediate corrective action in case the observed anomaly points to a safety-relevant incident.

> **Example:** It is an *anomaly* if a car keeps accelerating while the brake pedal is being pressed. In such a situation, an online integrity monitor should autonomously discontinue the acceleration.

In the area of self-driving cars, the ANSI/UL 4600 standard [ANS20] defines the use of safety performance indicators (SPIs) as a technique to collect information about anomalies detected during the operation of the vehicle. SPIs are an approach *to proactively monitor operations and to respond to all incidents, including near misses* [Koo20].

All detected anomalies should be documented in an *anomaly database* for further online or off-line analysis. The depth of investigation into an anomaly depends on the severity of the anomaly—the more severe the anomaly the more information about the occurrence of the anomaly should be recorded. The off-line analysis of the anomaly database can expose valuable information about *weak spots* of a system that can be corrected in a future version.

In a safety-critical system, every single observed anomaly must be scrutinized in detail until the final cause of the anomaly has been unambiguously identified.

6.3.2 Failure Detection

A failure can only be detected if the *observed behavior* of a component can be judged in relation to the *intended behavior*. Failure detection within a system is only possible if the system contains some form of redundant information about the *intended behavior*. The coverage of the *failure detector*, i.e., the probability that a failure will be detected if it is present, will increase if the information about the intended behavior becomes more detailed. In the extreme case, where every failure in the behavior of a component must be detected, a second component that provides the basis for the comparison—a *golden reference component*—is needed, i.e., the redundancy is 100%.

Knowledge about the regularity in the activation pattern of a computation can be used to detect temporal failures. If it is a priori known that a result message must arrive every second, the *non-arrival* of such a message can be detected within 1 second. If it is known that the result message must arrive *exactly at every full* second and a global time is available at the receiver, then the failure-detection latency is given by the precision of the clock synchronization. Systems that tolerate jitter do have a longer failure-detection latency than systems without jitter. The extra time gained from an earlier failure detection can be significant for initiating a mitigation action in a safety-critical real-time system.

In real-time systems, the *worst-case execution time* (WCET; see Sect. 10.2) of all real-time tasks must be known in advance in order to find a viable schedule for the task execution. This WCET can be used by the operating system to monitor the execution time of a task. If a task has not terminated before its WCET expires, a temporal failure of the task has been detected.

6.3.3 Error Detection

As mentioned before, an *error* is an incorrect data structure, e.g., an incorrect *state* or an incorrect *program*. We can only detect an error if we have some *redundant information* about the intended properties of the data structure under investigation. This information can be part of the data structure itself, such as a CRC field, or it can come from some other source, such as *a priori knowledge* expressed in the form of *assertions* or a *golden channel* that provides a result that acts as *golden reference data structure*.

Syntactic Knowledge About the Code Space The code space is subdivided into two partitions, one partition encompassing syntactically correct values, with the other containing detectably erroneous codewords. This a priori knowledge about the syntactic structure of valid codewords can be used for error detection. One plus the maximum number of bit errors that can be detected in a codeword is called the *Hamming distance* of the code. Examples of the use of error-detecting codes are error-detecting codes (e.g., parity bit) in memory, CRC polynomials in data transmission, and check digits at the man-machine interface. Such codes are very effective in detecting the corruption of a value.

> **Example:** Consider the scenario where each symbol of an alphabet of 128 symbols is encoded using a single byte. Because only seven bits ($2^7=128$) are needed to encode a symbol, the eighth bit can be used as a parity bit to be able to distinguish a *valid codeword* from an *invalid codeword* of the 256 codewords in the code space. This code has a *Hamming distance of 2*.

Duplicate Channels If two independent deterministic channels calculate two results using the same input data, we can compare the results to detect a failure but cannot decide which one of the two channels is wrong. Fault-injection experiments [Arl03] have shown that the duplicate execution of application tasks at different times is an effective technique for the detection of transient hardware faults. This technique can be applied to increase the failure-detection coverage, even if it cannot be guaranteed that *all* task instances can be completed twice in the available time interval.

There are many different possible combinations of hardware, software, and time redundancy that can be used to detect different types of failures by performing the computations twice. Of course, both computations must be *replica determinate*; otherwise, many more discrepancies are detected between the redundant channels than those that are actually caused by faults. The problems in implementing replica-deterministic fault-tolerant software have already been discussed in Sect. 5.6.

Golden Reference If one of the channels acts as a golden reference that is considered correct by definition, we can determine if the result produced by the other channel is correct or faulty. Alternatively, we need three channels with majority voting to find out about the single faulty channel, under the assumption that all three channels are synchronized.

> **Example:** David Cummings reports about his experience with error detection in the software for NASA's Mars Pathfinder spacecraft [Cum10]: *Because of Pathfinder's high reliability requirements and the probability of unpredictable hardware errors due to the increased radiation effects in space, we adopted a highly "defensive" programming style. This included performing extensive error checks in the software to detect the possible side effects of radiation-induced hardware glitches and certain software bugs. One member of our team, Steve Stolper, had a simple arithmetic computation in his software that was guaranteed to produce an even result (2, 4, 6 and so on) if the computer was working correctly. Many programmers would not bother to check the result of such a simple computation. Stolper, however, put in an explicit test to see if the result was even. We referred to this test as his "two-plus-two-equals-five check." We never expected to see it fail. Lo and behold, during software testing we saw Stolper's error message indicating the check had failed. We saw it just once. We were never able to reproduce the failure, despite repeated attempts over many thousands if not millions of iterations. We scratched our heads. How could this happen, especially in the benign environment of our software test lab, where radiation effects were virtually nonexistent? We looked carefully at Stolper's code, and it was sound.*

What can we learn from this example? We should never build a safety-critical system that relies on the results of a single channel only.

6.4 Fault Tolerance

The design of any fault-tolerant system starts with the precise specification of a *fault hypothesis*. The fault hypothesis states what types of faults must be tolerated by the fault-tolerant system and divides the fault space into two domains, the domain of *normal faults* (i.e., the faults that must be tolerated) and the domain of *rare faults*, i.e., faults that are outside the fault hypotheses and are assumed to be *rare events*. Figure 6.5 depicts the state space of a fault-tolerant system. In the center we see the correct states. A *normal failure* will bring the system into *a normal fault state* (i.e., a state that is covered by the fault hypothesis). A normal fault will be corrected by an available fault-tolerance mechanism that brings the system back into the domain of correct states. A *rare fault* will bring the system into a state that is outside the specified fault hypothesis and therefore will not be covered by the provided fault-tolerance mechanisms. Nevertheless, instead of giving up, *a never-give-up (NGU)* strategy should try to bring the system back into a correct state.

> **Example:** Let us assume that the fault hypothesis states that during a specified time interval, a fault of any single component must be tolerated. The case that two components fail simultaneously is thus outside the fault hypothesis, because it is considered to be a *rare fault*. If the simultaneous failure of two components is detected, then the NGU strategy kicks in. In the NGU strategy, it is assumed that the simultaneous faults are transient and a fast restart of the complete system will bring the system back into a correct state. In order

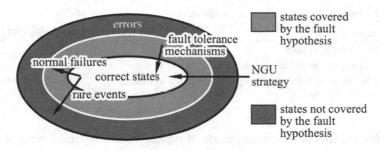

Fig. 6.5 State space of a fault-tolerant system

to be able to promptly activate the NGU strategy, we must have a detection mechanism inside the system that detects the violation of the fault hypothesis. A distributed fault-tolerant membership service, such as the membership protocol contained in the time-triggered protocol (TTP) [Kop93], implements such a detection mechanism.

6.4.1   Fault Hypotheses

Fault-Containment Unit (FCU)

The fault hypothesis begins with a specification of the *units of failure*, i.e., the fault-containment units (FCUs). It is up to quality engineering to ensure that FCUs fail independently. Even a small correlation of the failure rates of FCUs has a tremendous impact on the overall reliability of a system. If a fault can cause more than one FCU to fail, then the probability of such a correlated failure must be carefully analyzed and documented in the fault hypotheses.

Example: In a distributed system, a component, including hardware and software, can be considered to form an FCU. Given proper engineering precautions concerning the power supply and the electrical isolation of process signals have been made, the assumption that components of a distributed system that are physically at a distance will fail independently is realistic. On an MPSoC, an IP core that communicates with other IP cores solely by the exchange of messages can be considered to form an FCU. However, since the IP cores of an MPSoC are physically close together (potential for *spatial proximity faults*), having a common power supply and a common timing source, it is not justified to assume that the failures of IP cores are fully independent. For example, in the aerospace domain, a failure rate of 10 FITs is assumed for a total MPSoC failure, no matter what kind of MPSoC-internal fault-containment mechanisms are available.

Failure Modes and Failure Rates   In the next step, the assumed failure modes of the FCUs are described, and an estimated failure rate for each failure mode is documented. These estimated failure rates serve as an input for a reliability model to find out whether the calculated reliability of a design is in agreement with the required reliability. Later, after the system has been built, these estimated failure rates are compared with the failure rates that are observed in the field. This is to check whether the fault hypothesis is reasonable and the required reliability goals can be

Table 6.2 Order of magnitude of hardware failure rates

Failure mode	Failure rate (FIT)
Permanent hardware failures	10–100
Non-fail-silent permanent hardware failures	1–10
Transient hardware failures (strong dependence on environment)	1 000–1 000 000

met. Table 6.2 lists orders of magnitude of typical *hardware failure rates* of large VLSI chips [Pau98] that are used in the industrial and automotive domain.

In addition to the failure modes and failure rates, the fault hypothesis must contain a section that discusses the error and failure-detection mechanisms that are designed to detect a failure. This topic is discussed in Sect. 6.3.

Recovery Time The time needed to recover after a transient failure is an important input for a reliability model. In a state-aware design, the recovery time depends on the duration of the ground cycle (see Sect. 6.6) and the time it takes to restart a component.

6.4.2   Fault-Tolerant Unit

In order to tolerate the failure of a fault-containment unit (FCU), FCUs are grouped into *fault-tolerant units (FTUs)*. The purpose of an FTU is to mask the failure of a single FCU inside the FTU. If an FCU implements the fail-silent abstraction, then an FTU consists of two FCUs. If no assumptions can be made about the failure behavior of an FCU, i.e., an FCU can exhibit Byzantine failures, then four FCUs linked by two independent communication channels are needed to form an FTU. If we can assume that a fault-tolerant global time is existent at all FCUs and that the communication network contains temporal failures of an FCU, then it is possible to mask the failure of a non-fail-silent FCU by triplication, called *triple modular redundancy (TMR)*. TMR is the most important fault-masking method.

Although a failure of an FCU is masked by the fault-tolerant mechanism and is thus not visible at the user interface, a permanent failure of an FCU nevertheless reduces or eliminates any further fault-masking capability. It is therefore essential that *masked failures* are reported to a diagnostic system so that they can be repaired. Furthermore, special testing techniques must be provided to periodically check whether all FCUs and the fault-tolerance mechanisms are operational. The name *scrubbing* refers to testing techniques that are periodically applied to detect faulty units and avoid the accumulation of errors.

> **Example:** If the data words in memory are protected by an error-correcting code, then the data words must be accessed periodically in order to correct errors and thus avoid the accumulation of errors.

Fail-Silent FCUs   A *fail-silent FCU* consists of a computational subsystem and an error detector (Fig. 6.6) or of two FCUs and a self-checking checker to compare the results. A fail-silent FCU produces either correct results (in the value and time domain) or no results at all. In a time-triggered architecture, an FTU that consists of two deterministic fail-silent FCUs produces zero, one, or two correct result messages at about the same instant. If it produces no message, it has failed. If it produces one or two messages, it is operational. The receiver must discard redundant result messages. Since the two FCUs are *deterministic*, both results, if available, are correct and it does not matter which one of the two results is taken. If the result messages are *idempotent*, two replicated messages will have the same effect as a single message.

Triple Modular Redundancy   If a fault-containment unit (FCU) can exhibit value failures at its linking interface (LIF) with a probability that cannot be tolerated in the given application domain, then these value failures can be detected and masked in a *triple modular redundant (TMR)* configuration. In a TMR configuration, a fault-tolerant unit (FTU) must consist of three synchronized *deterministic* FCUs, where each FCU is composed of a *voter* and the *computational subsystem*. Any two successive FTUs must be connected by two independent real-time communication systems to tolerate a failure in any of the two communication systems (Fig. 6.7). All FCUs and the communication system must have access to a fault-tolerant global time base. The communication system must perform error containment in the temporal domain, i.e., it must have knowledge about the permitted temporal behavior of an FCU. In case an FCU violates its temporal specification, the communication

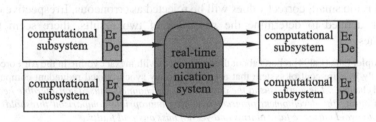

Fig. 6.6 FTU consisting of two fail-silent FCUs

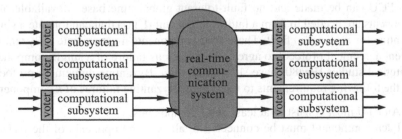

Fig. 6.7 Two FTUs, each one consisting of three FCUs with voters

system will discard all messages received from this FCU in order to protect itself from an overload condition. In the fault-free case, each receiving FCU will receive six physical messages, two (via the two independent communication systems) from each sending FCU. Since all FCUs are deterministic, the *correct FCUs* will produce identical messages. The voter detects an erroneous message and masks the error in one step by comparing the three independently computed results and then selecting the result that has been computed by the majority, i.e., by two out of three FCUs.

A TMR configuration that is set up according to the above specified rules will tolerate an *arbitrary failure* of any FCU and any communication system, provided that a fault-tolerant global time base is available.

Two different kinds of voting strategies can be distinguished: *exact voting* and *inexact voting*. In *exact voting*, a bit-by-bit comparison of the data fields in the result messages of the three FCUs forming an FTU is performed. If two out of the three available messages have exactly the same bit pattern, then one of the two messages is selected as the output of the triad. The underlying assumption is that correctly operating replica-determinate components produce exactly the same results. Exact voting requires that the input messages and the g-state of the three FCUs that form an FTU are *bit-identical*. If the inputs originate from redundant sensors to the physical environment, an agreement protocol must be executed to enforce bit-identical input messages.

In *inexact voting*, two messages are assumed to contain semantically the *same* result if the results are within some *application-specific* interval. Inexact voting must be used if the replica determinism of the replicated components cannot be guaranteed. The selection of an appropriate interval for an inexact voter is a delicate task: if the interval is too large, erroneous values will be accepted as correct; if the interval is too small, correct values will be rejected as erroneous. Irrespective of the criterion defined to determine the *sameness* of two results, there seem to be difficulties.

> **Example:** Lala [Lal94] reports about the experiences with inexact voting in the Air Force's F-16 fly-by-wire control system that uses four loosely synchronized redundant computational channels: *The consensus at the outputs of these channels caused considerable headaches during the development program in setting appropriate comparison thresholds in order to avoid nuisance false alarms and yet not miss any real faults.*

Byzantine-Resilient Fault-Tolerant Unit   If no assumption about the failure mode of an FCU can be made and no fault-tolerant global time base is available, four components are needed to form a fault-tolerant unit (FTU) that can tolerate a single Byzantine (or *malicious*) fault. These four components must execute a Byzantine-resilient agreement protocol to agree on a malicious failure of a single component. Theoretical studies [Pea80] have shown that these Byzantine agreement protocols have the following requirements to tolerate the Byzantine failures of k components:

(i) An FTU must consist of at least *3k+1* components.
(ii) Each component must be connected to all other components of the FTU by *k+1* disjoint communication paths.

(iii) To detect the malicious components, $k+1$ rounds of communication must be executed among the components. A round of communication requires every component to send a message to all the other components.

An example of an architecture that tolerates Byzantine failures of the components is given in [Hop78].

6.4.3 The Membership Service

The failure of an FTU must be reported in a consistent manner to all operating FTUs with a low latency. This is the task of the *membership service*. A point in real time when the membership of a component can be established is called a *membership point* of the component. A small temporal delay between the membership point of a component and the instant when all other components of the ensemble are informed in a consistent manner about the current membership is critical for the correct operation of many safety-relevant applications. The consistent activation of a never-give-up (NGU) strategy in case the fault hypothesis is violated is another important function of the membership service.

> **Example:** Consider an intelligent ABS (antilock braking system) braking system in a car with a node of a distributed computer system placed at each wheel. A distributed algorithm in each of the four nodes, one at each wheel, calculates the brake-force distribution to the wheels (Fig. 6.8), depending on the position of the brake pedal actuated by the driver. If a wheel node fails or the communication to a wheel computer is lost, the hydraulic brake force actuator at this wheel autonomously transits to a defined state, e.g., in which the wheel is free running. If the other nodes learn about the computer failure at this wheel within a short latency, e.g., a single control-loop cycle of about 2 msec, then the brake force can be redistributed to the three functioning wheels, and the car can still be controlled. If, however, the loss of a node is not recognized with such a low latency, then the brake-force distribution to the wheels, based on the assumptions that all four-wheel computers are operational, is wrong, and the car will go out of control.

ET Architecture In an ET architecture, messages are sent only when a significant event happens at a component. Silence of a component in an ET architecture means that *either* no significant event has occurred at the component *or* a fail-silent failure has occurred (the loss of communication or the fail-silent shutdown of the component). Even if the communication system is assumed to be perfectly reliable, it is not

Fig. 6.8 Example of an intelligent ABS in a car

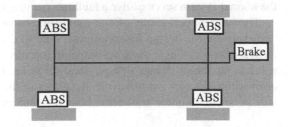

possible to distinguish when there is *no activity at the component* from the situation when a *silent component failure* occurs in an ET architecture. An additional time-triggered service, e.g., a periodic watchdog service (see Sect. 9.7.4), must be implemented in an ET architecture to solve the membership problem.

TT Architecture In a TT architecture, the periodic message-send times are the membership points of the sender. Let us assume that a failed component remains out of service for an interval with duration greater than the maximum time interval between two membership points. Every receiver knows a priori when a message of a sender is supposed to arrive and interprets the arrival of the message as a life sign at the membership point of the sender [Kop91]. It is then possible to conclude, from the arrival of the expected messages at two consecutive membership points, that the component was alive during the complete interval delimited by these two membership points (there is a tacit assumption that a transiently failed node does not recover within this interval). The membership of the FTUs in a cluster at any point in time can thus be established with a delay of one round of information exchange. Because the delay of one round of information exchange is known a priori in a TT architecture, it is possible to derive an a priori bound for the temporal accuracy of the membership service.

6.5   Robustness and Resilience

6.5.1   The Concept of Robustness

In the domain of embedded systems, we consider a system to be robust *if the severity of the consequences of a fault is inversely proportional to the probability of fault occurrence*, i.e., faults that are expected to occur frequently should have only a minor effect on the quality of service of the system. Irrespective of the concrete type and source of a fault, a robust embedded system will try to recover from the effects of a fault as quickly as possible in order to minimize the impact of the fault on the user. As noted above in Sect. 6.1, the immediate consequence of a fault is an error, i.e., an unintended state. If we detect and correct the error before it has had a serious effect on the quality of service, we have increased the robustness of the system. Design for robustness is not concerned with finding the detailed cause of a failure—this is the task of the *diagnostic subsystem*—but rather with the fast restoration of the normal system service after a fault has occurred.

The inherent periodicity of many real-time control systems and multimedia systems helps in the design for robustness. Due to the constrained physical power of most actuators, a single incorrect output in a control cycle will—in most cases—not result in an abrupt change of a physical set point. If we can detect and correct the error within the next control cycle, the effect of the fault on the control application will be small. Similar arguments hold for multimedia system. If a single frame contains some incorrect pixels, or even if a complete frame is lost, but the next frame in sequence is correct again, then the impact of a fault on the *quality of the multimedia experience* is limited.

A robust system consists of at least two subsystems (Fig. 6.9) implemented as independent FCUs, one *operational component* that performs the planned operations and controls the physical environment and a second *monitoring component* that *reflects* whether the results and the g-state of the operational component are in agreement with the *intentions* of the user [Tai03].

In a periodic application such as a control application, every control cycle starts with reading the g-state and the input data, then the control algorithm is calculated, and finally the new set points and the new g-state are produced (see Fig. 3.9). A transient fault in one control cycle can only propagate to the next control cycle if the g-state has been contaminated by the fault. In a robust system, the operational component must externalize its g-state in every control cycle such that the monitoring component can check the plausibility of the g-state and perform a corrective action in case a severe anomaly has been detected in the g-state. The corrective action can consist of resetting the operational component and restarting it with a repaired g-state.

In a safety-critical application, this *two-channel approach*—one channel produces a result and the other channel, the *safety monitor*, monitors whether the result is plausible—is absolutely essential. Even if the software has been proven correct, it cannot be assumed that there will be no transient faults during the execution of the hardware. The *IEC 61508 standard* on functional safety requires such a two-channel approach, one channel for the normal function and another independent channel to ensure the *functional safety* of a control system (see also Sect. 11.4).

In a fail-safe application, the safety monitor has no other authority than to bring the application to the safe state. A *fail-silent failure* of the safety monitor will result in a loss of the safety monitoring function, while a *non-fail-silent failure* of the safety monitor will cause a reduction of the availability but will not impact the safety.

In a fail-operational application, a *non-fail-silent failure* of the safety monitor has an impact on the safety of the application. Therefore, the safety monitor must be part of a holistic safety architecture.

> **Example:** [Kop21] presents a fail-operational architecture for self-driving cars. In this architecture, a monitor (the monitoring subsystem—MSS) continually checks the operational FCU (the computer-controlled driving subsystem—CCDSS). If the MSS or the CCDSS is faulty, a fallback FCU (the critical event handling subsystem—CEHSS) takes over. A fault-tolerant decision subsystem (FTDSS) ensures that all actuators consistently use either the output of the CCDSS+MSS pair or the CEHSS.

Fig. 6.9 Structure of a robust system

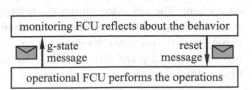

6.5.2   The Concept of Resilience

Resilience is defined as *the persistence of dependability when facing changes* [Lap08]. These changes can be classified in three dimensions: *nature (functional, environmental, or technological), prospect (foreseen or unforeseen),* and *timing (seconds to years)*. Resilience is, thus, a broad concept. Indeed, several concepts discussed in this book directly contribute to the resilience of a system.

> **Example:** System-of-systems design (see Sect. 4.7.3) is an approach to achieve resilience systematically.

> **Example:** A never-give-up strategy (see Sect. 6.4) is a technique to react to a *short-term, unforeseen* changes in the *environment* (i.e., a violation of the fault hypothesis).

> **Example:** Autonomic components (see Sect. 13.3.5) learn from their deployment and can adapt to changes.

Despite this broad definition of resilience, the close concept of *cyber-resiliency* has emerged in the area of security [NIS21]: *cyber resiliency is the ability to antici-pate, withstand, recover from, and adapt to adverse conditions, stresses, attacks, or compromises on systems that use or are enabled by cyber resources.* Cyber-resiliency is resilience concerning unforeseen changes in the environment, typically associated with security attacks.

6.6   Component Reintegration

Most computer system faults are transient, i.e., they occur sporadically for a very short interval, corrupt the state, but do not permanently damage the hardware. If the service of the system can be reestablished quickly after a transient fault has occurred, then in most cases the user will not be seriously affected by the consequences of the fault. In many embedded applications, the fast reintegration of a failed component is thus of paramount importance and must be supported by proper architectural mechanisms.

6.6.1   Finding a Reintegration Point

While a failure can occur at an arbitrary moment outside the control of the system designer, the system designer can plan the proper point of reintegration of a repaired component. The key issue during the reintegration of a component in a real-time system is to find a future point in time when the state of the component is in syn-chrony with the component's environment, i.e., the other components of the cluster and the physical plant. Because real-time data are invalidated by the passage of time, rolling back to a past checkpoint can be futile: it is possible and probable that the progression of time has already invalidated the checkpoint information (see also Table 4.1).

Reintegration is simplified if the state that must be reloaded into the reintegrating component is of *small size* and fits into a single message. Since the size of the state has a relative minimum immediately after the completion of an atomic operation, this is an ideal instant for the reintegration of a component. In Sect. 4.2.3 we have introduced the notion of the *g-state* (ground state) to refer to the state at the *reintegration instant*. In cyclic systems—many embedded control and multimedia systems are cyclic—an ideal reintegration instant of a component is at the beginning of a new cycle. The temporal distance between two consecutive reintegration instants, the reintegration cycle, is then identical to the duration of the control cycle. If the g-state is empty at the reintegration instant, then the reintegration of a repaired component is trivial at this moment. In many situations, however, there is no instant during the lifetime of a component when its g-state is completely empty.

6.6.2   Minimizing the Ground State

After a cyclic reintegration instant has been established, the g-state at this selected instant must be analyzed and minimized to simplify the reintegration procedure.

In a first phase, all system data structures within the component must be investigated to locate any hidden state. In particular, all variables that must be initialized must be identified, and the state of all semaphores and operating system queues at the reintegration instant must be checked. It is good programming practice to output the g-state of a task in a special output message when a task with g-state is detected and to re-read the g-state of the task when the task is reactivated. This identifies the g-state and makes it possible to pack all g-states of all tasks of a component into a *g-state message* particular to this component.

In a second phase, the identified g-state must be analyzed and minimized. Figure 6.10 displays a suggested division of the g-state information into three parts:

(i) The first part of the g-state consists of input data that can be retrieved from the instrumentation in the environment. If the instrumentation is state-based and sends the absolute values of the RT entities (state messages) rather than their relative values (event messages), a complete scan of all the sensors in the environment can establish a set of current images in the reintegrating component and thus resynchronize the component with the external world.

(ii) The second part of the g-state consists of output data that are in the control of the computer and can be enforced on the environment. We call the set of the output data a *restart vector*. In a number of applications, a restart vector can be defined at development time. Whenever a component must be reintegrated, this restart vector is enforced on the environment to achieve agreement with the outside world. If different process modes require different restart vectors, a set of restart vectors can be defined at development time, one for each mode.

(iii) The third part of the g-state contains g-state data that do not fall into category
       (i) or category (ii). This part of the g-state must be recovered from some
       component-external source: from a replicated component of a fault-tolerant
       system, from the monitoring component, or from the operator. In some situa-
       tions, a redesign of the process instrumentation may be considered to trans-
       form g-state of category (iii) into g-state of category (i).

> **Example:** When a traffic control system is restarted, it is possible to enforce a restart vector
> on the traffic lights that sets all cross-road lights first to yellow, and then to red, and finally
> turns the main street lights to green. This is a relatively simple way to achieve synchroniza-
> tion between the external world and the computer system. The alternative, which involves
> the reconstruction of the current state of all traffic lights from some log file that recorded
> the output commands up to the point of failure, would be more complicated.

In a system with replicated components in an FTU, the g-state data that cannot
be retrieved directly from the environment must be communicated from one component of the FTU to the other components of the FTU by means of a g-state message.
In a TT system, sending such a g-state message should be part of the standard component cycle.

6.6.3 Component Restart

The restart of a component after a failure has been detected by a monitoring component (Fig. 6.10) can proceed as follows: (i) The *monitoring component* sends a
trusted reset message to the TII interface of the operational component to enforce a
hardware reset. (ii) After the reset, the operational component performs a self-test
and verifies the correctness of its core image (the *job*) by checking the provided
signatures in the core image data structures. If the core image is erroneous, a copy
of the static core image must be reloaded from stable storage. (iii) The operational
component scans all sensors and waits for a cluster cycle to acquire all available
current information about its environment. After an analysis of this information, the

Fig. 6.10 Partitioning of
the g-state

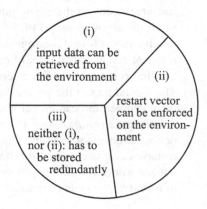

operational component decides the *mode* of the controlled object and selects the restart vector that must be enforced on the environment. (iv) Finally, after the operational component has received the g-state information that is relevant at the next reintegration instant from the monitoring component, the operational component starts its tasks in synchrony with the rest of the cluster and its physical environment. Depending on the hardware performance and the characteristics of the real-time operating system, the time interval between the arrival of the reset message and the arrival of the g-state information message can be significantly longer than the duration of a reintegration cycle. In this case, the monitoring component must perform a far-reaching state estimation to establish a relevant g-state at the proper reintegration point.

Points to Remember

- A *fault* is the adjudged cause of an error or failure.
- An *error* is that part of the state of a system that deviates from the *intended* (correct) state.
- A *failure* is an event that denotes a deviation of the actual service from the intended service, occurring at a particular point in real time.
- The failure rate for permanent failures of an industrial-quality chip is in a range between *10* and *100* FITs. The failure rate for transient failures is orders of magnitude higher.
- *Information security* deals with the *authenticity, integrity, confidentiality, privacy*, and *availability* of information and services that are provided by computer system. The main security concerns in embedded systems are the *authenticity and integrity of data.*
- A *vulnerability* is a deficiency in the design or operation of a computer system that can lead to a security incident. We call the successful exploitation of a vulnerability an *intrusion*.
- The typical attacker proceeds according to the following three phases: *access to the selected subsystem, search for and discovery of a vulnerability*, and finally *intrusion and control of the selected subsystem.*
- It is widely acknowledged in security research and practice that many security incidents are caused by human rather than technical failures.
- The basic cryptographic primitives that must be supported in any security architecture are *symmetric key encryption, public key encryption, hash functions*, and *random number generation.*
- An *anomaly* is a system state that lies in the gray zone between *correct* and *erroneous*. The detection of anomalies is important, since the occurrence of an anomaly is an indication that some atypical scenario that may require immediate corrective action is developing (e.g., the intrusion by an adversary).

- In a safety-critical system, every single observed anomaly must be scrutinized in detail until the final cause of the anomaly has been unambiguously identified.
- Failure detection within a system is only possible if the system contains some form of redundant information about the *intended behavior*.
- The fault hypothesis states what types of faults must be tolerated by a fault-tolerant system and divides the fault space into two domains, the domain of *normal faults* (i.e., the faults that must be tolerated) and the domain of *rare faults*, i.e., faults that are outside the fault hypotheses and are assumed to be *rare events*.
- A *rare fault* will bring the system into a state that is outside the specified fault hypothesis and therefore will not be covered by the provided fault-tolerance mechanisms. Nevertheless, instead of giving up, *a never-give-up (NGU)* strategy should be employed to try to bring the system back to a correct state.
- It is up to quality engineering to ensure that FCUs fail independently. Even a small correlation of the failure rates of FCUs has a tremendous impact on the overall reliability of a system.
- The purpose of a fault-tolerant unit (FTU) is to mask the failure of a single FCU inside the FTU. Although a failure of an FCU is masked by the fault-tolerant mechanism and is thus not visible at the user interface, a permanent failure of an FCU nevertheless reduces or eliminates any further fault-masking capability.
- In a triple modular redundant (TMR) configuration, a fault-tolerant unit (FTU) consists of three synchronized *deterministic* FCUs, where each FCU is composed of a *voter* and the *computational subsystem*.
- A *membership service* consistently reports *the operational state* of every FTU to all operating FTUs.
- In many embedded applications, the fast reintegration of a failed component is of paramount importance and must be supported by proper architectural mechanisms.
- Design for robustness is not concerned with finding the detailed cause of a failure—this is the task of the *diagnostic subsystem*—but rather with the fast restoration of the normal system service.
- In a safety-critical application, a *two-channel approach*, in which one channel produces a result and the other channel monitors whether the result is plausible, is absolutely essential.

Bibliographic Notes

The seminal paper by Avizienis, Laprie, Randell, and Landwehr [Avi04] introduces the fundamental concepts in the field of dependability and security. Anomaly detection is covered in the comprehensive survey by Chandola [Cha09], and online failure prediction is the topic of [Sal10]. Several industries have developed industry-specific cybersecurity standards, e.g., ISO/SAE 21434 for automotive [ISO21] or IEC 62443 for automation and control systems [IEC21]. The yearly

DSN conference (organized by the IEEE and the IFIP WG 10.4) is the most important forum for presenting research papers in the field of dependable and secure computing.

Review Questions and Problems

6.1 Give the precise meaning of the terms failure, error, and fault. What are the characteristics of an FCU?

6.2 What are typical permanent and transient failure rates of VLSI chips?

6.3 What is an anomaly? Why is anomaly detection important?

6.4 Why is a short recovery time from transient faults important?

6.5 What are the basic techniques for error detection? Compare ET systems and TT systems from the point of view of error detection.

6.6 What is the difference between *robustness* and *fault tolerance*? Describe the structure of a robust system!

6.7 What is the difference between a *Heisenbug* and a *Bohrbug*?

6.8 Describe the characteristics of a Byzantine failure! What is an SOS failure?

6.9 Give some examples of security threats! What is a *botnet*?

6.10 What is the difference between the *Biba model* and the *Bell-LaPadula model* for secure systems?

6.11 What steps must be taken in a systematic *security analysis*? What is a *vulnerability*? What is an *intrusion*?

6.12 What is a membership service? Give a practical example for the need of a membership service. What is the quality parameter of the membership service? How can you implement a membership service in an ET architecture?

6.13 Describe the contents of the fault hypothesis document? What is an NGU strategy?

6.14 Discuss the different types of faults that can be masked by the replication of components. Which faults cannot be masked by the replication of components?

6.15 What is required for the implementation of fault tolerance by TMR?

6.16 What is a restart vector? Give an example.

Chapter 7
Real-Time Communication

Overview

Many real-time systems are distributed computer systems. Thus, they require a real-time communication subsystem that ensures reliable and timely message exchange to allow the distributed computer system to operate as a coordinated whole. We call the real-time communication subsystem the *real-time network*. A real-time network is fundamentally different from communication networks found in common IT infrastructures. It must ensure timely message deliveries that are not required in IT networks. In the past many different real-time networks have been designed and deployed. Often, particular real-time networks are highly specialized to specific industries and only address niche markets. However, with standardization, more and more industries moved toward only a few solutions, and this adoption process of standard solutions is still ongoing. While standardized Ethernet-based real-time networks have been in use for some time now, only rather recently key real-time functions have been incorporated in the IEEE 802.1 set of standards by the time-sensitive networking (TSN) task group. However, although today's Ethernet is suitable as a real-time network for many industries, it is still a long way to reach market dominance and interoperability comparable to office equipment.

We start this chapter with the definition of a real-time network and discuss its typical requirements, timeliness being the most important one. Although many different real-time networks exist, they share common design principles. We discuss these based on a *simple model* that interprets the real-time network as a set of resources used for message exchange. Our *simple model* distinguishes three types of messages, event-triggered, rate-constrained, and time-triggered, and a real-time network may support one or many of these message types. We also mention general design limitations and pitfalls. The remainder of the chapter uses one section per message type, for which we first discuss general aspects, followed by concrete examples. For event-triggered messages, we discuss basic Ethernet and CAN. Next, Avionics Full-Duplex Switched Ethernet (AFDX) and Audio/Video Bridging (AVB) illustrate rate-constrained messages. Finally, we discuss the time-triggered protocol

© The Author(s), under exclusive license to Springer Nature Switzerland AG 2022   177
H. Kopetz, W. Steiner, *Real-Time Systems*,
https://doi.org/10.1007/978-3-031-11992-7_7

(TTP), time-triggered Ethernet (TTEthernet), and time-sensitive networking (TSN) as examples for real-time networks supporting time-triggered messages.

7.1 Requirements

Modern real-time systems often implement a distributed computer system where a real-time network interconnects spatially distributed nodes. A real-time network is necessary when:

  (i) The real-time system is inherently distributed, e.g., industrial automation in a factory floor, energy production plants, trains, large aircrafts.
 (ii) There is need to partition a real-time system into nearly independent *fault-containment units* to provide fault tolerance, e.g., spacecrafts, automobiles.
(iii) The real-time system's processing needs exceed the processing capability of a single node, e.g., self-driving cars.
(iv) The real-time system's instrumentation interface cannot be implemented by a single node alone. For example, in the case of an industrial robot, it would be impractical to directly connect all sensors and actuators to a single processing node.

The real-time network performs all real-time communication between distributed nodes as defined by the real-time system and may use heterogeneous technologies and paradigms. The real-time network also includes transmit and receive subsystems at the nodes. Real-time networks must maintain the properties of *real-time data* elaborated in the previous chapters. Thus, real-time networks must satisfy architectural requirements that substantially differ from the ones of non-real-time communication services.

7.1.1 Timeliness

The most important requirement of a real-time network is *guaranteed message-transport latency* with *bounded message-transport jitter*.

Guaranteed Message-Transport Latency A network is a real-time network, *if and only if a bound for the worst-case message-transport latency for time-critical messages can be determined by analysis, and this bound holds during operation with a sufficiently high probability.* In hard real-time systems, said probability evolves toward a requirement of a de facto guarantee. For example, a common reliability requirement for ultrahigh dependable systems (e.g., airplanes) requires a failure rate of 10^{-9} failures/hour. It is this guarantee that common IT network technologies do not meet. The target values for the latency and its bound, on the other hand, are determined by the needs of the specific real-time application. It is, thus, also the concrete real-time application that determines whether a particular network technology is suitable as a real-time network or not. In many real-time systems, minimizing the worst-case message-transport latency will directly yield quality improvements on the system level. For example, if the real-time system implements a *real-time transaction* (see Sect. 1.7.3) that starts with the acquisition of sensor values and terminates with providing output to an actuator, minimizing the message-transport latency reduces the *response time of* the real-time system's control loop and thereby improves control stability and control quality.

Bounded Message-Transport Jitter The jitter is the difference between the worst-case and the best-case message-transport latencies. A large jitter has a negative effect on the duration of the *action delay* (see Sect. 5.5.1) and the precision of the clock synchronization (see Sect. 3.1.3).

Time-Measurement Service A *real-time image* (see Sect. 5.4) must be *temporally accurate* at the *instant of use* (see Sect. 5.4), and a node using the real-time image must be able to check its temporal accuracy. The real-time network must, thus, support distributed *time measurement* because, in a distributed computer system, the observation of a real-time entity and the use of its associated real-time image will often happen at different nodes. A real-time network that implements a clock synchronization protocol, e.g., IEEE 1588, can use time-stamps to realize a time measurement service. The necessary quality of the clock synchronization protocol (precision, tolerance/robustness to failures and attacks) is determined by the needs of the real-time system.

7.1.2 Dependability and Security

Communication Reliability Messages in transport may be lost or corrupted in the time and value domain. Therefore, real-time networks often implement redundancy mechanisms. Lost messages are compensated by redundancy in space, i.e., transmission along disjoint communication paths, or time, i.e., sequential message retransmissions. In real-time systems that take benefit of low message-transport latencies, redundancy in space is preferable since redundant copies of a message can be communicated in parallel. Corrupted messages are typically addressed by data redundancy within the message itself. Data redundancy techniques to improve message integrity are error-detecting and error-correcting codes, time-stamps, and sequence numbers. We call *communication reliability* the probability of successful message transport by the real-time network after applying all implemented message-transport redundancy mechanisms.

Fault Containment The real-time network must implement *traffic-policing* methods when multiple nodes use the same network resource (e.g., they are connected through the same physical bus or share the same network switch). Otherwise, faulty nodes may use the resource beyond their specified quota and limit its use to non-faulty nodes. Traffic-policing methods can be implemented at the boundary of the real-time network as part of the distributed nodes themselves, called *fault containment at the source*. Here, an independent component local to the node monitors the node's communication behavior and intervenes if this behavior diverges from its specification. For example, it may shut down the complete node. On the other hand, traffic policing will often be implemented inside the real-time network, e.g., in a network switch. A common traffic-policing method, both *at the source* and *in the network*, is the *leaky bucket algorithm*, which only allows a node to transmit a specified amount of data within a specified interval.

Message Integrity and Confidentiality   While message integrity is essential for communication reliability, the confidentiality of real-time messages has been rarely required by real-time systems in the past. However, real-time message confidentiality must be reconsidered because new attacks are enabled as real-time systems more and more connect to IT networks and even to the Internet. While the monetary value of information content of single real-time messages may be limited, it scales up with the consolidation of multiple, different messages. For example, an attacker who consolidates real-time messages in an industrial production plant may learn confidential procedures in process automation. Therefore, a security architecture may require network-level cryptography in sensitive parts of the real-time network. Protocols using block ciphers such as MACsec (IEEE 802.1AE) can be a good match for real-time systems as they can cause acceptable additional message-transport latency.

Configuration Integrity   As we build more and more complex real-time systems, the complexity of their real-time networks increases as well. This complexity is also reflected in the growing configuration space of real-time networks. The *configuration integrity* of real-time networks must also be preserved in presence of failures and attacks. In particular, only trusted entities after successful authentication may reconfigure the real-time network.

7.1.3   Flexibility

A real-time system may inherently require an ability to add and remove nodes or even entire subsystems, e.g., adding and removing machines in a production hall (*plug-and-produce*), a maintenance procedure to replace a node while the system operates, or coupling a spacecraft to a space station. Furthermore, the requirements on a real-time system may change over time, causing hardware/software updates.

Changing Communication Demands   A real-time network can support intentional changes in the communication demands but must maintain the *worst-case message-transport latencies* of real-time messages from distributed applications that are not addressed by this change. Naturally, changes can only be supported as far as they do not exceed physical or technological limitations (e.g., the total communication bandwidth cannot be exceeded). A communication change can also require reconfiguring the message transport of non-affected applications. It depends on the concrete real-time system if such a change is permitted during system operation or a system maintenance phase or not permitted at all.

Network Engineering vs. Plug-and-Play   Different real-time networks provide different degrees of automation in their reconfiguration response to changing communication demands. *Network engineering* refers to a manual or computer-aided generation of a new communication configuration. On the other hand, *plug-and-play* is the ability of a real-time network to accommodate a change in communication demand without human interaction. While real-time networks like AVB (IEEE 802.1 Audio/Video Bridging) provide plug-and-play, its use in safety-critical systems is a risk.

7.1.4   Communication Bandwidth and Cost Efficiency

Communication Bandwidth   The required communication bandwidth for a real-time image is determined by its size and maximum update frequency. This required bandwidth can vary tremendously with the type of real-time entity and use case, e.g., from a few bit/sec for a room-temperature sensor at 10 Hz to multiple Gbit/sec for an ultrahigh-definition camera for self-driving cars.

The acronym *SWaP-C* stands for the system aspects: *size, weight, power, and cost*. While power and cost are important factors in basically every real-time system, weight and size are especially crucial in mobile real-time systems like automobiles, airplanes, or spacecrafts.

Cost Efficiency   A real-time network must satisfy the requirements discussed in this chapter cost-efficiently. There is a cross-industry trend toward *converged networks* in real-time systems: a single homogeneous network technology serves multiple, if not all, distributed applications in the real-time system. Converged networks typically increase network utilization and, therefore, also improve cost efficiency. The IEEE 802.1 Time-Sensitive Network (IEEE 802.1 TSN) is a real-time network technology for converged networks (see Sec. 7.5.3).

7.2   Design Principles and Pitfalls

There are many different ways to implement a real-time network, but the number of underlying design principles is quite limited. Following the discussion in Chap. 2, these design principles serve as a *simple model* that captures all system-level relevant communication aspects. In this simple model, we consider the real-time network as an abstract *set of resources* and a *basic message-transport service (BMTS)* that transports messages from a sending component to one or many receiving components by more or less coordinated usage of said resources. BMTS differ by the message types they support (see Sect. 7.2.2). On top of the BMTS, higher-layer protocols can be implemented (see Sect. 14.4.2).

7.2.1   Real-Time Network Model

In the simple model, the real-time network is a set of resources, and only one sender can use a resource at a time.

Individual Resource   There is only one sender in the real-time system for each individual resource. The resource is, for example, a direct communication link between a sender and a receiver.

Shared Resource   Multiple senders share the resource. The resource is, for example, a *physical bus*, a *network hub*, or *message buffers* in a network switch.

> **Example—Resources in Switched Ethernet:** Nodes connect with bidirectional communication links to Ethernet switches. The unidirectional portion of the link from the sender to a switch is an *individual resource*. All other communication links and the message buffers in the network switches are *shared resources*.

7.2.2   Message Types

Nodes use the resources by exchanging messages, and we can distinguish the following message types.

Event-Triggered Messages   The sender produces messages whenever a significant event occurs. These events can be outside the real-time system's sphere of control, and, thus, no minimum time in between messages can be established. Consequently, the total number of messages from all the senders may reach or exceed the resource capacity of the real-time network, causing long message-transport latencies and even message loss. A real-time network cannot provide temporal guarantees for message-transport latencies for event-triggered messages.

Rate-Constrained Messages   The sender produces messages and guarantees not to exceed a maximum message rate, defined a priori. Based on the total of the message rates, the real-time network can dimension the capacity of its resources accordingly, and bounds for the worst-case message-transport latencies can be calculated off-line. However, the message-transport jitter can be high because rate-constrained messages from different senders may come sparsely or in a *burst*.

Time-Triggered Messages The sender produces messages at a priori defined points in a synchronized global time. These points in time from different senders can be coordinated, allowing the real-time network to dimension its resource capacities. In contrast to rate-constrained messages, the real-time network can minimize its resources by this coordination. The real-time network can also a priori define points in time when to temporarily buffer and forward time-triggered messages as well as the points in time of reception of the final receiving nodes. Consequently, the message-transport jitter can be reduced to the precision of the synchronized global time.

The act of producing rate-constrained messages or time-triggered messages from any given input is called *traffic shaping*.

7.2.3 Flow Control

While the previous *message types* are defined based on the sender's behavior, in any communication scenario, the *receiver* and the limited resource capacities of the network determine the maximum speed of communication. *Flow control* is, thus, concerned with the control of the speed of information flow between one or a multitude of senders and a receiver (or the network—in this case, the term *congestion control* is sometimes used) in such a manner that the receiver (or the network) can keep up with the sender(s). While the a priori information of rate constrained and time-triggered messages allow *implicit flow control*, flow control for event-triggered messages requires *explicit flow-control* mechanisms.

Explicit Flow Control When a sender or multiple senders aim to exceed the resource capacities of the network or the receiver, messages will be lost. *Explicit flow-control* mechanisms prevent these situations by signaling all or some senders to pause and resume their message transmissions. Explicit flow control is also called *backpressure flow control*.

> **Example—Ethernet Pause Command:** When the receive buffers in a receiver node (or a switch) in an Ethernet network exceed a given threshold, the receiving node sends an Ethernet *pause* message to the sender to pause the message transmissions for a defined duration. Furthermore, Ethernet frames with VLAN tag distinguish eight priorities. *Data center bridging* also uses these priorities in the pause message and can selectively pause only the transmission of Ethernet frames with particular priorities.

Explicit flow control not always requires the exchange of dedicated messages but can also be established by network access protocols.

> **Example—CAN:** In a CAN system, a node cannot access the bus if a transmission by another sender is in progress, i.e., the access protocol at the sender exerts back pressure on the node that intends to send a message.

Implicit Flow Control The a priori knowledge of the real-time network's usage patterns inherently excludes overload scenarios at the receivers and the network.

Flow control is, thus, also established a priori. The real-time network can implement *traffic-policing* mechanisms that ensure that even faulty senders will not violate said usage patterns but stick to their worst-case resource usage quota.

From a complexity point of view, explicit flow control introduces control of the receiver over the sender. Thus, a faulty receiver can impact a correct sender. While the interface between a sender and receiver in implicit flow control is unidirectional, this interface is bidirectional—and thus more complex—in explicit flow control.

7.2.4 Design Limitations

Any physical communication channel is characterized by its *bandwidth* and its *propagation delay*. The *bandwidth* denotes the number of bits that can traverse the channel in unit time. The length of the channel and the transmission speed of the wave (electromagnetic, optical) within the channel determine the *propagation delay*, which is the duration it takes for a single bit to travel from one end of the channel to the other end. Because the transmission speed of a wave in a cable is approximately 2/3 of the transmission speed of light in a vacuum (about 300,000 km/s), it takes a signal about 5 μs to travel across a cable of 1 km length. The term *bit length of a channel* denotes the number of bits that can reside in the channel within one propagation delay.

> **Example:** If the channel bandwidth is 100 Mbit/s and the channel is 200 m long, the bit length of the channel is 100 bits since the propagation delay of this channel is 1 μs.

7.2.4.1 Bus-Based Real-Time Networks

In a bus, only one node may send a message at a time. Otherwise the multiple messages' signals *collide* and cause interference on the bus. This interference will lead to message loss, in some cases even asymmetric message loss: some nodes will receive a message correctly, while others will not. Two types of media-access protocols exist: *collision-detection protocols* detect and resolve message collisions as they occur, and *collision avoidance protocols* prevent collisions from happening. Both protocol types require the nodes to sense the bus and distinguish between *bus activity* and *bus idle* phases. The minimum duration of the *bus idle* phase equals the propagation delay because only then do all nodes reliably sense *bus idle*. From this lower bound on *bus idle* follows the *data efficiency* in a physical bus as a function of its message length m and bit length bl:

$$data\_efficiency < m / (m + bl)$$

Example: Consider a 1 km bus with a bandwidth equal to 100 Mbit/s and a message length of 100 bits. It follows that the bit length of the channel is 500 bits, and the data efficiency is 16.6%.

In general, it will be wasteful to send short messages in long busses and busses with high bandwidth. For example, if the message length is less than the bit length of the channel, then the data efficiency will be less than 50 %.

7.2.4.2   Switch-Based Real-Time Networks

In a switched network, the nodes connect to switches that serve as intermediate stations in the communication path between sending and receiving nodes. All connections are bidirectional and point to point, either between a node and a switch or between two switches. Thus, the connections do not require a media-access protocol. However, the bit length parameter in the data efficiency equation is replaced by a technology-dependent *inter-frame gap* parameter in switched networks. The receiver uses the inter-frame gap to resynchronize to the sender.

Example: In 1 Gbit/s Ethernet, the sending node will transmit 96-bit *idle symbols* in the inter-frame gap. Ethernet messages (Layer 1 Ethernet packets) are between 72 and 1530 octets long. Thus, a point-to-point 1 Gbit/s Ethernet link has a data efficiency between 86% and 99%.

While in bus-based networks, the shared resource is the physical bus, it is the message buffers and their management functions inside the switches in switched networks. These message buffers may be used in two modes: *store-and-forward*, in which the switch receives a message completely before forwarding, and *cut-through*, in which the switch starts forwarding messages before their complete reception. The *technology latency* denotes the minimum time it takes a switch to start forwarding a message assuming the forwarding queues for this message are empty, and a sufficient portion of the message has been received to start transmission.

Example: The *TTE-Switch Controller Space* is a *store-and-forward* switch that supports bandwidths of up to 1 Gbit/s and has a technology latency of about 10μs.

In *store-and-forward* mode, a lower bound for the *best-case message-transport latency* of a message from sending node to a receiving node is given by the sum of the message-transport latencies on the point-to-point links and the technology latencies of the switches in the communication path. An upper bound for the *worst-case message-transport latency* also considers the specific *traffic shaping* method used to establish a *message type* and the *queuing delays* in the switches that result from concurrent message receptions. *Cut-through* mode significantly reduces the best-case and worst-case message-transport latencies because (i) only a portion of a message needs to be received in a switch to start its forwarding transmission and (ii) the transmissions of the message on multiple point-to-point links in the communication path may overlap.

7.2.4.3   Wireless Real-Time Networks

Wireless real-time networks are also common in real-time systems. Wireless protocols are similar to the ones used in a physical bus. However, in the absence of a
wired transport medium, they are susceptible to noise, signal reflection, fading, and
shadowing, and their *communication reliability* is typically much lower than wired
solutions. Still, many low-criticality use cases for wireless communication and
wireless solutions exist that implement one or many of the message types outlined
above. In addition, some wireless protocols schedule communication time and frequency, for example, IEEE 802.15.4 *time-slotted channel hopping*.

7.2.5   Design Pitfalls

7.2.5.1   Misconception of Subsystem Scope

Subsystem Scope   The real-time network is only a subsystem of an overall real-
time system. It is, thus, crucial to understand its *subsystem scope*. A real-time network technology must implement the previously discussed requirements but may
provide services that go well beyond. Examples of higher-level network services are
group membership or file-transfer protocols. The *subsystem scope* may include such
services. On the other hand, the real-time system will implement mechanisms
clearly outside the *subsystem scope*. For example, monitoring whether a message
reception has a physical effect on the controlled object is out of the scope of a communication subsystem. The research problem of which functions to include in a
communication system and which not (also generally applicable to real-time networks) has been originally presented by Saltzer et al. [Sal84].

> **Example—Three Mile Island:** A misconception of a real-time network's subsystem scope
> can have serious consequences, as seen in the following quote [Sev81, p. 414] regarding the
> Three Mile Island nuclear reactor no. 2 accident on March 28, 1979: *Perhaps the single
> most important and damaging failure in the relatively long chain of failures during this
> accident was that of the Pressure Operated Relief Valve (PORV) on the pressurizer. The
> PORV did not close; yet its monitoring light was signaling green (meaning closed).* The
> fundamental design principle to *never trust an actuator* was violated in this system. The
> designers assumed that the acknowledged arrival of a control output signal that commanded
> the valve to close implied that the valve was closed. Since there was an electromechanical
> fault in the valve, this implication was not true. A proper system design that used an inde-
> pendent subsystem to mechanically sense the closed position of the valve would have
> avoided this catastrophic false information.

7.2.5.2   Incompatible Network Services

The design of individual network protocols to achieve the desired *emergent behavior* under protocol composition is generally nontrivial and error-prone.

Example—Redundancy Management and End-to-end Protocols: IEEE 802.1CB (part of the IEEE 802.1 TSN set of protocols) and ARINC 664-p7 define methods for a receiver to eliminate redundant copies of a message sent over disjoint paths in the real-time network. The intent of these protocols is to only forward one of the copies for further processing to off-load the receiving processor. End-to-end protocols such as an application-level CRC are intended to protect against failures in the network—a faulty Ethernet switch can modify a message's content and generate a correct Ethernet-level CRC (switches implement this functionality). However, it may only generate the application-level CRC with a much lower, negligible probability. Both IEEE 802.1CB and ARINC 664-p7 redundancy management are incompatible with end-to-end protocols. A faulty network switch may corrupt a message in the application-level CRC, resulting in it still appearing correct to the redundancy management (as the switch knows how to generate a correct Ethernet CRC) and selected by redundancy management as the one copy to be forwarded. The application will detect the message corruption and not receive another (correct) copy because redundancy management will eliminate them.

7.2.5.3   Design for Average Case Instead of Worst Case

Networks designed for the average case instead of the worst case will fail to deliver the timeliness requirements of a real-time network. An example effect of an average-case design is *thrashing*.

Thrashing   In an ideal system, growing demand for a shared resource is served as long as there are free resource capacities. In reality, an arbitration scheme that manages the resource to demand allocation can cause an *arbitration overhead* preventing the resource from being fully utilized. Some arbitration schemes suffer from *thrashing*: their *arbitration overhead* grows rapidly when the demand exceeds a certain level, called the *thrashing point*, causing even lower resource utilization than at lower demands. For example, the *carrier-sense multiple access collision-detection* protocol of Ethernet is susceptible to thrashing (see Sect. 7.3.2). Arbitration schemes with thrashing can be acceptable for noncritical systems if the average use case is below the *thrashing point*. However, hard real-time systems must be free of *thrashing* effects.

7.3   Event-Triggered Communication

The different message types (event-triggered, rate-constrained, and time-triggered) differ in which a priori system information is available (and how the real-time network uses it). Event-triggered messages may not have a priori system information at all. This is attractive for non-real-time systems as it minimizes the configuration burden and simplifies *plug-and-play*. However, such a priori system information is required to calculate bounds on the worst-case message-transport latencies in a real-time system. Thus, if a real-time system implements a shared network based on event-triggered messages, the analysis of their worst-case message-transport latencies *cannot be addressed as an isolated problem.* It is always integrated with a holistic real-time system analysis that couples the analysis of distributed real-time task executions with their communication. *It is **impossible** to provide temporal guarantees for event-triggered messages independently from the complete real-time system analysis.*

An event-triggered message transmission starts with a sender initiating the generation of an event-triggered message upon the occurrence of a *node-internal* or *node-external* event by providing the message's data to the sender's communication subsystem. If the network is a shared resource, it may already be in use by another message transmission. Thus, the sender's communication subsystem temporarily stores messages until it is this sender's turn to use the network. Typically, this storage is organized as one or many FIFO (first-in, first-out) message queues with priorities. Messages from higher-priority message queues are selected before messages from lower-priority queues.

In bus-based networks, the media-access protocol of the network determines the points in time when messages are selected for transmission from the multiple senders' communication subsystems. Once the network selects a message from a sender, it directly delivers it to the receiving nodes' communication subsystems.

In switch-based networks, a sender's communication subsystem may immediately transmit the event-triggered message to its directly connected switch unless the switch exerts *backpressure flow control.* Then, the switch will forward the message to further switches and the final receiving nodes' communication subsystems.

7.3.1   CAN

The CAN (control area network) protocol developed by Bosch [CAN90] is a bus-based CSMA/CA (carrier-sense multiple access/collision avoidance) protocol that exercises *backpressure flow control* on the sender. For this purpose, a CAN message contains an *arbitration field* of either 12 or 32 bits for *standard* or *extended* messages (messages are called *frames* in CAN). This arbitration field also serves as message identifier and message priority, where 0s indicate high priority and 1s low priority. The actual data is transported in the message's *data field*, which also contains a *CRC* field and an *end-of-frame* field.

In the *bus activity* phase, a bit is transmitted by setting the CAN bus in either the *dominant* state, if the bit is of value 0, or the *recessive* state if the bit is of value 1. It will frequently be the case that multiple senders will have messages ready in their communication subsystems when the bus becomes idle. Thus, CAN uses a *bit-wise arbitration* protocol to identify the sender of the next message: each candidate sender will put its first bit of the arbitration field on the bus and observe the bus state. There are three possible outcomes: (i) the candidate sender puts a 0 (dominant) and observes a 0, (ii) the candidate sender puts a 1 (recessive) and observes a

1, or (iii) the candidate sender puts a 1 (recessive) but observes a 0 (dominant). In cases (i) and (ii), the candidate sender is either the sole sender or one of many senders with an equal bit in this (first) position and will continue the bit arbitration process with the next bit. However, case (iii) means for the candidate sender that there must be another candidate sender present in this arbitration phase with a higher-priority bit in this position. In this case (iii), the candidate sender will back off and attempt to transmit its message in the subsequent arbitration phase. All other senders will continue this arbitration protocol with the next bit in the arbitration field. The unique message identifiers ensure that the arbitration protocol determines a single sender.

The bit-wise arbitration requires the length of a bit (called a *bit cell*) on the bus to be at least as long as its propagation delay. Only then all candidate senders can reliably observe case (iii).

CAN Flexible Data rate (FD)   Limiting the length of a bit cell to the propagation delay also limits the achievable bandwidth of the bus. CAN FD (Flexible Data rate) addresses this limitation by using different data rates. CAN uses the normal data rate for the bit-wise arbitration and a higher data rate for the transmission of the data in the message's data field.

CAN deployment in a real-time system requires the system integrator to specify a unique priority for each CAN message. Also, the analysis of worst-case message-transport latencies of CAN messages requires the system integrator to analyze the behavior of the tasks in the real-time system. If the analyzed latencies are too long, the message priorities may be reconfigured, causing a highly complex iterative design loop of analysis and reconfiguration. In a CAN system, a faulty node can inhibit the transmission of messages among healthy nodes.

7.3.2   Ethernet

Ethernet (IEEE 802.3) defines a *carrier-sense multiple access collision-detection* protocol (CSMA/CD) for networks that realize a physical bus. A node senses the bus and may start to transmit its message when it perceives the bus as idle. However, multiple nodes may perceive the bus as being idle at about the same time (bounded by the propagation delay of the bus and node-internal delays). Thus, it will frequently happen that multiple senders transmit their message to the bus concurrently, resulting in the physical overlay of their signals. The nodes will detect this *collision* on the bus and send a *jamming signal* to ensure that all nodes consistently classify the ongoing transmission as a collision. When the senders perceive this jamming signal, they will preempt their message transmission and back off for a random timeout. When this timeout expires and the bus is idle, a sender will retry its message transmission.

Switched Ethernet made this collision-detection protocol obsolete. As nodes directly connect to switches, collisions on the physical layer are excluded by design. However, *logical collisions* may still occur when multiple messages compete for transmission on a receiver's direct link or a direct link that connects two switches. Switches resolve logical collisions by message buffering.

While, in principle, a holistic analysis of the total real-time system could also derive bounds for the worst-case message-transport latencies for *basic* Ethernet networks, it is common to equip basic Ethernet with additional functionality as enabler for real-time use. Indeed, tens of real-time Ethernet variants are used for real-time systems. We will continue the discussion of some variants in the following sections.

7.4 Rate-Constrained Communication

We can calculate bounds for worst-case message-transport latencies in a real-time network when each node limits the number and size of these messages it sends per given time interval and makes this information a priori available for analysis. We call these messages *rate-constrained messages*. Still, these messages may collide, but we can adequately dimension the network's resources to resolve these collisions. These resources will be the shared message buffers in switched networks into which the messages are enqueued as a temporary buffer. In a physical bus, these resources are individual resources in the nodes' communication subsystems.

> **Example:** In a 100 Mbit/s Ethernet network of three nodes connected via a single store-and-forward switch, where two nodes send a 10,000-bit message each to the third node, the queuing delay for each message is at maximum the transmission time of the respective other message, which is 100 μs. If there are 100 sending nodes instead of 2, the queuing delay increases to 10 milliseconds.

Network Calculus Analyzing worst-case queuing delays for rate-constrained messages in practical real-time systems can become quite complex. Therefore, different mathematical analysis methods have been developed, the most prominent one being *network calculus* [Cru91a-a, Cru91b-b]. Rate-constrained message transmissions are formalized as *arrival curves* and *service curves* from which network calculus derives bounds for worst-case message queue lengths as well as worst-case end-to-end message transmission latencies. Network calculus returns provable latency bounds, but these bounds can become quite pessimistic. Thus, improvements to network calculus and new techniques such as *trajectory approach* [Bau09] have been developed which return results closer to the actual worst-case latencies. However, in safety-critical real-time systems, the pessimism of a mathematical framework can be acceptable when its certification cost is lower.

7.4.1 Avionics Full-Duplex Switched Ethernet (AFDX): ARINC 664-p7

When the avionics industry developed the *Integrated Modular Avionics* (IMA) para-digm, according to which compute nodes integrate multiple functions each, it also created a demand for more powerful real-time networks to interconnect these nodes. The *Avionics Full-Duplex Switched Ethernet* (AFDX) technology has been devel-oped for this purpose. AFDX is a 100 Mbit/s real-time Ethernet variant standardized as ARINC 664 part 7.

Virtual Links AFDX messages are standard Ethernet multicast messages. Each AFDX message has a unique sender and one or many receivers. An AFDX network is a switch-based network with statically configured communication paths for all AFDX messages. AFDX calls the sum of the communication paths for an AFDX message from its sender to all its receivers a *virtual link*. Indeed, each AFDX mes-sage is identified by its *virtual link identifier* (VL ID), which is encoded in the last 16 bits of its *destination address field*. An AFDX switch, upon reception of an AFDX message, can determine its ports to forward said message simply by a table lookup.

Bandwidth Allocation Gap (BAG) AFDX messages are rate-constrained messages where the rate is determined by 1/BAG. BAG is a configurable parameter from 1 ms to 128 ms (in steps of power of 2).

Traffic Shaping and Traffic Policing The sending node's communication subsys-tem *shapes* the outing traffic to establish a minimum time between transmitting two messages on the same *virtual link*. AFDX switches monitor these times and discard messages if they do not meet their inter-arrival time (*traffic policing*). Traffic shap-ing and traffic policing ensure that nodes adhere to their a priori message rates and sizes, and the aforementioned mathematical analysis frameworks return sound results.

Ordered Delivery AFDX messages carry a *sequence number*, which receiving nodes use to identify and drop messages received out of order (called *integrity checking*).

Redundancy Management AFDX can be configured to transport messages on two redundant networks. A receiving node's communication subsystem implements a redundancy management function that eliminates the redundant copy of the same message. However, as discussed in a previous example, this redundancy manage-ment is not composable with application-level end-to-end checks. Therefore, safety-critical real-time systems may choose to disable redundancy management and forward redundant copies to the application.

7.4.2 Audio/Video Bridging: IEEE 802.1 AVB

Many real-time extensions to Ethernet exist, but AVB has been the first one standardized within the IEEE Standardization Association (IEEE SA), followed by time-sensitive networking (see Sec. 7.5.3). It targets distributed audio and video applications, including professional studio equipment, event halls, and in-car infotainment networks. AVB can be deployed in switched Ethernet networks with 100 Mbit/s or higher bandwidth.

AVB Streams AVB messages are VLAN-tagged Ethernet multicast frames where the sending node of a message is called the *talker* and the receiving nodes are called the *listeners*. The related messages a talker sends to the listeners are called a *stream*, and a node may be talker of multiple streams. AVB establishes the communication between talkers and listeners at runtime using the *stream reservation protocol* (SRP). AVB streams are classified into *SR Class A* and *SR Class B* streams by the priority field in the VLAN tag of Ethernet frames. AVB messages are rate-constrained messages where the rate is defined by the amount of data a talker may transmit per *Class Measurement Interval* (CMI). The CMI is 125 µs for SR Class A and 250 µs for SR Class B. The amount of data a talker may transmit per CMI is determined at runtime by SRP.

Forwarding and Queuing Enhancements AVB switches generally prioritize AVB messages over non-AVB (SR Class A is prioritized over SR Class B). However, a traffic shaping mechanism called *credit-based shaping* at the egress of AVB switches occasionally pauses AVB message transport and lets the switch serve non-AVB messages.

Clock Synchronization IEEE 802.1AS Although not used for message communication, AVB also standardizes a clock synchronization protocol in IEEE 802.1AS, which resembles IEEE 1588 in an IEEE 802.1 conform manner.

Message-Transport Latency IEEE 802.1BA defines AVB profiles and gives a latency target of 2 ms for SR Class A traffic (50ms for SR Class B) in a 100 Mbit/s Ethernet network with a maximum of seven hops in between the talker and the listeners. While IEEE 802.1BA also gives example calculations that meet the target latencies, a more formal mathematical treatment is necessary. Such a treatment, based on *network calculus*, is provided in [Rui14].

7.5 Time-Triggered Communication

Time-triggered messages are *periodic messages* that maximize the use of a priori information about the real-time network to achieve optimal temporal predictability of message transports. In its *strong form*, all relevant communication actions of time-triggered messages happen at pre-planned instants in a synchronized global time. These instants are the transmission instant at the sender's communication subsystem, the reception instant at a receiver's communication subsystem, and all message forwarding instants (in case of a switched network). The collective configuration of these pre-planned instants is called the *communication schedule*, and nodes (and switches) will periodically cycle through this schedule in a coordinated manner based on the global time. The communication schedule specifies for each time-triggered message a *period* and a *phase* within this period in a way that excludes collisions of time-triggered message transmissions on shared resources per design.

Real-time networks can implement *strong* time-triggered messages, but they may choose a *weak form* because of design tradeoffs of *action delay* and *jitter* toward network utilization. In this weak form, the message-specific actions for time-triggered messages are also pre-planned. However, their effect (transmission, reception, forwarding) may happen only within a known interval after the pre-planned instant and not necessarily at the instant itself. Furthermore, the communication schedule for weak time-triggered messages may tolerate collisions of messages up to a defined limit.

> **Example:** A sender's communication subsystem may have a configured message transmission instant for a time-triggered message. However, the action at this instant may cause the time-triggered message to be placed into a transmit queue at the sender's communication subsystem instead of an immediate transmission. The effect of this time-triggered message being actually transmitted happens only after all preceding messages in the transmit queue are processed. Therefore, the number and size of such preceding messages must be bounded and known a priori.

Communication Schedule Planning Time-triggered messages give a system designer *full control* over the message-transport times, but this full control comes at the cost of generating the communication schedules. Schedule planning is a search problem that may become quite resource-intensive, but candidate solutions can easily be checked for correctness. Formally, in terms of computational complexity, this means that the schedule planning problem is in the *NP* complexity class. Indeed, schedule planning for time-triggered networks has been expressed as SMT (satisfiability modulo theories) problem in [Ste10]: the phases of all time-triggered messages are treated as integer variables, and different types of linear inequalities (called *constraints*) relate these integer variables to each other.

> **Example:** A basic type of constraint is the *collision-free* constraint which expresses that the phases of any two time-triggered messages must be selected such that their transmissions on the same communication link will not coincide. In the formalization, communication bandwidth is split up into repeating cycles of sequential communication slots of equal size. Messages are assigned to unique slots and, thus, a message's phase is also given by a slot number. The *collision-free* constraint, therefore, simply requires that the slot numbers of different messages on the same link must be different.

Schedule planning is realized by a search for concrete evaluations of all the integer variables (i.e., assign message phases) such that all the constraints are satisfied. As this search is complex, quite sophisticated tools have been developed or adopted, like SMT solvers. Complex tooling may be seen as an issue for safety-critical systems. However, since it is easy to analyze whether a given candidate solution is correct, we can take advantage of simple *schedule verification tools* to analyze the output of the complex planning tool: given a candidate solution (from the complex planning tool), check whether it satisfies all the constraints.

Schedule Planning and Optimization   Schedule planning is a highly active research direction (particularly stimulated by the standardization of time-triggered messages in IEEE 802.1 TSN). Various research works closely relate schedule planning to optimization. On the one hand, optimization tools like ILP (integer linear programming) are used for schedule planning. On the other hand, the schedule planning problem itself is extended to an optimization problem, e.g., optimization for resource usage or optimization in the context of integration of time-triggered messages with rate-constrained messages.

7.5.1   TTP

The time-triggered protocol (TTP) is a bus-based protocol that integrates *strong* time-triggered messages, temporal error detection, a fault-tolerant clock synchronization service, and a membership service in a *single protocol* with small protocol overhead [Kop93]. The system integrator must set up the parameters for the transmission slots of the nodes a priori. Event-triggered communication can be implemented by an overlay protocol on top of TTP.

Fault-tolerant clock synchronization is achieved by taking the difference between the *measured* and *specified* arrival time of every message as a measure for the difference of the clocks of the sender and receiver and by applying the *fault-tolerant average algorithm* (see Sect. 3.4.3) on these differences to calculate a correction factor for each local clock.

The *membership service of TTP* informs all connected nodes about the *health state* of every cluster node and of the *violation of the fault hypothesis* (should it occur) such that a never-give-up (NGU) strategy can be activated quickly. The membership is encoded in a *membership vector* that contains as many bits as there are nodes in a cluster. A specified bit position of the membership vector is assigned to each node. When this bit is set to *TRUE*, a node is operating; if this bit is set to *FALSE*, this node is not operating correctly. The membership instant of a node is the periodic send instant of a message by this node. The *state* of a TTP controller (*C-state*) consists of the current time and the node membership vector. To enforce agreement on the C-state of all nodes of an ensemble, TTP calculates the CRC at the sender over the message contents concatenated with the C-state of the sender. The CRC at the receiver is calculated from the received message contents concatenated with the C-state of the receiver. If the result of the CRC check at the receiver is negative, then either the message was corrupted during transmission or there is a disagreement between the C-states of the sender and receiver. In both cases, the message is discarded, and the receiving node assumes that the sender has been faulty. We call this assumption the *self-confidence principle*. The self-confidence principle ensures that a single faulty node cannot kill a correct node in a system that is assumed to contain at most one faulty node.

If, in the above scenario, the sender has been correct—all other working nodes have received the message correctly—then the receiving node must have been faulty. An algorithm must tolerate that a faulty node makes further faulty decisions. It will send a message with a wrong membership vector and will be taken out of the membership by all other working nodes. If the receiving node had been right, a correct decision would have been taken, and the sender of the original message would have been taken out of the membership. TTP operates on two physical channels and has an independent bus guardian at every node that protects the bus from babbling idiots—even a faulty node can send a message during its assigned time slot only. It fails silently outside its time slot.

Considering the services TTP provides, it is a very data-efficient protocol well suited for applications that require a frequent update of short real-time data elements. Examples of such applications are industrial controls or the control of robot movements.

TTP has been formally certified for use in airborne systems [Rus02]. It is deployed in the A380 and the Boeing 787 aircraft and other aerospace and industrial control applications. TTP is standardized as SAE AS6003.

7.5.2 TTEthernet

TTEthernet implements time-triggered, rate-constrained, and event-triggered messages (called *traffic classes*). It has been designed as a cross-industry real-time network for safety-critical and mixed-criticality systems. Most use cases of TTEthernet can be found in the space and aerospace markets.

TTEthernet Traffic Classes TTEthernet typically implements 100 Mbit/s and 1 Gbit/s switched Ethernet. It uses standard Ethernet frames and the AFDX *virtual link* address scheme for time-triggered and rate-constrained messages (i.e., these message types are identified by pre-configured multicast addresses). Indeed, rate-constrained messages are backward compatible to AFDX, while event-triggered messages are *basic* Ethernet messages (i.e., their transmission may fail, for example, due to message queue overflows). A TTEthernet communication schedule plans all message-transport instants (transmission, forwarding, reception). It supports both *strong* and *weak* time-triggered messages via configuration. The *weak* form follows the previously presented example where the sender can be the sending node or a forwarding switch, and TTEthernet ensures that there may be only one preceding message in the queue. TTEthernet achieves the *strong* form by ensuring that the transmit queue is empty at the planned message transmit instant. Thus, although rate-constrained or event-triggered messages may be ready to transmit, TTEthernet *timely blocks* their transmission by not adding them to the transmit queue in favor of timely transmission of the planned time-triggered message. However, if this time-triggered message is not ready for transmission in time, TTEthernet reallocates its bandwidth to other message types. A time-triggered message may not be ready because of a decision in the sending node or because of message loss/corruption.

Fault Tolerance   TTEthernet supports network replication where each node can attach to up to three isolated switched networks. The TTEthernet switches implement *traffic policing* for time-triggered and rate-constrained messages. Event-triggered messages have the lowest priority and are dropped if necessary. As TTEthernet switches realize different message buffers for the different traffic types, memory violations of lower-priority messages on higher-priority messages are excluded by design. Furthermore, TTEthernet node and switch designs can follow a *self-checking pair* paradigm, which allows arguing their failure mode to be benign and close to a *fail-silent* failure mode.

Fault-Tolerant Clock Synchronization   TTEthernet implements a distributed fault-tolerant clock synchronization algorithm to establish a global time. It distinguishes *Synchronization Masters* (SMs), *Compression Masters* (CMs), and *Synchronization Clients* (SCs). Typically, SMs are realized by nodes and CMs are realized by switches, and at least one switch per network will be configured as CM. Every node and switch that is not an SM nor a CM operates as SC and only passively synchronizes to the fault-tolerant time base. Clock synchronization happens in two steps. In the first one, the SMs provide clock synchronization messages, called *Integration Frames* (IN), to the CMs, which use the IN timing as an input to a fault-tolerant averaging function. In the second step, the CMs provide a *compressed Integration Frame* (cIN) back to the SMs (and the SCs) at a point in time determined from the output of the fault-tolerant averaging function in the first step. The SMs and SCs calculate an average from the received cIN frames and use this output to correct their local clocks. This algorithm has been formally verified and tolerates single failures and coinciding SM and CM failures [Ste11a]. TTEthernet also defines protocols for initial synchronization of the distributed system and clique detection and restart procedures.

TTEthernet has been standardized as SAE AS6802 and is embedded in the ECSS-E-ST-50-16C standard. It has been selected as the communication backbone system of the NASA Orion Program [Bag10] and for several elements of NASA's Space Gateway. TTEthernet has also been selected for Ariane 6 [Cla14].

7.5.3   Time-Sensitive Networking: IEEE 802.1 TSN

IEEE 802.1 Time-Sensitive Networking (TSN) continues the IEEE AVB standardization activities. It is the first time that time-triggered messages are introduced into the IEEE 802.1 set of standards. The main target areas for TSN are industrial automation and automotive in-vehicle networking, but use cases go well beyond, for example, into the aerospace market or telecom. TSN, like AVB, is a set of standard amendments (primary amendments to IEEE 802.1Q—indicated by lowercase letters like IEEE 802.1Qbv) and some stand-alone standards (indicated by uppercase letters like IEEE 802.1CB). The standardization activities can be classified into (i) low-latency message transmission; (ii) enhancements of the IEEE 802.1AS clock synchronization protocol; (iii) improvements toward message transmission reliability, e.g., through message replication; and (iv) network configuration and resource management. A detailed discussion of the TSN protocols and use cases is given in [LoB19]. This section focuses on two low-latency aspects: time-triggered messages and message preemption. We will also limit the discussion to TSN in switched Ethernet real-time networks with VLAN-tagged Ethernet frames (the typical use case).

In IEEE 802.1 switches (the standard uses the term *bridges*), we can distinguish three main phases of operation: (i) ingress operation, (ii) message switching, and (iii) egress operation. A switch handles the frame reception at the ingress port, which may also execute *traffic policing*. Message switching determines the target egress ports and places the message in egress queues. Ethernet switches may have up to eight message queues per egress port, and the message switching function selects the queue based on the VLAN tag's 3-bit *priority code point* field (IEEE 802.1CB also specifies the use of other frame fields for queue selection). The *transmission selection* function at the egress port selects from the (up to eight) message queues the next message for transmission. The *transmission selection* may follow different policies, the simplest one being a *strict priority* policy, in which messages from higher-priority queues are always selected prior to messages from lower-priority queues.

IEEE 802.1Qbv—Scheduled Traffic Traffic shaping to achieve time-triggered message transport is also part of the egress operation. For this, IEEE 802.1Qbv defines a *gate* for each egress queue. The gates can be either *open* or *closed*, but *transmission selection* may only select messages from queues with their gate in the *open* state. TSN uses a *gate control list* (GCL) as a communication schedule that defines instants in a global time base when the state of gates shall change.

> **Example:** Given a switch with two gated queues (Q1 and Q2) at an egress port, we can achieve a time-triggered message transport from Q1 by configuring the GCL to set the gates of both Q1 and Q2 in the closed state at an instant t0 and to set the gate of Q1 into the open state at (t0 + max(message length)). This ensures that the communication link is idle at the Q1 gate open instant and the *transmission selection* function selects a message from Q1 for transmission.

The queue-gating mechanism allows realizing different types of effects through the appropriate configuration of the GCL in the switches and the message transmission instants in the sending nodes. For example, Craciunas et al. [Cra16] achieve *strong* time-triggered message transport by appropriately configuring the messages' phases in the sending nodes such that only one message will be in a respective switch queue at a time. The gate open event for this queue is, thus, also uniquely associated with this particular message. On the other hand, sending nodes can be configured such that their messages do queue up in the switch, for example, to ease schedule planning. Then setting the queue in the open state achieves a *weak form* of time-triggered messages. Sending nodes may even be unsynchronized, and the queue-gating mechanism converts these unsynchronized messages (e.g., rate-constrained messages) into *weak* time-triggered messages.

IEEE 802.1Qbu—Frame Preemption Many real-time networks are *converged networks* that transport real-time and non-real-time messages, which may become quite long.

> **Example:** Some implementations of Ethernet allow *jumbo* frames that are up to 9,000 bytes long. The message-transport time of such a jumbo frame on *one* communication link in 100 Mbit/s is 720 μs (72 μs in 1 Gbit/s).

Methods to limit the delay of real-time messages are necessary, and *frame preemption* is such a method. Frame preemption distinguishes *preemptable* from *express* frames by allocating the frames to different egress queues. *Express* frames preempt the transmissions of *preemptable* frames. The preemption mechanism ensures that the preempted frame is a well-formed Ethernet frame of at least 64 bytes with a CRC (called *mCRC*) that represents all the message fragments of the preempted frame transmitted so far (i.e., a receiver continues the calculation of the CRC with each fragment of a frame it receives). After preemption, the message transmission of the preempted frame is continued (unless there are more express frames ready), and the receiving node reassembles the original full frame. Frame preemption can also be scheduled with respect to the global time by configuration in the GCL.

TSN Profiles TSN specifies quite a high number of protocols, mechanisms, and features. Thus, different industries are in the process of defining specific TSN profiles. For example, the automotive industry standardizes a *TSN Profile for Automotive In-Vehicle Ethernet Communications* in P802.1DG and an aerospace profile as *TSN for Aerospace Onboard Ethernet Communications* in P802.1DP.

Points to Remember

- A network is a real-time network, *if and only if a bound for the worst-case message-transport latency for time-critical messages can be determined by analysis, and this bound holds during operation with a sufficiently high probability.*

- In hard real-time systems, the required message-transport latency of time-critical messages is challenging, often in the order of a few milliseconds or below. At the same time, the probability that the calculated bounds on the worst-case message-transport latencies hold must be very high. For example, a common reliability requirement for ultrahigh dependable systems (e.g., airplanes) requires a failure rate of 10^{-9} failures/hour.
- Commonly used IT networks are not hard real-time networks (see Sect. 1.5.1). They do not meet the combined requirements of low message-transport latency and high probability for bounds on worst-case message latency.
- Typical requirements of real-time networks are timeliness, dependability and security, flexibility, and SWaP-C (*size, weight, power, and cost*).
- The required bandwidth of a real-time network can vary highly, e.g., from a few bit/sec for a room-temperature sensor at 10 Hz to multiple Gbit/sec for an ultrahigh-definition camera for self-driving cars.
- In a *simple model*, the real-time network is a set of resources, and only one sender can use a resource at a time. Nodes use the resources by exchanging messages.
- In a real-time system, the following message types are distinguished: event-triggered messages, rate-constrained messages, and time-triggered messages.
- The different message types (event-triggered, rate-constrained, and time-triggered) differ in which *a priori* system information is available (and how the real-time network uses it).
- Event-triggered messages may not have *a priori* system information at all.
- *It is **impossible** to provide temporal guarantees for event-triggered messages independently from the complete real-time system analysis.*
- CAN and basic Ethernet are examples of event-triggered real-time networks.
- Rate-constrained messages are limited in number and size, and, thus a bound for their worst-case latency can be calculated.
- Network calculus is the most important tool to calculate latency bounds for rate-constrained networks.
- Avionics Full-Duplex Switched Ethernet (AFDX) and Audio/Video Bridging are examples of rate-constrained networks.
- Time-triggered messages are *periodic messages* that maximize the use of *a priori* information about the real-time network to achieve optimal temporal predictability of message transports.
- Time-triggered messages are either strong or weak.
- For strong time-triggered messages, the visible communication effects follow the communication plan closely.
- Weak time-triggered messages allow some bounded latency between the visible effect and the plan and may account for such uncertainty in the plan as well.
- TTP, TTEthernet, and TSN are examples of time-triggered networks.

Bibliographic Notes

The requirements for distributed safety-critical real-time systems onboard vehicles are analyzed in the SAE report J20056/1 "Class C Application Requirements" [SAE95]. An interesting report about a *Comparison of Bus Architectures for Safety Critical Embedded Systems* has been published by NASA [Rus03]. A rationale for the design of time-triggered Ethernet is published in [Kop08].

Review Questions and Problems

7.1 Compare the requirements of *real-time* communication systems with those of *non-real-time* communication systems. What are the most significant differences?

7.2 Why are application-specific end-to-end protocols needed at the interface between the computer system and the controlled object?

7.3 Describe the different flow-control strategies. Which subsystem controls the speed of communication if an *explicit* flow-control schema is deployed?

7.4 Compare the efficiency of event-triggered and time-triggered communication protocols at low load and at peak load.

7.5 What mechanisms can lead to thrashing? How should a system react if thrashing is observed?

7.6 Given a bandwidth of *500* Mbits/sec, a channel length of *100* m, and a message length of *80* bits, what is the limit of the protocol efficiency that can be achieved at the media-access level of a bus system?

7.7 Give an example of incompatible network services!

7.8 What is the Ethernet PAUSE command?

7.9 How does the CAN protocol determine the next sender?

7.10 Can rate-constrained communication protocols guarantee a worst-case communication latency? Discuss!

7.11 What are the differences between TTP, TTEthernet, and IEEE TSN 802.1Qbv time-triggered messages? Give examples!

Chapter 8
Power and Energy Awareness

Overview

The increasing growth of energy-aware and power-aware computing is driven by the following concerns:

- The widespread use of mobile battery-powered devices, where the available *time-for- use* depends on the power consumption of the device.
- The power dissipation within a large system-on-chip that leads to high internal temperatures and hot spots that have a negative impact on the chip's reliability, possibly physically destroying the chip.
- The high cost of the energy for the operation and cooling of large data centers, and finally.
- The general concern about the carbon emissions of the ICT industry, which is of about the same magnitude as the carbon emissions of the air transport industry.

In the past, the number of instructions executed by a computer system in a unit of time (e.g., *MIPS* or *FLOPS*) was the important indicator for measuring the performance of a system. In the future, the number of instructions executed per unit of energy (e.g., a Joule) will be of equal importance.

It is the objective of this chapter to establish a framework for developing an understanding for energy-efficient embedded computing. In the first section, we introduce some basic concepts from physics and a simple model to estimate the energy consumption of different computing tasks. This gives the reader an indication of where energy is dissipated and what are the mechanisms of energy dissipation. Since energy consumption depends very much on the considered technology, we assume a hypothetical 100 nm CMOS VLSI technology as the reference for the estimation. The next section focuses on hardware techniques for energy saving, followed by a discussion about the impact of system architecture decisions on the energy consumption. Software techniques that help save energy are treated in the following section.

8.1   Power and Energy

Computers require electric energy for the processing and transmission of data. According to a report from the International Energy Agency [IEA21]: "Global data center electricity use in 2020 was 200–250 TWh, or around 1% of global final electricity demand. This excludes energy used for cryptocurrency mining, which was ~100 TWh in 2020. Globally, data transmission networks consumed 260–340 TWh in 2020, or 1.1–1.4% of global electricity use." If we add to this number the energy required for the operation of the end-user equipment, we arrive at a worldwide electric energy requirement of about 800–1000 TWh for the entire ICT sector in the year 2020. This is about 4% of the electricity production of about 25,000 TWh worldwide. The climate impact of the ICT industry [Fri20] is about of the same magnitude as the climate impact of the airline travel industry [Ric20].

8.1.1   Basic Concepts

The concept of energy, initially introduced in the field of mechanics, refers to a scalar quantity that describes the *capability of work* that can be performed by a *system*. The *intensity of the work*, which is the energy used up per unit of time, is called *power*. Energy is thus the integral of power over time. There exist different forms of energy such as *mechanical energy (potential energy* and *kinetic energy)* and *electrical energy* on the one hand and *thermal energy* and *chemical energy* on the other hand.

The *first law of thermodynamics*, relating to the *conservation of energy*, states that in a closed system, the total sum of all forms of energy is constant. The transformation of one form of energy to another form of energy is governed by the *second law of thermodynamics* that states that thermal energy can only be partially transformed to other forms of energy.

It is *impossible* to convert thermal energy to electrical energy with an efficiency of 100%, while the converse, the transformation of electrical energy to thermal energy can be performed with 100% efficiency. Furthermore, heat-pumps upgrade low thermal energy from the environment to a higher level of temperature for use in heating. As a rule of thumb, the conversion efficiency of the chemical energy stored in gasoline to mechanical energy by an automotive engine is between 20% and 40%. However, the efficiency of the conversion of electrical energy into mechanical energy, e.g., by an electric motor, or the conversion of mechanical energy to electrical energy by a generator is better than 95%.

There exist many different—and confusing—units to measure the amount of energy. The unit of energy in the *SI* system is the *Joule (J)* which is defined as the amount of work done by moving a mass of one *kg* for a distance of one *meter* against a force of one *newton*. One J corresponds also to the power of one *Watt* lasting for one second (*Ws*). A widely used energy measure is the *kWh* which is

1000 × 3600 Ws or 3.6 10^6 *J* (since an hour has 3600 s). A *MWh* is 1000 kWh, a *GWh* is 1000 MWh, and a *TWh* is 1000 GWh. Thermal energy, i.e., heat, is often measured in *calories or BTUs* (*British Thermal Units*). One *kilo-calorie* (kcal) is defined as the heat needed to increase the temperature of one kg of water at room temperature by one degree *Celsius*. One kcal is about 1.16 Wh, while one *BTU* corresponds to about 0.29 Wh. In this chapter, we use the *Watt* as a *unit of power* and the *Watt.hour* as a unit of energy if the amount of energy is large (>1 Wh) and the *Joule* (or *Watt.second*) if the amount of energy is small (<1 J).

> **Example**: A liter of gasoline contains chemical energy in the amount of about 9.6 kWh. An automotive engine converts about one fourth of this chemical energy to mechanical energy, i.e., one liter of gasoline provides about *2.4 kWh* of mechanical energy. The rest of the chemical energy is converted to heat. If a car requires 18 kWh of mechanical energy for 100 km, this amounts to about 7.5 liters of gasoline.

The *potential (mechanical) energy E_p* required to lift an object on earth by *h* meters is given by

$$E_p = m \times g \times h$$

where *m* is the mass of the object expressed in *kg*, *g* is the gravitational constant (9.8 m sec^{-2} on earth), and *h* is the difference in altitude, expressed in meters (*m*).

> **Example**: If a car with a weight of 1800 kg is driven from the ground to a mountain that is 1000 m above ground, the potential energy increase is E_p = *1800 × 9.8 × 1000 = 17,640,000 J* or 4.9 kWh. This is the mechanical energy provided by an automotive engine when it burns about 2 liters of gasoline.

The *kinetic (mechanical) energy E_k* required to accelerate an object of mass *m* from 0 to the speed *v* is given by

$$E_k = m \times v^2 / 2$$

> **Example**: If a car with a weight of 1800 kg is accelerated from 0 m/second to a speed of 30 m/second (which corresponds to 108 km/hour), the required energy is 810,000 J or 225 Wh, corresponding to about *0.1 l* of gasoline. This kinetic energy heats up the brakes when the car is stopped.

Batteries A battery is a storage device for electric energy. The voltage that exists at the terminals of a battery can drive electric current through a resistor. In the resistor, electric energy is converted to heat. If an *electric current I* flows through a wire with a *resistance R*, the voltage drop on the wire will be, according to Ohm's law

$$V = I \times R$$

and the dissipated electric power is given by

$$W = V \times I$$

Table 8.1 Energy content of batteries

Battery type	Voltage (V)	Energy density (J/g)	Mass (g)	Energy (Wh)
AA (disposable)	1.5	670	23	4.3
AAA (disposable)	1.5	587	11.5	1.9
Button cell CR2032 (disposable)	3	792	3	0.7
NiCd (rechargeable)	1.2	140		
Lead acid (rechargeable)	2.1	140		
Lithium ion (rechargeable)	3.6	500		
Ultra-capacitor		Up to 100		
(Gasoline)		44,000		

where W denotes the dissipated electric power (in *watt*), I the current (in *ampere*), and R the resistance (in *ohm*). If a constant voltage of V is applied to a system with a resistance of R over a time of t seconds, the energy that is dissipated equals.

$$E = t \times V^2/R \text{ or } E = t \times I^2 \times R$$

where E is the dissipated energy in *joule or Ws*, t denotes the time in *seconds*, and V, R, and I denote the voltage, the resistance, and the current, respectively.

Table 8.1 depicts the nominal energy content of different disposable and rechargeable batteries. The actual amount of energy that can be retrieved from a battery depends on the discharge pattern. If the requested battery power is highly irregular, the battery efficiency is reduced and the actual energy that can be drawn out of a battery can be less than half of the nominal energy [Mar99]. The discharge level of a rechargeable battery has an influence on the lifetime of the battery, i.e., the number of times the battery can be recharged before it breaks down.

The efficiency of a rechargeable battery, i.e., the relation *energy-input-for-charging* and *energy-output-for-use* of well-managed batteries, is between 75% and 90% under the assumption that during the *charging phase* and the *use phase*, the power that flows into and out of the battery is carefully controlled and matched to the electrochemical parameters of the battery. If an application has highly irregular power characteristics, then an intermediate energy storage device, such as an *ultra-capacitor*, must be provided to smoothen the power flow in and out of the battery. An ultra-capacitor is an electro-chemical capacitor that has a high energy density when compared to conventional capacitors. It can be used as an intermediate energy storage device to buffer large power-spikes over a short time.

8.1.2   Energy Estimation

The electric energy that is needed by a computing device to execute a program can be expressed as the sum of the following four terms:

$$E_{total} = E_{comp} + E_{mem} + E_{comm} + E_{IO}$$

where E_{total} is the total energy needed, E_{comp} denotes the energy needed to perform the computations in a CMOS VLSI circuit, E_{mem} denotes the energy consumed by the memory subsystem, E_{comm} denotes the energy required for the communication, and E_{IO} is the energy consumed by I/O devices, such as a screen. In the following, we investigate each of these terms in some detail by presenting simplified energy models. The numerical values of the parameters depend strongly on the technology that is used. We give some approximate value ranges for the parameters for a *hypothetical 100 nm CMOS* technology which is operated with a supply voltage of 1 V in order to enable the reader to develop an understanding for the relative importance of each one of the terms.

Computation Energy   The energy E_{comp} required for the computations consists of two parts: (i) the dynamic energy $E_{dynamic}$ that denotes the energy needed to perform the switching functions in a CMOS VLSI circuit and (ii) the static energy E_{static} that is dissipated due to a leakage current drawn from the power supply irrespective of any switching activity.

The dynamic energy consumption of a switching action of a transistor can be modeled by a first-order RC network as shown in Fig. 8.1. In such a network, consisting of a power supply with voltage V (measured in volt), a capacitor C (measured in farad), and a resistor R (measured in ohm), the voltage on the resistor R as a function of time t (measured in seconds) after the switching event is given by

$$V_{res}(t) = Ve^{-t/\tau}$$

where $\tau = RC$, which has the dimension of *second*, is called the *time-constant* of the RC network. This time-constant characterizes the speed of the switching operation.

The total energy required for one switching operation is given by the integral of the product *voltage $V(t)$* and *current $I(t)$* through the resistor over the time period from zero to infinity and amounts to

$$E = \frac{1}{2}CV^2$$

The current that flows into the capacitor builds up an electric charge in the capacitor and does not directly contribute to the energy dissipation. But indirectly, the current flow produces heat caused by the resistance of the wire.

Let us set up a hypothetical effective capacitance C_{eff} such that the term $C_{eff}\,V^2$ covers the energy needed to switch all output transitions of an average instruction of

Fig. 8.1 First-order RC network

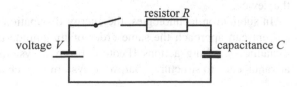

voltage V resistor R capacitance C

a processor and the energy that is dissipated by the brief current flow from the voltage source.

Then the dynamic energy needed for the execution of a single instruction can be expressed by the simple equation

$$E_{\text{dynamic}-\text{instruction}} = C_{\text{eff}}\, V^2$$

and the dynamic energy needed to run a program with N instructions is given by

$$E_{\text{dynamic}-\text{program}} = C_{\text{eff}}\, V^2 N$$

In the following, the term *IP-core* (intellectual property core) is introduced to refer to a well- specified functional unit of a system-on-chip (SoC). An IP-core can be a complete processor with local scratchpad memory or some other functional unit, such as an MPEG decoder. Let us now assume that an IP-core executes on the average one instruction per clock cycle and this IP-core is driven with a frequency of f clock cycles per second. Then, the dynamic power dissipation per second of the IP-core equals

$$P_{\text{dynamic}} = C_{\text{eff}}\, V^2 f$$

Example: Assume a program that causes the execution of 1000 million machine instructions per second is executed on an IP-core with a voltage of 1 V and where an average instruction is characterized by an effective capacitance C_{eff} of *1 nF*. The execution of one instruction requires an energy of *1 nJ* (*1 nano J*, i.e., $10^{-9}\ J$). The dynamic energy that is needed for the execution of this program is then *1 J*.

There are different mechanisms that contribute to a steady flow of current (the *leakage current*) between the terminals of a transistor, even if the transistor is not performing any switching action. According to [Ped06], the most important contribution to the leakage current in a technology which is below *100 nm* is the *subthreshold leakage current* I_{sub}, which increases with decreasing threshold voltage and with increasing temperature. Another important effect is the *tunnel current* flowing from the gate into the channel, although these are separated by the non-conducting gate oxide. A quantum mechanical effect causes electrons to *tunnel* through the very thin isolation, and this current grows exponentially as the oxide becomes thinner with smaller feature sizes. The exponential growth of the leakage current with rising temperature is of major concern in a submicron device with a low threshold voltage. This increase of the leakage current can lead to run-away thermal effects that, if not properly controlled, end with the thermal destruction of the device.

In submicron technologies, the energy dissipation caused by the static leakage current can approach the same order of magnitude as the energy needed for the dynamic switching actions [Ped06]. From a system perspective, it is therefore advantageous to structure a hardware system in such a way that subsystems can be switched off completely if their services are not needed during short intervals. This is the topic of power gating discussed in Sect. 8.3.3.

Communication The energy requirement for the transmission of a bit stream from a sender to a receiver is highly asymmetric. While the sender needs to generate signals that are strong enough to reach the intended receiver, the receiver must only sense the (weak) incoming signals. In addition to the energy required for the reception of a message, a receiver needs energy during the standby period of waiting for a message to arrive. If the *receive/wait ratio* is small, the standby energy consumption can be substantial.

Let us call the energy needed to transmit a single bit E_{tr}. This term is different for wired and wireless communication. For a wired interconnect, this term depends on wire length d, the effective capacitance of the interconnect per unit length $C_{effunit}$, and the square of voltage V:

$$E_{tr} = dC_{effunit}V^2$$

A typical value for the effective capacitance $C_{effunit}$ per unit length of an interconnect on a die is in the order of 1 pF/cm (*1 pF is 10^{-12} F*) [Smi97, p. 858].

For wireless communication, the transmit energy per bit can be approximated by

$$E_{tr} = d^2b$$

where d is the distance between the sender and the receiver in m and b is a constant. This square relationship between distance and energy is a gross approximation and must be adapted to the energy characteristics of the concrete antenna. In many portable devices, such as mobile phones, the sender adjusts the transmit energy dynamically in order to find the most appropriate energy level for the actual distance to the receiver. A typical value for the parameter b is given by [Hei00] as 100 pJ/(bit/m²). If we send a 32 byte message from a sender to a receiver which is *10 m* away, then we need a transmit energy in the order of about *10 nJ/bit* or about *2.5 µJ* per message.

The asymmetry of the energy requirements of a sender and a receiver has a decisive influence on the design of communication protocols between a base station that is connected to the electric utility and an energy-constrained mobile battery-powered device.

In a wireless sensor network, the low amount of energy required by the sensor can be harvested from the physical environment, e.g., from light, sound, or the incoming electric energy of the sender [Sah20].

Memory In memory-intensive applications, the energy that is required to access the memory subsystems can be larger than the energy that is needed for the computations. The energy that is consumed by a memory subsystem consists of two terms. The first term is the product of the power that is consumed by an idle memory subsystem and the time the memory subsystem is powered up. This term depends on the type and size of the memory and is different for SRAM, DRAM, and nonvolatile FLASH memory. The second term consists of the product of the number of memory accesses and the energy requirement for a single memory access. In an MPSoC that consists of a number of IP-cores that are connected by a network-on-chip (NoC), three types of memory accesses can be distinguished:

- Scratchpad memory that is within an IP-core.
- Shared on-chip memory in a separate IP-core of the SoC. It is accessed via the NoC.
- Off-chip memory, e.g., a large DRAM, that is accessed via the NoC and a memory gateway to the memory chip.

The energy requirements for accessing each of these three types of memory are substantially different. Whereas the access for the instruction and data in the scratchpad memory is considered in the effective capacitance C_{eff} of an instruction, the access to the on-chip and off-chip memory requires the exchange of two messages via the NoC: a *request message* containing the address to the memory and a *response message* with the data from the memory. In our model, we make the gross estimate that a memory access for a 32 byte block of on-chip memory requires an energy in the order of 200 *nJ* and access to the off-chip memory requires about hundred times as much, i.e., 20 *μJ*.

I/O Devices The energy consumed by I/O devices such as screens, sensors, and actuators is application specific (e.g., size of a screen). It can be substantial.

8.1.3   Thermal Effects and Reliability

The dissipation of electric energy in a device causes a heating of the device. The temperature rise of the device depends on the amount of energy dissipated, the heat capacity of the device, and the heat flow between the device and its environment.

> **Example**: Let us assume an MPSoC with an area of *100 mm²* housing *32* IP-cores connected by a network-on-chip. One of the IP-cores executes the program of the above example. It is physically contained in a block of *1* mm × *1* mm of a *0.5*-mm-thick silicon die. If we assume that this block of silicon, measuring *0.5 mm³*, is thermally isolated and no heat can be transferred to the block's environment, then the temperature of our silicon block will increase by 1 *°C* if about *815 μJ* of energy are dissipated in this block. (Silicon has density of *2.33 g/cm³* and a specific heat of *0.7 J/g°C*.) The execution of the program of the above hypothetical example—generating a heat of 1 joule—would thus lead to a temperature rise in the order of *1000 °C* and result in a *hot spot* on the die. In reality, such a temperature rise will not occur, since any temperature difference between two bodies forces a heat flow from the warmer body to the colder body that reduces the temperature difference.

In a VLSI device, the heat flow from a hot spot to the chip's environment takes place in two steps. In the first step, the heat of the hot spot flows to the die and increases the temperature of the entire die. In the second step, the heat flows from the die through the package to the environment. The heat flow from the hot spot to the entire die and the exact temperature profile of the die can be calculated by solving the partial differential equation for heat transfer.

Hot spots will develop if a substantial amount of power is dissipated in a very small area. For example, let us assume that a temporary low impedance path between the power rails of a transistor, a *latch-up* (which can be corrected by a power-cycle) develops in a small part of a circuit due to a fault caused by a neutron from ambient cosmic radiation. The current that dissipates in this path will result in a hot spot that can physically destroy the circuit. It is therefore expedient to monitor the current drawn by a device and switch off the power quickly (e.g., within less than 10 µsec) if an unexpected current surge is observed.

The temperature difference that develops between the die and the environment is determined by the power dissipation in the die and the thermal conductivity $H_{package}$ of the package. This thermal conductivity $H_{package}$ of a typical chip package is between 0.1 and 1 W/°C depending on the package geometry, size, and material. Plastic packages have a significantly lower thermal conductivity than ceramic packages. If the heat flow through the package is more than 10 Watt, then a fan should cool the package. The introduction of fans has a number of disadvantages, such as the additional energy required to operate the fan, the noise of the fan, and the reliability of the mechanical fan. If a fan fails, overheating might destroy the circuit.

A high substrate temperature has a negative effect on the reliability of a device and can cause transient and permanent failures. High substrate temperatures change the timing parameters of the transistors and the circuits. If the specified timing patterns are violated, transient and data-dependent device failures will occur.

The *Arrhenius equation* gives a gross estimate for the acceleration of the failure rate caused by an increase of the temperature of the silicon substrate [Vig10].

> **Example:** If the temperature of the substrate of a device increases from 50 °C (i.e., 323 K) to 100 °C (i.e., 373 K), and a failure mechanism with an activation energy of 0.5 eV is assumed, then the failure rate of the device will increase by a factor of about 11.

8.2 Hardware Power Reduction Techniques

8.2.1 Device Scaling

The most effective way to reduce the power consumption of CMOS devices is the scaling of the device parameters, i.e., making the transistors smaller [Fra01]. Table 8.2 depicts the effect of *ideal scaling* on the different parameters of a CMOS device. The scaling factor α from one microelectronic generation to the next is normally $1/\sqrt{2}$, i.e., about 0.7, such that the area of a scaled version of a design is reduced by a factor of 2, the power requirement is reduced by a factor of 2, the speed is increased by a factor $\sqrt{2}$, and the energy needed for the execution of an instruction (the *energy performance*) is reduced by $2\sqrt{2}$. Note from Table 8.2 that *ideal device scaling* has no effect on the power density that is dissipated in a given area of the die. It follows that ideal scaling will not result in a temperature increase of the die.

Example: Let us assume that an IP-core scales down ideally by a factor of $1/\sqrt{2}$ every 2 years. At the start, the IP-core has a size of 16 mm^2 and executes *125* MIPS, consuming a power of *16* Watt. Eight years later, after four generations of shrinking, this IP-core has a size of *1* mm^2, executes *500* MIPS, and consumes a power of *1* Watt. The energy needed for the execution of an instruction has been reduced by a factor of *64*, while the time performance has increased by a factor of 4.

Device scaling has made it possible to place up to one billion transistors on a single die. It is thus within the capabilities of the semiconductor industry to place a complete system, including processor, memory, and input/output circuitry on a single die, resulting in a system-on-chip (SoC). Spatial and temporal closeness of subsystems that are involved in a computation leads to a significant improvement of the energy efficiency. Spatial locality reduces the effective capacitances of the switching actions, which implies lower energy needs and faster operations. Temporal locality reduces the number of cache misses. If subsystems residing on different chips are integrated on a single die, the significant amount of energy needed to exchange data and control among chips can be saved.

Example: According to Intel [Int09], the 1996 design of the first *teraflop super computer*, consisting of 10,000 Pentium Pro Processors, operated with an energy efficiency of *2 MegaFlops/Joule* or *500 nJ* per instruction. Ten years later, in 2006, a *teraflop research chip* of Intel containing *80* IP-cores on a single die connected by a network-on-chip achieved an energy efficiency of *16,000 MegaFlops/Joule* or *62 pJ/instruction*. This is an increase in the energy performance by a factor of *8000* within 10 years. If we assume that in 10 years five generations of scaling are taking place, the increase in the energy performance in one generation is not only factor of $2\sqrt{2}$, the value stipulated by ideal scaling, but by a factor of more than *4*. This additional improvement is caused by the integration of all subsystems on a single die.

Over the last 25 years, device scaling has also had a very beneficial effect on device reliability. The failure rates of transistors have been reduced even *faster* than the increase in the number of transistors on a die, resulting in an increase in chip

Table 8.2 The effect of ideal device scaling on device parameters

Physical parameter	Scaling factor
Channel length, oxide thickness, wiring width	α
Electric field in device	1
Voltage	α
Capacitance	α
RC delay	α
Power dissipation	α^2
Power density	1
Time performance in MIPS	$1/\alpha$
Energy performance	$1/\alpha^3$

reliability despite the fact that many more transistors are contained in a scaled chip. The MTTF w.r.t. permanent failures of industrial state-of-the-art chips are significantly lower than *100* FIT [Pau98].

Scaling cannot continue indefinitely because there are limits due to the discrete structure of matter and quantum mechanical effects, such as *electron tunneling*. The reduction in the number of dopants in a transistor increases the statistical variations. The thermal energy of noise limits the reduction of the supply voltage. If the supply voltage is higher than stipulated by *ideal scaling*, then scaling will lead to an increased thermal stress. In submicron technologies, we have reached the point where these effects cannot be neglected any more. The International Technology Roadmap for Semiconductors 2009 [ITR09, p. 15] summarizes these challenges in a single sentence: *The ITRS is entering a new era as the industry begins to address the theoretical limits of CMOS scaling.*

8.2.2 Low-Power Hardware Design

Over the past few years, a number of hardware design techniques have been developed that help to reduce the power needs of VLSI circuits [Kea07]. In this section, we give a very short overview of some of these techniques.

Clock Gating: In many highly integrated chips, a significant fraction of the total power is consumed in the distribution network of the execution clock. One way to reduce power in the execution clock network is to turn execution clocks off when and where they are not needed. Clock gating can reduce the power consumption of chips by 20% or more.

Transistor Sizing and Circuit Design: Energy can be saved if the transistor and circuit design is optimized with the goal to save power instead with the goal to get the optimal speed. Special sizing of transistors can help to reduce the capacitance that must be switched.

Multi-threshold Logic: With present-day microelectronic design tools, it is possible to build transistors with different threshold voltages on the same die. High-threshold transistors have a lower leakage current but are slower than low-threshold transistors. It is thus possible to save dynamic and static power by properly combining these two types of transistors.

8.2.3 Voltage and Frequency Scaling

It has been observed that, within limits characteristic for each technology, there is a nearly linear dependency of frequency on voltage. If the frequency for the operation of the device is reduced, the voltage can be reduced as well without disturbing the functionality of the device [Kea07]. Since the power consumption grows linearly with frequency, but with the square of the voltage, combined voltage and frequency scaling causes not only a reduction of power, but also a reduction of energy required to perform a computation.

Example: The Intel XScale® processor can dynamically operate over the voltage range of *0.7–1.75 V* and at a frequency range of *150–800 MHz*. The highest energy consumption is *6.3* times the lowest energy consumption.

Voltage scaling can be performed in the interval $< V_{threshold}, V_{normal}>$. Since V_{normal} is reduced as a device is scaled to lower dimensions (see Sect. 8.2.1), the range that is available for voltage scaling in submicron devices is reduced and voltage scaling becomes less effective.

The additional circuitry that is needed to perform software-controlled dynamic voltage and frequency scaling is substantial. In order to reduce this circuitry, some designs support only two operating modes: a *high-performance operating mode* that maximizes performance and an *energy-efficient operating mode* that maximizes energy efficiency. The switchover between these two modes can be controlled by software. For example, a laptop can run in the high-performance operating mode if it is connected to the power grid and in the energy-efficient operating mode if it runs on battery power.

Given that the hardware supports voltage and frequency scaling, the operating system can integrate power management with real-time scheduling of time-critical tasks to optimize the overall energy consumption. If the *Worst-Case Execution Time* (the WCET) of a task is known on a processor running at a given frequency and the task has some slack until it must finish, then the frequency and voltage can be reduced to let the task complete just in time and save energy. This integrated real-time and power management scheduling has to be supported at the level of the operating system.

8.2.4   Sub-threshold Logic

There are an increasing number of applications where ultralow power consumption with reduced computational demands is desired. Take the example of the billions of standby circuits in electronic devices (e.g., television sets) that are continuously draining power while waiting for a significant event to occur (e.g., a start command from a remote console or a significant event in a sensor network). The technique of sub-threshold logic uses the (normally unwanted) sub-threshold leakage current of a submicron device to encode logic functionality. This novel technique has the potential to design low time performance devices with a very low power requirement [Soe01].

8.3   System Architecture

Next to device scaling, the following system architecture techniques are most effective in reducing the energy requirement significantly.

8.3.1   Technology-Agnostic Design

At a high level of abstraction, an application requirement can be expressed by a platform-independent model (PIM) (see also Sect. 4.4). A PIM describes the functional and temporal properties of the requested solution without making any reference to the concrete hardware implementation. For example, when we specify the functionality and timing of the braking system of a car, we demand that the proper braking action will start within 2 milliseconds after stepping on the brake pedal. We say that such a high-level description of an application is *technology agnostic*. A PIM can be expressed in a procedural language, e.g., System C, augmented by the required timing information, e.g., by UML-MARTE [OMG08]. The system implementer has then the freedom to select the implementation technology that is most appropriate for his/her purpose.

In the second step, the PIM must be transformed into a representation that can be executed on the selected target hardware, resulting in the platform-specific model (PSM). The target hardware can be either a specific CPU with memory, a field-programmable gate array (FPGA), or a dedicated application-specific integrated circuit (ASIC). Although the functional and temporal requirements of the PIM are satisfied by all of these implementation choices, they differ significantly in their non-functional properties, such as energy requirements, silicon real-estate, or reliability. Figure 8.2 gives a gross indication of the energy required to execute a given computation in the three mentioned technologies. CPU-based computations have a built-in power overhead for instruction fetch and decoding that is not present in hardwired logic.

More recently, GPUs (graphics processing units) and TPUs (tensor processing units) have also become available for use in real-time systems, e.g., in self-driving cars. Experiments in embedded computer vision applications show that the energy efficiency of GPUs is typically in between the ones of CPUs and FPGAs [Qas19]. TPUs implement a *domain-specific architecture* for machine learning applications. *The first-generation tensor processing unit (TPU) runs deep neural network (DNN) inference 15–30 times faster with 30–80 times better energy efficiency than contemporary CPUs and GPUs in similar semiconductor technologies* [Jou18]. Today, MPSoCs for real-time systems may include CPUs, GPUs, and TPUs, e.g., the NVIDIA Xavier.

The *technology-agnostic design* makes it possible to change the target hardware of a single component, e.g., replace a CPU-based component by an ASIC without having to revalidate the complete system. Such an implementation flexibility is of

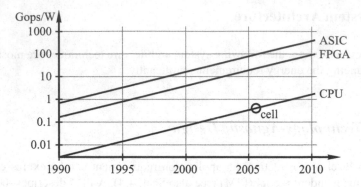

Fig. 8.2 Power requirement of different implementation technologies. (Adapted from [Lau06, slide 7])

particular importance for battery-operated mass-market devices where the initial test version of the functionality of a component can be realized and tested on a CPU-based implementation and later transferred to an ASIC for the mass-market production.

The *technology-agnostic design* makes it possible to address the *technology-obsolescence problem* as well: In long-lived applications, such as the control system on an airplane, the services of the control system must be provided for a long time-span, e.g., 50 years. During this time-span, the original hardware technology becomes outdated. Technology-agnostic design makes it possible to change the hardware and the related transformation of the PIM to the PSM, without having to change the interfaces to the other subsystems.

8.3.2 Pollack's Rule

Over the past 20 years, we have seen a tremendous performance increase of single processor systems. New architectural mechanisms, such as pipelining, out-of-order execution, speculative branching, and many levels of caching, have made it possible to significantly reduce the execution time needed for a sequential program without having to invest in alternative system and software architectures that support a highly parallel execution environment. However, this performance increase of the sequential processor has its (energy) price. Fred Pollack from Intel looked at the integer performance increase of a new micro-architecture against area and power of the previous micro-architecture, implemented in the same process technology [Bor07]. Pollack found that over a number of Intel architectures, starting with the i386 in 1986, the performance of every subsequent micro-architecture increased only with the square root of the power or silicon area. This relationship is normally referred to as *Pollack's rule*.

Embedded systems are characterized by an *application-inherent parallelism*, i.e., they consist of many concurrent, nearly independent processes. In order to establish a viable software execution environment for these nearly independent parallel processes, a complex operating system that provides spatial and temporal partitioning must be implemented on top of a sequential processor. From the energy perspective, this is yet another setback. At first, energy is wasted by the execution of the powerful sequential machine, and then energy is wasted again to provide the encapsulated parallel execution environments for the support of the parallel processes running on this sequential machine.

Example: According to Pollack's rule, the speed improvement of an IP-core achieved by advanced micro-architectural mechanisms scales by the square root of two, while the required energy and the silicon area increase by factor of 2. After four generations of micro-architecture evolutions, an IP-core would have grown *16* times its original size, would consume *16* times as much energy as the original, and achieve a time performance improvement of four. The micro-architecture evolution has degraded the energy efficiency by *400%*.

The introduction of multi-core systems on chip (MPSoC), where simple IP-cores are connected by a network-on-chip (NoC), will thus revolutionize the execution environment for embedded systems. Considering the size of the embedded system market, it can be expected that in the future energy-efficient multi-core systems that focus on this market will become dominant. The potential for energy savings of these systems is significant.

An important issue in the design of MPSoCs is related to the structure of the interconnect among the IP-cores. There are basically two alternatives: (i) a message-based communication infrastructure and (ii) the provision of a large shared memory. Poletti et al. [Pol07] have investigated the energy efficiency of these two alternatives and come to the conclusion that message-based systems are preferable if there is a high computation/communication ratio while shared memory outperforms message passing if the computation/communication ratio is low. The comparative analysis of memory models for chip multiprocessors comes to a similar conclusion [Lev08]. In many industrial embedded systems, the high computation/communication ratio suggests that message passing is the preferred alternative.

Example: In a premium car of today, one can find up to one hundred *electronic control units* (ECU) that are connected by a number of low-bandwidth CAN buses. Aggregating some of these ECUs on a single die of an MPSoC will put very little load on the high-bandwidth NoC.

Message passing has many other advantageous properties over shared memory, such as function encapsulation, fault containment, the support of *implementation-agnostic design methods*, and the support of *power gating*.

8.3.3   Power Gating

In a multi-core SoC, consisting of a set of heterogeneous IP-cores interconnected by a real-time network-on-chip, a well-defined application functionality can be implemented in a dedicated IP-core, i.e., a component (see Chap. 4). Examples for such an application functionality are resource management, security, MPEG processing, input-output controllers, external memory manager, etc. If the components interact with each other via message passing only and do not access a shared memory, then it is possible to encapsulate a component physically and logically in a small area of silicon that can be powered down when the services of the component are not needed, thus saving the dynamic and static power of the component. Normally, the *state* (see Sect. 4.2) that is contained in the component will be lost on power-down. It is therefore expedient to select a power/down power/up point when the state of the component is empty. Otherwise, the state of the component must be saved. There are two ways of saving the state, either by hardware techniques or by sending a message containing the state to another component such that this other component can save and update the state.

The hardware effort for saving the state transparently can be substantial [Kea07]. On the other hand, a distributed architecture that supports robustness must support the dynamic restart of components in case a component has failed due to a transient fault (see Chap. 6). This restart must be performed with a temporally accurate component state. This software-based state restoration mechanism can be used to support the state restoration required by power gating as well without any additional overhead.

In an MPSoC architecture that consists of a plurality of components that are connected by a network-on-chip (NoC), power gating is a very effective technique to save power. A component that is not in use can be shut down completely, thus not only saving dynamic power but also the static power. Since static power is increasing substantially as we deploy below 100 nm technology, power gating becomes an extremely important power-saving technology.

In many devices, it is useful to distinguish between two major modes of operation: *service mode* and *sleep mode*. In service mode, the full set of device services is provided and the dynamic power dissipation is dominant. In sleep mode, only the minimal functionality for activating the device upon arrival of a wake-up signal is required. In sleep mode, the static (leakage) power is of major concern. Power gating can be very effective in reducing the power demand in sleep mode. Alternatively, the sleep mode can be implemented in a completely different technology, e.g., in sub-threshold logic, that starts up the service mode as soon as a relevant wake-up signal is recognized. In this case, all components that are involved in the service mode can be shut down completely while in sleep mode, thus not consuming any power at all.

8.3.4 Real Time Versus Execution Time

It is important to stress the fundamental difference between *real time* and *execution time* in a distributed real-time system. There is no close relation between these two time bases. In an MPSoC, the granularity of the real time will be one or two orders of magnitude larger (and correspondingly, the frequency lower) than the granularity of the execution time at the chip level. Since the power consumption is proportional to the frequency—see Sect. 8.1.2—the global *real-time clock distribution network* will only consume a small fraction of the power that would be needed for a global *execution time clock distribution network*. Establishing a single global real-time base for the whole MPSoC, but many local asynchronous execution time bases, one in each IP-core of an MPSoC, can itself save a significant amount of energy and furthermore increase the energy savings potentials of a chip, as explained in the following paragraphs.

The *real-time base* makes the nodes of a distributed system aware of the progression of real time and provides the basis for the generation of temporal control signals (see also Sect. 4.1.3). The local real-time clocks should be incremented according to the international standard of time TIA. If no external clock synchronization is available, real time is established by a real-time reference clock that forms the source of the distributed real-time base. The granularity of the global real time depends on the precision of the clock synchronization and will be different at different integration levels. For example, at the chip level, where the IP-cores of a SoC communicate via a NoC, the local real-time clocks in the IP-cores will have a better precision (and consequently a smaller granularity) than the global real time at the device level, where devices communicate via a local area network (see Chap. 3 on clock synchronization). At the chip level, the establishment of the global real time can be realized by a stand-alone real-time clock distribution network or it can be integrated into the NoC.

The *execution time base* drives the algorithmic computations of the nodes and thus determines the speed of the computation (logical control—see Sect. 4.1.3). In a large SoC, the energy dissipation of a global execution time clocking system of a SoC can form a large part of the chip's energy consumption. In addition, the location-dependent delay of the central high-frequency timing signals results in a clock skew that is difficult to control. Furthermore, the individual control of the voltage and frequency of an IP-core is not possible if all IP-cores operate with the same clock signal. It makes therefore sense to design each IP-core and the NoC as an *island of synchronicity* that generates its clocking signal for the execution time locally. If the voltage of an IP-core can also be controlled locally, then the IP-core is an encapsulated subsystem with the capability for local voltage-frequency scaling and power gating. In the architecture model outlined in Chap. 4, clock-domain crossing occurs in the message-interface between an IP-core and the NoC, which must be carefully designed in order to avoid meta-stability problems.

8.4   Software Techniques

The equation $E = C_{\text{eff}} V^2 N$ of Sect. 8.1.2 gives the dynamic energy E required for the execution of a program. There are three parameters in this equation, the effective capacitance C_{eff}, the supply voltage V, and the number of instructions N. Reducing the effective capacitance C_{eff} and reducing the number of instructions per task reduces the time needed to complete a computational task. There is thus no inherent conflict at the software level between designing for energy performance and designing for time performance.

The voltage depends primarily on the hardware technology and can be controlled by software if dynamic voltage and frequency scaling is supported by the hardware. The effective capacitance C_{eff} can be reduced by *spatial and temporal locality*, particularly in the memory system. The number of instructions, the *instruction count N*, needed to achieve the intended result depends entirely on the software. The instruction count is the sum of the instructions executed by the *system software* and the *application software*.

8.4.1   System Software

System software consists of the operating system and the middleware. The objectives of a flexible system-software infrastructure versus minimal energy consumption drive the design process in different directions. Many operating systems of the past have considered flexibility as the key design driver, ignoring the topic of energy performance. In these systems, a long sequence of system-software instructions must be executed in order to finalize a single application command, such as the sending of a message.

In battery-operated embedded systems, the energy efficiency of the system software can be improved by off-line tailoring of the operating system functions to the specific requirements of the given application. Furthermore, an integrated resource management strategy that considers timeliness requirements and energy requirements in a holistic manner is followed.

Whenever hardware resources (processors, memory, caches) are shared among nearly independent processes, the implicit interactions (e.g., arbitration, cache reloading, processor switching) increase the energy consumption and make time and energy estimation more difficult.

The non-availability of a generally accepted architectural style for embedded systems leads to many property mismatches at the interfaces among subsystems (e.g., *big-endian* vs. *little-endian*) that must be reconciled by the system software, leading to necessary but unproductive glue software which consumes valuable energy when executed.

8.4.2 Application Software

Algorithm Design An algorithm that has been developed from the point of view of optimal time performance will be different from an algorithm that is developed from the point of view of optimal energy performance. In contrast to best-effort systems, where the average execution time of an algorithm is optimized, in real-time systems, the worst-case execution time of an algorithm is of relevance.

In many real-time applications, such as multimedia or control applications, there is not always a need for precise results—good approximate results will suffice. Finding algorithms that give a good approximate result under energy constraints is a relevant research topic (see also Sect. 10.2.3 on *anytime algorithms*).

In *cloud computing*, where some tasks are processed in the cloud (the servers in the cloud run on power available from the electric utility) and some tasks are computed locally in a battery-operated mobile device, the trade-off between the energy requirement of the algorithms for the local computations and the energy requirement for the transmission of data to and from the cloud is an important architectural design issue. Appropriate design tools must support the design exploration of different task allocation strategies to the cloud or to the mobile device.

Algorithm Analysis Many embedded applications contain a computationally intensive algorithmic segment, called a *computational kernel*. Profiling the execution of a program that implements an algorithm can identify the computational kernel. If a computational kernel is isolated into a self-contained component that puts all elements of the computational kernel (e.g., processing engine, memory) physically and temporally close together in an IP-core, then the effective capacity C_{eff} of the execution environment can be reduced, thus saving energy. If an identified computational kernel is mature, does not change, and is used by many applications, then it can be transformed to an ASIC IP-core of an MPSoC, resulting in orders of magnitude energy savings.

Data Structures The significant energy cost for memory access can be reduced by the design of data structures that are optimized from the point of view of energy efficiency for the specified use cases. For example, a strictly binary time-format, which can be implemented with a binary counter, requires much less energy to operate than a time-format that is based on the Gregorian calendar.

8.4.3 Software Tools

Next to the system software, compilers have an important role to play in the design of low-energy systems. Energy-aware compilers can select the instructions for the target code on the basis of their energy requirements. The energy-aware allocation of registers is another important issue, since the register file of modern processors is quite energy intensive.

System designers need tools to estimate the power consumption at an early stage of the design. These tools must be flexible to support different target architectures in order that *design explorations* can be carried out. They should be smoothly integrated in the design environment.

Points to Remember

- *Energy* is defined as the capability of performing work. *Power* refers to the intensity of work. Energy is the integral of power over time. Although power and energy savings are closely related, they are not the same.
- Energy that is needed by a computing device to execute a program can be expressed as the sum of the following four terms: $E_{total} = E_{comp} + E_{mem} + E_{comm} + E_{IO}$ where E_{total} is the total energy needed, E_{comp} denotes the energy needed to perform the computations in a CMOS VLSI circuit, E_{mem} denotes the energy consumed by the memory subsystem, E_{comm} denotes the energy required for the communication, and E_{IO} is the energy consumed by I/O devices, such as a screen.
- The dynamic energy needed to run a program with N instructions is given by $E = C_{eff} V^2 N$ where C_{eff} is the effective capacitance of an instruction, V is the supply voltage, and N denotes the number of instructions that must be executed.
- The exponential growth of the leakage current with rising temperature is of major concern in a submicron device with a low threshold voltage.
- The asymmetry of the energy requirements of a sender and a receiver has a decisive influence on the design of communication protocols between a base station connected to the power grid and an energy-constrained mobile battery-powered device.
- In memory-intensive applications, the energy that is required to access the memory subsystems can be larger than the energy needed for the computations.
- A high substrate temperature has a negative effect on the reliability of a VLSI device and can cause transient and permanent failures. Around half of all device failures are caused by thermal stress.
- The most effective way to reduce the power consumption of CMOS devices is the scaling of the device parameters, i.e., making the transistors smaller.
- The scaling factor α from one microelectronic generation to the next is normally 0.7 such that the area of a scaled version of a design is reduced by a factor of 2, the power requirement is reduced by a factor of 2, the speed is increased by a factor $\sqrt{2}$, and the energy performance increases by $2\sqrt{2}$.
- The positioning of all subsystems onto a single die (SoC) leads to a significant reduction of the distances between the transistors in the diverse subsystems (and in consequence to a reduction of the capacities of the signal lines) which results in major energy savings.

- Device scaling cannot go on indefinitely because there are limits that have their cause in the discrete structure of matter and quantum mechanical effects, such as *electron tunneling*.
- If the frequency for the operation of the device is reduced, the voltage can be reduced as well without disturbing the functionality of the device, resulting in substantial energy savings.
- Real time and execution time are two different time bases that are not closely related.
- A technology-agnostic design methodology makes it possible to move function-ality from software to hardware with a substantial gain in energy efficiency.
- Pollack's rule states that micro-architectural advances from one generation to the next increase the performance of sequential processors only with the square root of the increase in power or silicon area.
- Spatial locality of computational subsystems reduces the effective capacitance and thus increases energy efficiency.
- The most important contribution of software to energy efficiency is a reduction of the number and types of statements that must be executed to achieve the desired results.

Bibliographic Notes

The tutorial survey by Benini and Micheli [Ben00] gives an excellent overview of system- level design methods that lead to energy-efficient electronic systems. Pedram and Nazarian [Ped06] provide models to investigate the thermal effects in submicron VLSI circuits. The limits of device scaling are the topic of [Fra01]. Pollack's rule and the future of SoC architectures are the topic of [Bor07]. Issues of energy awareness in systems-on-chip are discussed in [Pol07]. The review paper by Sah and Amgoth [Sah20] contains a good survey of energy harvesting.

Review Questions and Problems

8.1 Explain the difference between power and energy.
8.2 How many Joule are contained in a calorie or in a kWh?
8.3 Calculate the dynamic energy of a program execution if the program contains 1,000,000 instructions, the supply voltage is 1 V, and the effective capacitance of an average instruction is 1 nF.
8.4 What is *static energy*? How does static energy change with temperature?
8.5 How much energy is required for access to the scratchpad memory, the on-chip memory, and the off-chip memory in the reference architecture introduced in this chapter?

8.6 A sensor node executes 100,000 instructions per second (supply voltage is 1 V and effective capacitance on an instruction is 1 nF) and sends every second a message with a length of 32 bytes to its neighbor node, which is 10 m away. The voltage of the transmitter is 3 V. How much power is needed to drive the sensor node? How many hours will the sensor node operate if the power supply contains two AAA batteries?

8.7 A processor has two operating modes, a time performance optimized mode characterized by a voltage of 2 V and a frequency of 500 MHz and an energy-optimized mode characterized by a voltage of 1 V and a frequency of 200 MHz. The effective capacity of an instruction is 1 nF. What is the power requirement in each of the two modes?

8.8 A lithium-ion laptop battery weighs 380 g. How long will a battery-load last if the laptop has a power demand of 10 W?

Chapter 9
Real-Time Operating Systems

Overview

In a component-based distributed real-time system, we distinguish two levels of system administration, the coordination of the message-based communication and resource allocation among the components and the establishment, coordination, and control of the concurrent tasks within each one of the components. The focus of this chapter is on the operating system and middleware functions within a component.

In case the software core image is not permanently residing in a component (e.g., in read-only memory), mechanisms must be provided for a *secure boot* of the component software via the technology-independent interface. Control mechanisms must be made available to *reset*, *start*, and *control* the execution of the component software at run time. The software within a component will normally be organized in a set of concurrent tasks. Task management and inter-component task interactions have to be designed carefully in order to ensure *temporal predictability* and *determinism*. The proper handling of time and time-related signals is of special importance in real-time operating systems. The operating system must also support the programmer in establishing new message communication channels at run time and in controlling the access to the message-based interfaces of the components. Domain-specific higher-level protocols, such as a simple *request-reply protocol*, that consist of a sequence of rule-based message exchanges should be implemented in the middleware of a component. Finally, the operating system must provide mechanisms to access the local process input/output interfaces that connect a component to the physical plant. Since the value domain and the time domain of the RT entities in the physical plant are *dense*, but the representation of the values and times inside the computer is *discrete*, some inaccuracy in the representation of values and times inside the computer system cannot be avoided. In order to reduce the effects of these representation inaccuracies and to establish a consistent (but not fully faithful) model of the physical plant inside the computer system, agreement protocols must be executed at the interface between the physical world and cyberspace to

© The Author(s), under exclusive license to Springer Nature Switzerland AG 2022
H. Kopetz, W. Steiner, *Real-Time Systems*,
https://doi.org/10.1007/978-3-031-11992-7_9

create a *consistent digital image* of the external world inside the distributed computer system.

A real-time operating system (OS) within a component must be temporally predictable. In contrast to operating systems for personal computers, a real-time OS should be deterministic and support the implementation of fault tolerance by active replication. In safety-critical applications, the OS must be certified. Since the certification of the behavior of a dynamic control structure is difficult, dynamic mechanisms should be avoided wherever possible.

9.1   Inter-Component Communication

The information exchange of a component with its environment, i.e., other components and the physical plant, is realized exclusively across the four message-based interfaces introduced in Sect. 4.4. It is up to the generic middleware and the component's operating system to manage the access to these four message interfaces for inter-component communication. The TII, the LIF, and the TDI are discussed in this section, while the *local interface* is discussed in the section on process input/output.

9.1.1   Technology-Independent Interface (TII)

In some sense, the TII is a meta-level interface that brings a new component out of the core image of the software, the *job*, and the given embodiment, the *component hardware* into existence. The purpose of the TII is the configuration of the component and the control of the execution of the software within a component. The component hardware must provide a dedicated TII port for the secure download of a new software image onto a component. Periodically, the g-state (see Sect. 4.2.3) of the component should be published at the TII in order to be able to check the contents of the g-state by a dedicated diagnostic component. A further TII port directly connected to the component hardware must allow the resetting of the component hardware and the restart of the component software at the next reintegration point with a relevant g-state that is contained in the reset message. The TII is also used to control the voltage and frequency of the component hardware, provided the given hardware supports voltage-frequency scaling. Since malicious TII messages have the potential to destroy the correct operation of a component, the authenticity and integrity of all messages that are sent to the TII interface must be assured.

9.1.2   Linking Interface (LIF)

The linking interface of a component is the interface where the services of the component are provided during normal operation. It is the most important interface from the point of view of operation and of composability of the components. The LIF has been discussed extensively in Sect. 4.6.

9.1.3   Technology-Dependent Interface (TDI)

In the domain of VLSI design, it is common practice to provide a dedicated interface port for testing and debugging, known as the JTAG port that has been standardized in IEEE standard 1149.1. Such a debugging port, the TDI, supports a detailed view inside a component that is needed by a component-designer to monitor and change the internal variables of a component that are not visible at any one of the other interfaces. The component-local OS should support such a testing and debugging interface.

9.1.4   Generic Middleware (GM)

The software structure within a component is depicted in Fig. 9.1. Between the local hardware-specific real-time operating system and the application software is the *generic middleware* (GM). The *execution control messages* that arrive at the TII (e.g., *start task, terminate task*, or *reset the component hardware* and r*estart the component* with a relevant g-state) or are produced at the TII (e.g., *periodic publication of the g-state*) are interpreted inside a component by the standardized generic middleware (GM). The application software, written in a high-level language, accesses the operational message-based interfaces (the LIF and the local interface)

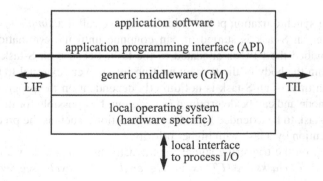

Fig. 9.1 Software structure within a component

by API system calls. The GM and the task-local operating system must manage the API system calls and the messages that arrive at the LIF and the commands that arrive via the TII messages. While the task-local operating system may be specific to a given component hardware, the GM layer provides standardized services, processes the standardized system control messages, and implements higher-level protocols.

> **Example**: A high-level time-monitored *request-reply protocol* that is a unique concept at the level of the API requires two or more independent messages at the basic message transport service (BMTS) level and a set of local timer and operating system calls for its implementation. The GM implements this high-level protocol. It keeps track of all relevant messages and coordinates the timeouts and operating system calls.

9.2   Task Management

In our model, the component software is assumed to be a *unit of design*, and a *whole component* is the smallest unit of fault containment. The concurrent tasks within a component are *cooperative* and not *competitive*. Since the whole component is a unit of failure, it is not justified to design and implement resource-intensive mechanisms to protect the component-internal tasks from each other. The component-internal operating system is thus a lightweight operating system that manages the task execution and the resource allocation inside a component.

A *task* is the execution of a sequential program. It starts with reading of input data and of its internal state and terminates with the production of the results and updated internal state. A task that does not have an internal state at its point of invocation is called a *stateless task*; otherwise, it is called a *statefull task*. *Task management* is concerned with the initialization, execution, monitoring, error handling, interaction, and termination of tasks.

9.2.1   Simple Tasks

If there is no synchronization point within a task, we call it a *simple task* (*S-task*), i.e., whenever an S-task is started, it can continue until its termination point is reached, provided the CPU is allocated to the task. Because an S-task cannot be blocked within the body of the task by waiting for an event external to the S-task, the execution time of an S-task is not directly dependent on the progress of other tasks in the node and can be determined *in isolation*. It is possible for the execution time of an S-task to be extended by indirect interactions, such as the preemption of the task execution by a task with higher priority.

Depending on the triggering signal for the activation of a task, we distinguish *time-triggered (TT) tasks* and *(ET) event-triggered tasks*. A *cycle* (see Sect. 3.3.4) is

Fig. 9.2 Task descriptor
list (TADL) in a TT
operating system

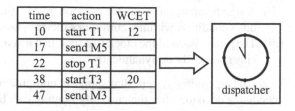

time	action	WCET
10	start T1	12
17	send M5	
22	stop T1	
38	start T3	20
47	send M3	

dispatcher

assigned to every TT-task, and the task execution is started whenever the global time
reaches the start of a new cycle. Event-triggered tasks are started whenever a *start-
event* for the task occurs. A start event can be the completion of another task or an
external event that is relayed to the operating system by an incoming message or by
the interrupt mechanism.

In an entirely time-triggered system, off-line scheduling tools establish the temporal control structure of all tasks a priori. This temporal control structure is encoded
in a *Task-Descriptor List (TADL)* that contains the cyclic schedule for all activities
of the node (Fig. 9.2). This schedule considers the required precedence and mutual
exclusion relationships among the tasks such that an explicit coordination of the
tasks by the operating system at run time to guarantee mutual exclusion is not
necessary.

Whenever the time reaches an entry point of the TADL, the dispatcher is activated. It performs the action that has been planned for this instant. If a task is started,
the operating system informs the task of its activation time, which is synchronized
within the cluster. After task termination, the operating system makes the results of
the task available to other tasks.

The application program interface (API) of an S-task in a TT system consists of
three data structures and two operating system calls. The data structures are the
input data structure, the *output data structure*, and the *g-state* data structure of the
task (which is empty, in case the task is *stateless*). The system calls are TERMINATE
TASK and ERROR. The TERMINATE TASK system call is executed by the task
whenever the task has reached its normal termination point. In the case of an error
that cannot be handled within the application task, the task terminates its operation
with the ERROR system call.

In an event-triggered system, the evolving application scenario determines the
sequence of task executions dynamically. Whenever a significant event happens, a
task is released to the *ready* state, and the dynamic task scheduler is invoked. It is up
to the scheduler to decide at run time, which one of the ready tasks is selected for
the next service by the CPU. Different dynamic algorithms for solving the scheduling problem are discussed in the following chapter. The *WCET* (*Worst-Case
Execution Time*) of the scheduler contributes to the *WCAO* (*Worst-Case
Administrative Overhead*) of the operating system.

Significant events that cause the activation of an ET task can be:

(i) An event from the node's environment, i.e., the arrival of a message or an interrupt from the controlled object, or

(ii) A significant event inside the component, i.e., the termination of a task or some
      other condition within a currently executing task, or
(iii) The progression of the clock to a specified instant. This instant can be specified
      either statically or dynamically

An ET operating system that supports non-preemptive S-tasks will take a new
scheduling decision after the currently running task has terminated. This simplifies
task management for the operating system but severely restricts its responsiveness.
If a significant event arrives immediately after the longest task has been scheduled,
this event will not be considered until this longest task has completed.

In an RT operating system that supports task preemption, each occurrence of a
significant event can potentially activate a new task and cause an immediate interruption of the currently executing task. Depending on the outcome of the dynamic
scheduling algorithm, the new task will be selected for execution or the interrupted
task will be continued. Data conflicts between concurrently executing S-tasks can
be avoided if the operating system copies all input data required by this task from
the global data area into a private data area of the task at the time of task invocation.
If components are replicated, care must be taken that the preemption points at all
replicas is at the *same statement*; otherwise replica determinism may be lost.

The API of an operating system that supports event-triggered S-tasks requires
more system calls than an operating system that only supports time-triggered tasks.
Along with the data structures and the already introduced system calls of a TT system, the operating system must provide system calls to ACTIVATE (make ready) a
new task, either immediately or at some future point in time. Another system call is
needed to DEACTIVATE an already activated task.

9.2.2   Trigger Tasks

In a TT system, control always remains within the computer system. To recognize
significant state changes outside the computer, a TT system must regularly monitor
the state of the environment. A *trigger task* is a time-triggered S-task that evaluates
a *trigger condition* on a set of temporally accurate state variables that reflect the
current state of the environment. The result of a trigger task can be a control signal
that activates another application task. Since the states, either external or internal,
are sampled at the frequency of the trigger task, only those states with a duration
longer than the sampling period of the trigger task are guaranteed to be observed.
Short-lived states, e.g., the push of a button, must be stored in a memory element
(e.g., in the interface) for a duration that is longer than the sampling period of the
trigger task. The periodic trigger task generates an administrative overhead in a TT
system. The period of the trigger task must be smaller than the laxity (i.e., the difference between deadline and execution time) of an RT transaction that is activated
by an event in the environment. If the laxity of the RT transaction is very small
(<1 msec), the overhead associated with a trigger task can become intolerable and
the implementation of an interrupt is needed.

9.2.3 Complex Tasks

A task is called a *complex task (C-Task)* if it contains a blocking synchronization statement (e.g., a *semaphore wait operation*) within the task body. Such a *wait* operation may be required because the task must wait until a condition outside the task is satisfied, e.g., until another task has finished updating a common data structure or until input from a terminal has arrived. If a common data structure is implemented as a protected shared object, only one task may update the data at any particular moment (mutual exclusion). All other tasks must be delayed by the *wait* operation until the currently active task finishes its critical section. The worst-case execution time of a complex task in a node is therefore a *global* issue because it depends directly on the progress of the other tasks within the node or within the environment of the node.

The WCET of a C-task cannot be determined independently of the other tasks in the node. It can depend on the occurrence of an event in the node environment, as seen from the example of waiting for an input message. The timing analysis is not a local issue of a single task anymore; it becomes a global system issue. It is impossible to give an upper bound for the WCET of a C-task by analyzing the task code only.

The application programming interface of a C-task is more complex than that of S-tasks. In addition to the three data structures already introduced, i.e., the *input data structure*, the *output data structure*, and the *g-state* data structure, the global data structures that are accessed at the blocking point must be defined. System calls must be provided that handle a WAIT-FOR-EVENT and a SIGNAL-EVENT. After the execution of the WAIT-FOR-EVENT, the task enters the blocked state and waits in the queue. The event occurrence releases the task from the blocked state. It must be monitored by a time-out task to avoid permanent blocking. The time-out task must be *deactivated* in case the awaited event occurs within the time-out period, otherwise the blocked task must be *killed*.

9.3 The Dual Role of Time

A real-time image must be *temporally accurate* at the *instant of use* (see Sect. 5.4). In a distributed system, the temporal accuracy can only be checked if the duration between the *instant of observation* of a RT entity, observed by the sensor node, and the *instant of use*, determined by the actuator node, can be measured. This requires the availability of a global time base of proper precision among all involved nodes. If fault tolerance is required, two independent self-checking channels must be provided to link an end system to the fault-tolerant communication infrastructure. The clock synchronization messages must be provided on both channels in order to tolerate the loss of any one of the channels.

Every I/O signal has two dimensions, the value dimension and the temporal dimension. The value dimension relates to the value of the I/O signal. The temporal dimension relates to the instant when the value was captured from the environment or released to the environment.

> **Example**: In the context of hardware design, the value dimension is concerned with the contents of a register and the temporal dimension is concerned with the *trigger signal*, i.e., the control signal that determines when the contents of an I/O register are transferred to another subsystem.

An event that happens in the environment of a real-time computer can be looked upon from two different timing perspectives:

(i) It defines the *instant* of a value change of an RT entity in the domain of time. The precise knowledge of this *instant* is an important input for the *later* analysis of the consequences of the event (*time as data*).

(ii) It may demand *immediate* action by the computer system to react as soon as possible to this event (*time as control*).

It is important to distinguish these two different roles of time. In the majority of situations, it is sufficient to treat *time as data*, and only in the minority of cases, an immediate action of a computer system is required (*time as control*).

> **Example:** Consider a computer system that must measure the time interval between *start* and *finish* during a downhill skiing competition. In this application, it is sufficient to treat time as data and to record the precise time of occurrence of the *start event* and *finish event*. The messages that contain these two instants are transported to another computer that later calculates the difference. The situation of a train-control system that recognizes a red alarm signal, meaning the train should stop immediately, is different. Here, an immediate action is required as a consequence of the event occurrence. The occurrence of the event must initiate a control action without delay.

9.3.1   Time as Data

The implementation of *time as data* is simple if a global time base of known precision is available in the distributed system. The observing component must include the time-stamp of event occurrence in the observation message. We call a message that contains the time-stamp of an event a *timed message*. The timed message can be processed at a later time and does not require any dynamic data-dependent modification of the temporal control structure. Alternatively, if a field bus communication protocol with a known constant delay is used, the time of message arrival, corrected by this known delay, can be used to establish the send time of the message.

The same technique of timed messages can be used on the output side. If an output signal must be invoked on the environment at a precise *instant* with a precision much finer than the jitter of the output messages, a *timed output message* can be sent to the node controlling the actuator. This node interprets the time in the message and acts on the environment precisely at the intended instant.

In a TT system that exchanges messages at a priori known instants with a fixed period between messages, the representation of time in a timed message can take advantage of this a priori information. The time value can be coded in fractions of the period of the message, thus increasing the data efficiency. For example, if an observation message is exchanged every *100 msec*, a 7 bit time representation of time relative to the start of the period will identify the event with a granularity of better than 1 *msec*. Such a 7-bit representation of time, along with the additional bit to denote the event occurrence, can be packed into a single byte. Such a compact representation of the instant of event occurrence is very useful in alarm monitoring systems, where thousands of alarms are periodically queried by a cyclic trigger task. The cycle of the trigger task determines the maximum delay of an alarm report (*time as control*), while the resolution of the time-stamp informs about the exact occurrence of the alarm event (*time as data*) in the last cycle.

> **Example:** In a single periodic TTEthernet message with a data field of 1000 bytes and cycle time of *10 msec*, 1000 alarms can be encoded in a single message with a worst-case reaction time of *10 msec* and an alarm resolution time of better than *100 µsec*. In a 100 Mbit/sec Ethernet system, these periodic alarm messages will generate a (background) system load of less than 1% of the network capacity. Such an alarm reporting system will not cause any increase in load if all *1000* alarms occur at the same instant. If, in an event-triggered system, a *100* byte Ethernet message is sent whenever an alarm occurs, then the peak-load of *1000* alarm messages will generate a load of 10 % of the network capacity and a worst-case reaction time of *100 msec*.

9.3.2   *Time as Control*

Time as control is more difficult to handle than *time as data*, because it may sometimes require a dynamic data-dependent modification of the temporal control structure. It is prudent to scrutinize the application requirements carefully to identify those cases where such a dynamic rescheduling of the tasks is absolutely necessary.

If an event requires immediate action, the worst-case delay of the message transmission is a critical parameter. In an event-triggered protocol such as CAN, the message priorities are used to resolve access conflicts to the common bus that result from nearly simultaneous events. The worst-case delay of a particular message can be calculated by taking the peak-load activation pattern of the message system into account [Tin95].

> **Example**: The prompt reaction to an emergency shutdown request requires time to act as control. Assume that the emergency message is the highest priority message in a CAN system. In a CAN system, the worst-case delay of the highest priority message is bounded by the transmission duration of the longest message (which is about 100 bits), because a message transmission cannot be preempted.

9.4   Inter-Task Interactions

Inter-task interactions are needed to exchange data among concurrently executing tasks inside a component such that progress toward the common goal can be achieved. There are two principal means to exchange data among a set of concurrently executing tasks: (i) by the exchange of messages and (ii) by providing a shared region of data that can be accessed by more than one task.

Within a component, shared data structures are widely used since this form of inter-task interaction can be implemented efficiently in a single component where the tasks cooperate. However, care must be taken that the *integrity of data* that is read or written concurrently by more than one task is maintained. Figure 9.3 depicts the problem. Two tasks, T1 and T2, access the same critical region of data. We call the interval during the program execution during which the critical region of data is accessed in the *critical section* of a task. If the critical sections of tasks overlap, bad things may occur. If the shared data is read by one task while it is modified by another task, then the reader may read inconsistent data. If the critical sections of two or more writing tasks overlap, the data may be corrupted.

The following three techniques can be applied to solve the problem:

(i)   Coordinated task schedules.
(ii) The non-blocking write protocol.
(iii) Semaphore operations.

9.4.1   Coordinated Static Schedules

In a time-triggered system, the task schedules can be constructed in such a way that critical sections of tasks do not overlap. This is a very effective way to solve the problem, because.

(i) The overhead of guaranteeing mutual exclusion is minimal and predictable.
(ii) The solution is deterministic.

Wherever possible, this solution should be selected.

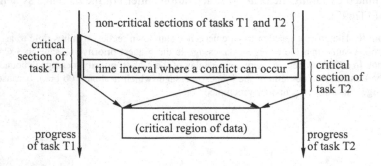

Fig. 9.3 Critical task sections and critical data regions

9.4.2   The Non-blocking Write (NBW) Protocol

If, however, the tasks with the critical sections are event-triggered, we cannot design conflict-free coordinated task schedules a priori. The *non-blocking write* (NBW) protocol is an example for a lock-free real-time protocol [Kop93a] that ensures data integrity of one or more readers if only a single task is writing into the critical region of data.

Let us analyze the operation of the NBW for the data transfer across the interface from the communication system to the host computer. At this interface, there is one writer, the communication system, and many readers, the tasks of the component. A reader does not destroy the information written by a writer, but a writer can interfere with the operation of the reader. In the NBW protocol, the real-time writer is never blocked. It will thus write a new version of the message into the critical data region whenever a new message arrives. If a reader reads the message while the writer is writing a new version, the retrieved message will contain inconsistent information and must be discarded. If the reader is able to detect the interference, then the reader can retry the read operation until it retrieves a consistent version of the data. It must be shown that the number of retries performed by the reader is bounded.

The protocol requires a concurrency control field, CCF, for every critical data region. Atomic access to the CCF must be guaranteed by the hardware. The concurrency control field is initialized to zero and incremented by the writer before the start of the write operation. It is again incremented by the writer after the completion of the write operation. The reader starts by reading the CCF at the start of the read operation. If the CCF is odd, then the reader retries immediately because a write operation is in progress. At the end of the read operation, the reader checks whether the writer has changed the CCF during the read operation. If so, it retries the read operation again until it can read an uncorrupted version of the data structure (see Fig. 9.4).

It can be shown that an upper bound for the number of read retries exists if the time between write operations is significantly longer than the duration of a write or read operation. The worst-case extension of the execution time of a typical real-time task caused by the retries of the reader is only a few percent of the original worst-case execution time (WCET) of the task [Kop93a].

initialization: CCF := 0;

writer:
start: CCF_old := CCF;
     CCF := CCF_old + 1;
     <write to data structure>
     CCF := CCF_old + 2;

reader:
start: CCF_begin := CCF;
     if CCF_begin = *odd*
     then goto *start*;
     <read data structure>
     CCF_end := CCF;
     if CCF_end ≠ CCF_begin
     then goto *start*;

Fig. 9.4 The non-blocking write (NBW) protocol

Non-locking synchronization has been implemented in other real-time systems, e.g., in a multimedia system [And95]. It has been shown that systems with non-locking synchronization achieve better performance than systems that lock the data.

9.4.3   Semaphore Operations

The *classic* mechanism to avoid data inconsistency is to enforce mutual exclusive execution of the critical task sections by a WAIT operation on a semaphore variable that protects the resource. Whenever one task is in its critical section, the other task must wait in a queue until the critical section is freed (*explicit synchronization*).

The implementation of a semaphore-initialize operation is expensive, both regarding memory requirements and operating system processing overhead. If a process runs into a blocked semaphore, a context switch must be made. The process is put into a queue and is delayed until the other process finishes its critical section. Then, the process is dequeued and another context switch is made to reestablish the original context. If the critical region is very small (this is the case in many real-time applications), the processing time for the semaphore operations can take *significantly* longer than the actual reading or writing of the common data.

Both the NBW protocol and semaphore operation can lead to a loss of replica determinism. The simultaneous access to CCF or a semaphore variable leads to a *race condition* that is resolved in an unpredictable manner in the replicas.

9.5   Process Input/Output

A *transducer* is a device that forms the interface between the plant (the physical world) and the computer (the cyber world). On the input side, *a sensor* transforms a mechanical or electrical quantity to a digital form, whereby the discreteness of the digital representation leads to an unavoidable error if the domain of the physical quantity is dense. The last bit of any digital representation of an analog quantity (both in the domain of value and time) is *non-predictable*, leading to potential inconsistencies in the cyber world representation if the same quantity is observed by two independent sensors. On the output side, a digital value is transformed to an appropriate physical signal by an *actuator*.

9.5.1   Analog Input/Output

In the first step, many sensors of analog physical quantities produce analog signals in the standard *4–20 mA* range (*4 mA* meaning *0%* of the value range and *20 mA* meaning *100%* of the value range) that is then transformed to its digital form by an analog-to-digital (AD) converter. If a measured value is encoded in the *4–20 mA* range, it is possible to distinguish a broken wire, where no current flows (*0 mA*), from a measured value of *0% (4 mA)*.

Without special care, the electric-noise level limits the accuracy of any analog control signal to about 0.1%. Analog-to-digital (AD) converters with a resolution of more than 10 bits require a carefully controlled physical environment that is not available in typical industrial applications. A 16-bit word length is thus more than sufficient to encode the value of an RT entity measured by an analog sensor.

The time interval between the occurrence of a value in the RT entity and the presentation of this value by the sensor at the sensor/computer interface is determined by the transfer function of the particular sensor. The step response of a sensor (see Fig. 1.4), denoting the *lag time* and the *rise time* of the sensor, gives an approximation of this transfer function. When reasoning about the temporal accuracy of a sensor/actuator signal, the parameters of the transfer functions of the sensors and the actuators must be considered (Fig. 9.5). They reduce the available time interval between the occurrence of a value at the RT entity and the use of this value for an output action by the computer. Transducers with short *lag times* increase the length of the temporal accuracy interval that is available to the computer system.

In many control applications, the instant when an analog physical quantity is observed (sampled) is in the sphere of control of the computer system. In order to reduce the dead time of a control loop, the instant of sampling, the transmission of the sampled data to the control node, and the transmission of the set point data to the actuator node should be *phase-aligned* (see Sect. 3.3.4).

9.5.2 Digital Input/Output

A digital I/O signal transits between the two states TRUE and FALSE. In many applications, the length of the time interval between two state changes is of semantic significance. In other applications, the moment when the transition occurs is important.

If the input signal originates from a simple mechanical switch, the new stable state is not reached immediately but only after a number of random oscillations (Fig. 9.6), called the *contact bounce*, caused by the mechanical vibrations of the switch contacts. This contact bounce must be eliminated either by an analog low-pass filter or, more often, within the computer system by software tasks, e.g., debouncing routines. Due to the low price of a microcontroller, it is cheaper to

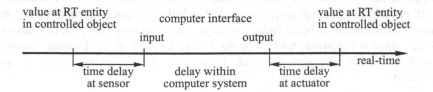

Fig. 9.5 Time delay of a complete I/O transaction

Fig. 9.6 Contact bounce of a mechanical switch

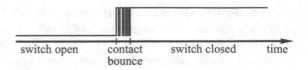

switch open   contact   switch closed   time
                    bounce

debounce a signal by software techniques than by hardware mechanisms (e.g., a low pass filter).

A number of sensor devices generate a sequence of pulse inputs, where each pulse carries information about the occurrence of an event. For example, distance measurements are often made by a wheel rolling along the object that must be measured. Every rotation of the wheel generates a defined number of pulses that can be converted to the distance traveled. The frequency of the pulses is an indication of the speed. If the wheel travels past a defined *calibration point*, an additional digital input is signaled to the computer to set the pulse counter to a defined value. It is good practice to convert the relative event values to absolute state values as soon as possible.

Time Encoded Signals Many output devices, e.g., power semiconductors such as IGBTs (insulated-gate-bipolar transistors), are controlled by pulse sequences of well-specified shape (pulse width modulation—PWM). A number of microcontrollers designed for I/O provide special hardware support for generating these digital pulse shapes.

9.5.3 Interrupts

The interrupt mechanism empowers a device outside the sphere of control of the computer to govern the temporal control pattern inside the computer. This is a powerful and potentially dangerous mechanism that must be used with great care. Interrupts are needed when an external event requires a reaction time from the computer (*time as control*) that cannot be implemented efficiently with a trigger task.

A trigger task extends the response time of an RT transaction that is initiated by an external event by at most one period of the trigger task. Increasing the trigger-task frequency can reduce this additional delay at the expense of an increased overhead. [Pol95b] has analyzed this increase in the overhead for the periodic execution of a trigger task as the required response time approaches the WCET of the trigger task. As a rule of thumb, only if the required response time is less than ten times the WCET of the trigger task, the implementation of an interrupt should be considered.

If information about the precise instant of arrival of a message is required, but no immediate action has to be taken, an interrupt-controlled time-stamping mechanism implemented in hardware should be used. Such a mechanism works autonomously and does not interfere with the control structure of tasks at the operating system level.

Example: In the hardware implementation of the IEEE 1588 clock synchronization protocol, a hardware mechanism autonomously generates the time-stamp of an arriving synchronization message [Eid06].

In an interrupt-driven software system, a transient error on the interrupt line may upset the temporal control pattern of the complete node and may cause the violation of important deadlines. Therefore, the time interval between the occurrence of any two interrupts must be continuously monitored and compared to the specified minimum duration between interrupting events.

Monitoring the Occurrence of an Interrupt   There are three tasks in the computer associated with every monitored interrupt [Pol95b] (Fig. 9.7). The first and second one are dynamically planned TT tasks that determine the interrupt window. The first one enables the interrupt line and thus opens the time window during which an interrupt is allowed to occur. The third task is the interrupt service task that is activated by the interrupt. Whenever the interrupt has occurred, the interrupt service task closes the time window by disabling the interrupt line. It then deactivates the scheduled future activation of the second task. In case the third task was not activated before the start of the second task, the second task, a dynamic TT task scheduled at the end of the time window, closes the time window by disabling the interrupt line. The second task then generates an error flag to inform the application of the missing interrupt.

The two time-triggered tasks are needed for error detection. The first task detects a sporadic interrupt that should not have occurred. The second task detects a missing interrupt that should have occurred. These different errors require different types of error handling. The more we know about the regularity of the controlled object, the smaller we can make the time window in which an interrupt may occur. This leads to better error detection coverage.

Example: An engine controller of an automotive engine has such a stringent requirement regarding the point of fuel injection relative to the position of the piston in the cylinder that the implementation must use an interrupt for measuring the position [Pol95b]. The position of the piston and the rotational speed of the crankshaft are measured by a number of sensors that generate rising edges whenever a defined section of the crankshaft passes the position of the sensor. Since the speed and the maximum angular acceleration (or deceleration) of the engine is known, the next correct interrupt must arrive within a small dynamically defined time window from the previous interrupt. The interrupt logic is only enabled during this short window and disabled at all other times to reduce the impact of sporadic interrupts

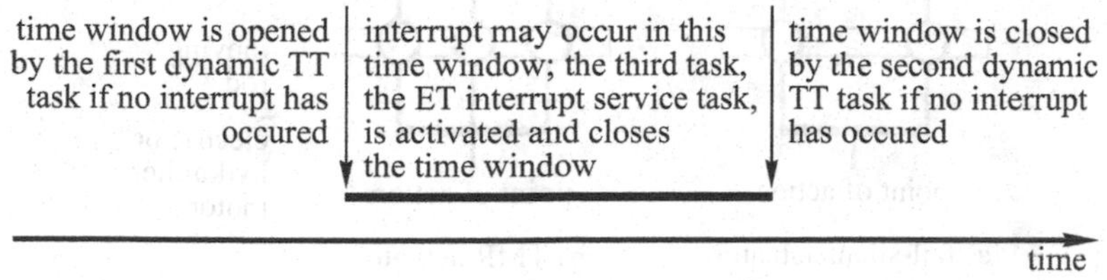

time window is opened by the first dynamic TT task if no interrupt has occured	interrupt may occur in this time window; the third task, the ET interrupt service task, is activated and closes the time window	time window is closed by the second dynamic TT task if no interrupt has occured

time →

Fig. 9.7 Time window of an interrupt

on the temporal control pattern within the host software. Such a sporadic interrupt, if not detected, may cause a mechanical damage to the engine.

9.5.4   Fault-Tolerant Actuators

An actuator must transform the signal generated at the output interface of the computer into some physical action in the controlled object (e.g., opening of a valve). The actuators form the last element in the chain between sensing the values of an RT entity and realizing the intended effect in the environment. In a fault-tolerant system, the actuators must perform the *final voting* on the output signals received on the replicated channels. Figure 9.8 shows an example where the intended action in the environment is the positioning of a mechanical lever. At the end of the lever, there may be any mechanical device that acts on the controlled object, e.g., there may be a piston of a control valve mounted at the point of action.

In a replica-determinate architecture, the correct replicated channels produce identical results in the value and in the time domains. We differentiate between the cases where the architecture supports the fail-silent property (Fig. 9.8a), i.e., all failed channels are silent, and where the fail-silence property is not supported (Fig. 9.8b), i.e., a failed channel can show an arbitrary behavior in the value domain.

Fail-Silent Actuator   In a fail-silent architecture, all subsystems must support the fail-silence property. A fail-silent actuator will either produce the intended (correct) output action or no result at all. In case a fail-silent actuator fails to produce an output action, it may not hinder the activity of the replicated fail-silent actuator. The fail-silent actuator of Fig. 9.8a consists of two motors where each one has enough power to move the point of action. Each motor is connected to one of the two replica-determinate output channels of the computer system. If one motor fails at any location, the other motor is still capable to move the point of action to the desired position.

Triple Modular Redundant Actuator   The triple modular redundant (TMR) actuator (Fig. 9.8b) consists of three motors, each one connected to one of the three replica-determinate output channels of the fault-tolerant computer. The force of any

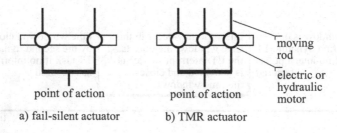

       point of action   point of action

    a) fail-silent actuator   b) TMR actuator

Fig. 9.8   Fault-tolerant actuators

two motors must be strong enough to override the force of the third motor; however, any single motor may not be strong enough to override the other two. The TMR actuator can be viewed as a *mechanical* voter that will place the point of action into a position that is determined by the majority of the three channels, outvoting the disagreeing channel.

Actuator with a Dedicated Stateless Voter In many applications where redundant actuators are already in place, a voting actuator can be constructed by combining the physical actuator with a small microcontroller that accepts the three input channels from the three lanes of a TMR system and votes on the messages received from the three lanes. This voter can be stateless, i.e., after every cycle the circuitry of the voter is reset in order to eliminate the accumulation of state errors caused by transient faults (Fig. 9.9).

> **Example**: In a car, a stateless voter can be placed at the brake actuator at each one of the four wheels. The voter will mask the failure in any one of the TMR channels. A stateless voter is an example for an intelligent instrumentation.

9.5.5 Intelligent Instrumentation

There is an increasing tendency to encapsulate a sensor/actuator and the associated microcontroller into a single physical housing to provide a standard abstract message interface to the outside world that produces *measured values* at a field bus, e.g., a CAN bus (Fig. 9.10). Such a unit is called an *intelligent instrument*.

The intelligent instrument hides the concrete sensor interface. Its single chip microcontroller provides the required control signals to the sensor/actuator, performs signal conditioning, signals smoothing and local error detection, and presents/takes a meaningful RT image in standard measuring units to/from the field bus message interface. Intelligent instruments simplify the connection of the plant equipment to the computer.

> **Example**: A MEMS acceleration sensor, micro machined into silicon, mounted with the appropriate microcontroller and network interface into a single package, forms an intelligent sensor.

To make the measured value fault-tolerant, a number of independent sensors can be packed into a single intelligent instrument. Inside the intelligent instrument, an agreement protocol is executed to arrive at an agreed sensor value, even if one of the

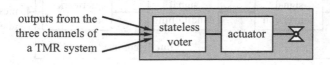

outputs from the
three channels of → stateless voter — actuator ⧗
a TMR system

Fig. 9.9 Stateless voter associated with an actuator

sensors has failed. This approach assumes that independent measurements can be taken in close spatial vicinity.

The integration of a field bus node with an actuator produces an intelligent actuator device.

> **Example**: An actuator of an airbag in an automobile must ignite an explosive charge to release the gas of a high-pressure container into the airbag at the appropriate moment. A small explosive charge, placed directly on the silicon of a microcontroller, can be ignited on-chip. The package is mounted at the proper mechanical position to open the critical valve. The microcontroller including the explosive charge forms an intelligent actuator.

Because many different field bus designs are available today, and no generally accepted industry wide field bus standard has emerged, the sensor manufacturer must cope with the dilemma to provide a different intelligent instrument network interface for different field buses.

9.5.6   Physical Installation

It is beyond the scope of this book to cover all the issues that must be considered in the physical installation of a sensor-based real-time control system. These complex topics are covered in books on computer hardware installation. However, a few critical issues are highlighted.

Power Supply Many computer failures are caused by power failures, i.e., long power outages, short power outages of less than a second also called sags, and power surges (short overvoltage). The provision of a reliable and clean power source is thus of crucial importance for the proper operation of any computer system.

Grounding The design of a proper grounding system in an industrial plant is a major task that requires considerable experience. Many transient computer hardware failures are caused by deficient grounding systems. It is important to connect all units in a tree-like manner to a high quality true ground point. Loops in the ground circuitry must be avoided because they pick up electromagnetic disturbances.

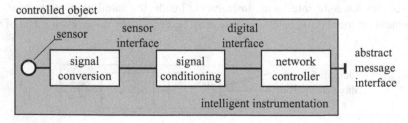

Fig. 9.10 Intelligent instrumentation

Electric Isolation   In many applications, complete electric isolation of the computer terminals from the signals in the plant is needed. Such isolation can be achieved by opto couplers for digital signals or signal transformers for analog signals.

9.6   Agreement Protocols

Sensors and actuators have failure rates that are considerably higher than those of single-chip microcomputers. No critical output action should rely on the input from a single sensor. It is necessary to observe the controlled object by a number of different sensors and to relate these observations to detect erroneous sensor values, to observe the effects of actuators, and to get an *agreed* image of the physical state of the controlled object. In a distributed system, *agreement* (also called *consensus* in [Bar93]) typically requires an information exchange among the agreeing partners. However, some information exchange can also be avoided by an appropriate system architecture.

> **Example:** The fail-operational architecture of [Kop21] discussed in the example of Sect. 6.5.1 avoids information exchange between the actuators in self-driving cars. This is achieved by a fault-tolerant decision subsystem (FTDSS). The FTDSS is composed of two fault-containment units (FCUs) that operate as intermediate stations in between the FCUs that generate the driving trajectories for the car and the actuators that use the trajectories. Thus, the required multiple rounds of communication (see Sect. 6.4.2, *Byzantine Resilient Fault-Tolerant Unit*) to achieve consensus are replaced by *multiple stages* of communication, a concept introduced by Miner [Min02].

In general, the number of rounds (or stages) of information exchange needed depends on the type of agreement and the assumptions about the possible failure modes of the partners.

9.6.1   Raw Data, Measured Data, and Agreed Data

In Sect. 1.2.1, the concepts of raw data, measured data, and agreed data have been introduced: *raw data* are produced at the digital hardware interface of the physical sensor. *Measured data*, presented in standard engineering units, are derived from one or a sequence of raw data samples by the process of *signal conditioning*. Measured data that are judged to be a correct image of the RT entity, e.g., after the comparison with other measured data elements that have been derived by diverse techniques, are called *agreed data*. Agreed data form the inputs to control actions. In a safety-critical system where no single point of failure is allowed to exist, an agreed data element may not originate from a *single* sensor. The challenge in the development of a safety-critical input system is the selection and placement of the redundant sensors and the design of the agreement algorithms. We distinguish two types of agreement, *syntactic agreement* and *semantic agreement*.

9.6.2 Syntactic Agreement

Assume that two independent sensors measure a single RT entity. When the two observations are transformed from the domain of analog values to the domain of discrete values, a slight difference between the two raw values caused by a measurement error and digitalization error is unavoidable. These different raw data values will cause different measured values. A digitalization error also occurs in the time domain when the time of occurrence of an event in the controlled object is mapped into the discrete time of the computer. Even in the fault-free case, these different measured values must be reconciled in some way to present an agreed view of the RT entity to the possibly replicated control tasks. In syntactic agreement, the agreement algorithm computes the agreed value without considering the context of the measured values. For example, the agreement algorithm always takes the average of a set of measured data values. If a Byzantine failure of one of the sensors must be tolerated, three additional sensors are needed (see Sect. 6.4.2).

9.6.3 Semantic Agreement

If the meanings of the different measured values are related to each other by a process model based on a priori knowledge about the relationships and the physical characteristics of the process parameters of the controlled object, we speak of *semantic agreement*. In semantic agreement, it is not necessary to duplicate or triplicate every sensor. Different redundant sensors observe different RT entities. A model of the physical process relates these redundant sensor readings to each other to find a set of plausible agreed values and to identify implausible values that indicate a sensor failure. Such an erroneous sensor value must be replaced by a calculated estimate of the most probable value at the given point in time, based on the inherent semantic redundancy in the set of measurements.

> **Example**: A number of *laws of nature* govern a chemical process: the conservation of mass, the conservation of energy, and some known maximum speed of the chemical reaction. These fundamental laws of nature can be applied to check the plausibility of the measured data set. In case one sensor reading deviates significantly from all other sensors, a sensor failure is assumed and the failed value is replaced by an estimate of the correct value at this instant, to be able to proceed with the control of the chemical process.

Semantic agreement requires a fundamental understanding of the applied process technology. It is common that an interdisciplinary team composed of process technologists, measurement specialists, and computer engineers cooperates to find the RT entities that can be measured with good precision at a reasonable cost. Typically, for every output value, about *three to seven* input values must be observed, not only to be able to diagnose erroneous measured data elements, but also to check the proper operation of the actuators. Independent sensors that observe the intended effect of the actuator (see Sect. 7.2.5.1) must monitor the proper operation of every actuator.

In engineering practice, semantic agreement of measured data values is more important than syntactic agreement. As a result of the agreement phase, an agreed (and consistent) set of digital input values is produced. These agreed values, defined in the value domain and in the time domain, are then used by all (replicated) tasks to achieve a replica-determinate behavior of the control system.

9.7 Error Detection

A real-time operating system must support error detection in the temporal domain and error detection in the value domain by generic methods. Some of these generic methods are described in this section.

9.7.1 Monitoring Task Execution Times

A tight upper bound on the worst-case execution time (WCET) of a real-time task must be established during software development (see Sect. 10.2). This WCET must be monitored by the operating system at run time to detect transient or permanent hardware errors. In case a task does not terminate its operation within the WCET, the execution of the task is terminated by the operating system. It is up to the application to specify which action should be taken in case of an error.

9.7.2 Monitoring Interrupts

An erroneous external interrupt has the potential to disrupt the temporal control structure of the real-time software within the node. At design time, the minimum inter-arrival periods of interrupts must be known to be able to estimate the peak load that must be handled by the software system. At run time, this minimum inter-arrival period must be enforced by the operating system by disabling the interrupt line to reduce the probability of erroneous sporadic interrupts (see Sect. 9.5.3).

9.7.3 Double Execution of Tasks

Fault-injection experiments have shown that the double execution of tasks and the subsequent comparison of the results is a very effective method for the detection of transient hardware faults that cause undetected errors in the value domain [Arl03].

The operating system can provide the execution environment for the double execution of application tasks without demanding any changes to the application task per se. It is thus possible to decide at the time of system configuration which tasks should be executed twice and for which tasks it is sufficient to rely on a single execution.

9.7.4   Watchdogs

A fail-silent node will produce correct results or no results at all. The failure of a fail-silent node can only be detected in the temporal domain. A standard technique is the provision of a watchdog signal (*heart-beat*) that must be periodically produced by the operating system of the node. If the node has access to the global time, the watchdog signal should be produced periodically at known absolute points in time. An outside observer can detect the failure of the node as soon as the watchdog signal disappears.

A more sophisticated error detection mechanism that also covers part of the value domain is the periodic execution of a *challenge-response protocol* by a node. An outside error detector provides an input pattern to the node and expects a defined response pattern within a specified time interval. The calculation of this response pattern should involve as many functional units of the node as possible. If the calculated response pattern deviates from the a priori known correct result, an error of the node is detected.

Points to Remember

- We distinguish two levels of system administration in a component-based distributed real-time system: (i) the coordination of the message-based communication and resource allocation among the components and (ii) the establishment, coordination of, and control of the concurrent tasks within each one of the components.
- Since the component software is assumed to be a *unit of design*, and a *whole component* is the smallest unit of fault containment, the concurrent tasks within a component are *cooperative* and not *competitive*.
- In an entirely time-triggered system, the static temporal control structure of all tasks is established a priori by off-line scheduling tools. This temporal control structure is encoded in a *Task-Descriptor List (TADL)* that contains the cyclic schedule for all activities of the node. This schedule considers the required precedence and mutual exclusion relationships among the tasks such that an explicit coordination of the tasks by the operating system at run time is not necessary.

- In a RT operating system that supports task preemption, each occurrence of a significant event can potentially activate a new task and cause an immediate interruption of the currently executing task. If components are replicated, care must be taken that the preemption points at all replicas is at the *same statement*; otherwise replica determinism may be lost.
- The timing analysis of a C-task is not a local issue of a single task anymore; it becomes a global system issue. In the general case, it is impossible to give an upper bound for the WCET of a C-task.
- It is important to distinguish *time as data* and *time as control*. *Time as control* is more difficult to handle than *time as data*, because it may sometimes require a dynamic data-dependent modification of the temporal control structure.
- Care must be taken that the integrity of data that is read or written concurrently by more than one task is maintained. In a time-triggered system, the task schedules can be constructed in such a way that critical sections of tasks do not overlap.
- In order to reduce the *dead time* of a control loop, the instant of sampling, the transmission of the sampled data to the control node, and the transmission of the set point data to the actuator node should be *phase-aligned* in a time-triggered system.
- In an interrupt-driven software system, a transient error on the interrupt line may upset the temporal control pattern of the complete node and may cause the violation of important deadlines.
- A voting actuator may be constructed by assigning a small microcontroller to the physical actuator that accepts the three input channels of the three lanes of a TMR system and votes on the messages received from the three lanes.
- Typically, for every output value, about *three to seven* input values must be observed, not only to be able to diagnose erroneous measured data elements, but also to check the proper operation of the actuators.

Bibliographic Notes

Many of the standard textbooks on operating systems contain sections on real-time operating systems, e.g., the textbook by Stallings [Sta18]. The ARINC 653 standard is used for operating systems in safety-critical avionics systems. AUTOSAR is an operating system for automotive applications, including safety-critical ones. Some commercial POSIX™ certified or mostly POSIX™ compliant real-time operating systems are used for safety-critical systems. The open-source operating system Linux is also mostly POSIX™ compliant. There are ongoing efforts toward a safety-critical Linux [All21]. However, due to its SW complexity, it is unclear if these efforts will be successful in the future. On the other hand, several Linux-based operating systems are used for non-safety-critical soft real-time systems. The most recent research contributions on real-time operating systems can be found in the annual Proceedings of *the IEEE Real-Time System Symposium* and the Journal *Real-Time Systems* from Springer Verlag.

Review Questions and Problems

9.1 Explain the difference between a standard operating system for a personal computer and an RT operating system within the node of a safety-critical real-time application!

9.2 What is meant by a *simple task*, a *trigger task*, and a *complex task*?

9.3 What is the difference between *time as data* and ti*me as control*?

9.4 Why is the classical mechanism of *semaphore operations* sub-optimal for the protection of critical data in a real-time OS? What alternatives are available?

9.5 How is *contact bounce* eliminated?

9.6 When do we need interrupts? What is the effect of spurious interrupts? How can we protect the software from spurious interrupts?

9.7 A node of an alarm monitoring system must monitor 50 alarms. The alarms must be reported to the rest of the cluster within 10 msec by a 100kbit/second CAN bus. Sketch an implementation that uses periodic CAN messages (time-triggered with a cycle of 10 msec) and an implementation that uses sporadic event-triggered messages, one for every occurring alarm. Compare the implementations from these points of view: generated load under the conditions of no alarm and all alarms occurring simultaneously, guaranteed response time, and detection of a crash failure of the alarm node.

9.8 Let us assume that an actuator has a failure rate of 10^6 FITs. If we construct a voting actuator by adding a microcontroller with a failure rate of 10^4 FITs to this actuator, what is the resultant failure rate of the voting actuator?

9.9 What is the difference between *raw data*, *measured data*, and *agreed data*?

9.10 What is the difference between *syntactic agreement* and *semantic agreement*? Which technique is more important in the design of real-time applications?

9.11 List some of the generic error detection techniques that should be supported by a real-time OS!

9.12 Which types of failures can be detected by the double execution of tasks?

Chapter 10
Real-Time Scheduling

Overview

Many thousands of research papers have been written about how to schedule a set of tasks in a system with a limited amount of resources such that all tasks will meet their deadlines. This chapter tries to summarize some important results of scheduling research that are relevant to the designer of real-time systems. The chapter starts by introducing the notion of a schedulability test to determine whether a given task set is schedulable or not. It distinguishes between a sufficient, an exact, and a necessary schedulability test. A scheduling algorithm is *optimal* if it will find a schedule whenever there is a solution. The adversary argument shows that generally it is not possible to design an optimal online scheduling algorithm. A prerequisite for the application of any scheduling technique is knowledge about the worst-case execution time (WCET) of all time-critical tasks. Section 10.2 presents techniques to estimate the WCET of simple tasks and complex tasks. Modern processors with pipelines and caches make it difficult to arrive at tight bounds for the WCET. *Anytime algorithms* that contain a root segment that provides a result of sufficient (but low) quality and an optional periodic segment that improves on the quality of the previous result point to a way out of this dilemma. They use the interval between the actual execution time of the root segment of a concrete task execution and the deadline, i.e., the worst execution time of the root segment, to improve the quality of the result. Section 10.3 covers the topic of static scheduling. The concept of the schedule period is introduced and an example of a simple search tree that covers a schedule period is given. A heuristic algorithm has to examine the search tree to find a feasible schedule. If it finds one, the solution can be considered a constructive schedulability test. Section 10.4 elaborates on dynamic scheduling. It starts by looking at the problem of scheduling a set of independent tasks by the rate-monotonic algorithm. Next, the problem of scheduling a set of dependent tasks is investigated. The priority ceiling protocol is introduced and a schedulability test for the priority ceiling protocol is sketched. Finally, the scheduling problem in distributed systems

H. Kopetz, W. Steiner, *Real-Time Systems*,
https://doi.org/10.1007/978-3-031-11992-7_10

is touched and some ideas about alternative scheduling strategies such as feedback scheduling are given.

10.1   The Scheduling Problem

A hard real-time system must execute a set of concurrent real-time tasks in such a way that all time-critical tasks meet their specified deadlines. Every task needs computational, data, and other resources (e.g., input/output devices) to proceed. The scheduling problem is concerned with the allocation of these resources to satisfy all timing requirements.

10.1.1   Classification of Scheduling Algorithms

The following diagram (Fig. 10.1) presents a taxonomy of real-time scheduling algorithms [Che87].

Static Scheduling A scheduler is called static (or pre-run time) if it makes its scheduling decisions prior to run time. It generates a dispatching table for the run time dispatcher off-line. For this purpose, it needs complete prior knowledge about the task set characteristics, e.g., maximum execution times, precedence constraints, mutual exclusion constraints, and deadlines. The dispatching table (see Fig. 9.2) contains all information the dispatcher needs at run time to decide at every point of the sparse time base which task is to be scheduled next. The run time overhead of the dispatcher is small. The system behavior is deterministic.

A scheduler is called *dynamic* (or *online*) if it makes its scheduling decisions at run time, selecting one out of the current set of *ready* tasks. Dynamic schedulers are flexible and adapt to an evolving task scenario. They consider only the *current* task requests. The run time effort involved in finding a schedule can be substantial. In general, the system behavior is non-deterministic.

Static Versus Dynamic Scheduling At its core, scheduling is a *resource allocation problem in time and space*, where tasks have demands for computing and communication resources. This resource allocation problem often becomes computation-

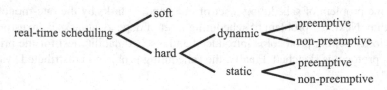

Fig. 10.1 Taxonomy of real-time scheduling algorithms

ally complex in industrial real-time systems, even NP-complete. Besides the operational differences discussed above, there is also a fundamental difference in how this computational complexity is handled by static and dynamic scheduling. The static scheduling strategy treats the resource allocation problem as a *search problem* for valid configurations (see Sect. 10.3.1), e.g., finding a valid dispatching table. Thus, the computationally intensive part of problem-solving is handled entirely at design time. The typical strategy of dynamic scheduling, on the other hand, is to put assumptions on the task set's characteristics, i.e., to *reduce the problem*, thereby *reducing* the necessity for intensive computation or avoiding it entirely. One typical reduction technique is the assumption of task independence (see Rate-Monotonic Scheduling in Sect. 10.4.1). Another reduction technique is limiting the resource utilization but allowing certain task dependencies (see the Priority Ceiling Protocol in Sect. 10.4.2). Many industrial real-time systems may not allow adequate reduction techniques or impose even stronger forms of dependencies between tasks (e.g., phase-aligned transactions, see Sect. 10.5.1) and may only be realized efficiently by static scheduling.

Non-preemptive and Preemptive Scheduling In non-preemptive scheduling, the currently executing task will not be interrupted until it decides on its own to release the allocated resources. Non-preemptive scheduling is reasonable in a task scenario where many short tasks (compared to the time it takes for a context switch) must be executed. In preemptive scheduling, the currently executing task may be preempted, i.e., interrupted, if a more urgent task requests service.

Centralized Versus Distributed Scheduling In a dynamic distributed real-time system, it is possible to make all scheduling decisions at one central site or to develop cooperative distributed algorithms for the solution of the scheduling problem. The central scheduler in a distributed system is a critical point of failure. Because it requires up-to-date information on the load situations of all nodes, it can also contribute to a communication bottleneck.

10.1.2 Schedulability Test

A test that determines whether a set of ready tasks can be scheduled such that each task meets its deadline is called a *schedulability test*. We distinguish between *exact*, *necessary*, and *sufficient* schedulability tests (Fig. 10.2).

A scheduler is called *optimal* if it will always find a feasible schedule whenever it exists. Alternatively, a scheduler is called *optimal*, if it can find a schedule whenever a clairvoyant scheduler, i.e., a scheduler with complete knowledge of the future request times, can find a schedule. Garey and Johnson [Gar75] have shown that in nearly all cases of task dependency, even if there is only one common resource, the complexity of an exact schedulability test algorithm belongs to the class of NP-complete problems and is thus computationally intractable. *Sufficient*

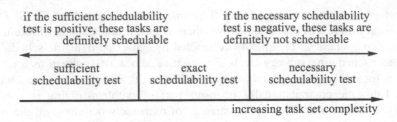

Fig. 10.2 Necessary and sufficient schedulability test

schedulability test algorithms can be simpler at the expense of giving a negative result for task sets that are, in fact, schedulable. A task set is definitely not schedulable if a *necessary schedulability* test gives a negative result. If a necessary schedulability test gives a positive result, there is still a probability that the task set may not be schedulable. The *task request time* is the instant when a request for a task execution is made. Based on the request times, it is useful to distinguish between two different task types: *periodic* and *sporadic* tasks. This distinction is important from the point of view of schedulability.

If we start with an initial request, all future request times of a periodic task are known a priori by adding multiples of the known period to the initial request time. Let us assume that there is a task set $\{T_i\}$ of periodic tasks with periods p_i, deadline interval d_i, and execution time c_i. The deadline interval is the duration between the deadline of a task and the task request instant, i.e., the instant when a task becomes ready for execution. We call the difference $d_i - c_i$ the *laxity* l_i of a task. It is sufficient to examine schedules of length of the least common multiples of the periods of these tasks, the *schedule period*, to determine schedulability. A necessary schedulability test for a set of periodic tasks states that the sum of the utilization factors.

$$\mu = \Sigma c_i / p_i \leq n,$$

must be less or equal to n, where n is the number of available processors. This is evident because the utilization factor of task T_i, μ_i, denotes the percentage of time task T_i requires service from a processor.

The request times of *sporadic* tasks are not known a priori. To be schedulable, there must be a minimum interval between any two request times of sporadic tasks. Otherwise, the necessary schedulability test introduced above will fail. If there is no constraint on the request times of task activations, the task is called an *aperiodic* task.

10.1.3   The Adversary Argument

Let us assume that a real-time computer system contains a dynamic scheduler with full knowledge of the past but without any knowledge about future request times of tasks. Schedulability of the current task set may depend on when a sporadic task will request service in the future.

The *adversary argument* [Mok83, p.41] states that, in general, it is not possible to construct an optimal totally online dynamic scheduler if there are mutual exclusion constraints between a periodic and a sporadic task. The proof of the adversary argument is relatively simple.

Consider two mutually exclusive tasks, task *T1* is periodic and the other task *T2* is sporadic, with the parameters given in Fig. 10.3. The necessary schedulability test introduced above is satisfied, because

$$\mu = 2/4 + 1/4 = 3/4 \leq 1.$$

Whenever the periodic task is executing, an *adversary* requests service for the sporadic task. Due to the mutual exclusion constraint, the sporadic task must wait until the periodic task is finished. Since the sporadic task has a laxity of 0, it will miss its deadline.

The clairvoyant scheduler knows all the future request times of the sporadic tasks and at first schedules the sporadic task and thereafter the periodic task in the gap between two sporadic task activations (Fig. 10.3).

The adversary argument demonstrates the importance of information about the future behavior of tasks for solving the scheduling problem. If the online scheduler does not have any further knowledge about the request times of the sporadic task, the dynamic scheduling problem is not solvable, although the processor capacity is more than sufficient for the given task scenario. The design of predictable hard real-time systems is simplified if *regularity assumptions* about the future scheduling requests can be made. This is the case in cyclic systems that restrain the points in time at which external requests are recognized by the computing system.

10.2   Worst-Case Execution Time

A deadline for completing an RT transaction can only be guaranteed if the worst-case execution times (WCET) of all application tasks and communication actions that are part of the RT transaction are known a priori. The WCET of a task is a guaranteed upper bound for the time between task activation and task termination. It must be valid for all possible input data and execution scenarios of the task and should be a tight bound.

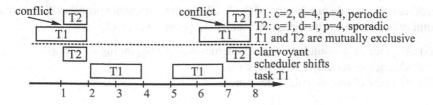

Fig. 10.3 The adversary argument

In addition to the knowledge about the WCET of the application tasks, we must find an upper bound for the delays caused by the administrative services of the operating system, the worst-case administrative overhead (WCAO). The WCAO includes all administrative delays that affect an application task but are not under the direct control of the application task (e.g., those caused by context switches, scheduling, cache reloading because of task preemption by interrupts or blocking, and direct memory access).

This section starts with an analysis of the WCET of a non-preemptive simple task. We then proceed to investigate the WCET of a preemptive simple task before looking at the WCET of complex tasks, and, finally, we discuss common techniques regarding the timing analysis of real-time programs.

10.2.1   WCET of Simple Tasks

The simplest task we can envision is a single sequential S-task that runs on dedicated hardware without preemption and without requiring any operating system services. The WCET of such a task depends on

 (i) The source code of the task.
 (ii) The properties of the object code generated by the compiler.
(iii) The characteristics of the target hardware.

In this section, we investigate the analytical construction of a tight worst-case execution time bound of such a simple task on hardware, where the execution time of an instruction is context independent.

Source Code Analysis   The first problem concerns the calculation of the WCET of a program written in a higher-level language, under the assumption that the maximum execution times of the basic language constructs are known and context independent. In general, the problem of determining the WCET of an arbitrary sequential program is unsolvable and is equivalent to the halting problem for Turing machines. Consider, for example, the simple statement that controls the entry to a loop:

  S: **while** (*exp*)
         do *loop*;

It is not possible to determine a priori after how many iterations, if at all, the Boolean expression *exp* will evaluate to the value FALSE and when statement S will terminate. For the determination of the WCET to be a tractable problem, there are a number of constraints that must be met by the program [Pus89]:

 (i) Absence of unbounded control statements at the beginning of a loop.
 (ii) Absence of recursive function calls.
(iii) Absence of dynamic data structures.

The WCET analysis concerns only the temporal properties of a program. The temporal characteristics of a program can be abstracted into a WCET bound for every program statement using the known WCET bound of the basic language constructs. For example, the WCET bound of a conditional statement

S: **if** (*exp*)
>> **then** S$_1$
>> **else** S$_2$;

can be abstracted as

$$T(S) = \max\left[T(\exp) + T(S_1), T(\exp) + T(S_2)\right]$$

where $T(S)$ is the maximum execution time of statement S, with $T(exp)$, $T(S_1)$, and $T(S_2)$ being the WCET bounds of the respective constructs. Such a formula for reasoning about the timing behavior of a program is called a *timing schema* [Sha89].

The WCET analysis of a program which is written in a high-level language must determine which program path, i.e., which sequence of instructions, will be executed in the worst-case scenario. The longest program path is called the *critical path*. Because the number of program paths normally grows exponentially with the program size, the search for the critical path can become intractable if the search is not properly guided and the search space is not reduced by excluding infeasible paths.

Compiler Analysis The next problem concerns the determination of the maximum execution time of the basic language constructs of the source language under the assumption that the maximum execution times of the machine language commands are known and context independent. For this purpose, the code generation strategy of the compiler must be analyzed, and the timing information that is available at the source code level must be mapped into the object code representation of the program such that an object-code timing analysis tool can make use of this information.

Execution Time Analysis The next problem concerns the determination of the worst-case execution time of the commands on the target hardware. If the processor of the target hardware has fixed instruction execution times, the duration of the hardware instructions can be found in the hardware documentation and can be retrieved by an elementary table look-up. Such a simple approach does not work if the target hardware is a modern RISC processor with pipelined execution units and instruction/data caches. While these architectural features result in significant performance improvements, they also introduce a high level of unpredictability. Dependencies among instructions can cause pipeline hazards, and cache misses will lead to a significant delay of the instruction execution. To make things worse, these two effects are not independent. A significant amount of research deals with the execution time analysis on machines with pipelines and caches. The excellent survey article by Wilhelm et al. [Wil08] presents WCET analysis in research and industry and describes many of the tools available for the support of WCET analysis.

Fig. 10.4 Worst-case
administrative overhead
(WCAO) of a task
preemption

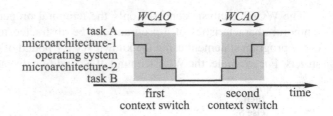

Preemptive S-Tasks If a simple task (S task) is preempted by another independent task, e.g., a higher priority task that must service a pending interrupt, the execution time of the S-task under consideration is extended by three terms:

  (i) The WCET of the interrupting task (task B in Fig. 10.4)
 (ii) The WCET of the operating system required for context switching
(iii) The time required for reloading the instruction cache and the data cache of the processor whenever the context of the processor is switched

We call the sum of the worst-case delays caused by the context switch (ii) and the cache reloading (iii) the Worst-Case Administrative Overhead (WCAO) of a task preemption. The WCAO is an unproductive administrative operating system overhead that is avoided if task preemption is forbidden.

The additional delay caused by the preemption of task A by task B is the WCET of the independent task B and the sum of the two WCAOs for the two context switches (shaded area in Fig. 10.4). The times spent in Microarchitecture-1 and Microarchitecture-2 are the delays caused by cache reloading. The Microarchitecture-2 time of the first context switch is part of the WCET of task B, because task B is assumed to start on an empty cache. The second context switch includes the cache reload time of task A, because in a non-preemptive system, this delay would not occur. In many applications with modern processors, the micro-architecture delays can be the significant terms that determine the cost of task pre-emption because the WCET of the interrupting task is normally quite short. The problem of WCAO analysis in operating systems is studied in [Lv09].

10.2.2   WCET of Complex Tasks

We now turn to the WCET analysis of a preemptive complex task (C-task) that accesses protected shared objects. The WCET of such a task depends not only on behavior of the task itself, but also on the behavior of other tasks and the operating system of the node. WCET analysis of a C-task is therefore not a local problem of a single task, but a global problem involving all the interacting tasks within a node.

In addition to the delays caused by the task preemption (which was analyzed in the previous section), an additional delay that originates from the direct interactions caused by the intended task dependencies (mutual exclusion, precedence) must be considered. There also exist techniques for coping with the direct interactions caused by the intended task dependencies—e.g., access to protected shared objects controlled by the priority ceiling protocol [Sha94]. This topic will be investigated in Sect. 10.4.2 on the scheduling of dependent tasks.

10.2.3   Anytime Algorithms

In practice, the time difference between the best-case execution time (BCET) and a guaranteed upper bound for the worst-case execution time (WCET) of a task can be substantial. *Anytime algorithms* are algorithms that use this time difference to improve the quality of the result as more execution time is provided [Chu90]. Anytime algorithms consist of a *root segment* that calculates a first approximation of the result of sufficient quality and a *periodic segment* that improves the quality of the previously calculated result. The periodic segment is executed repeatedly until the deadline, i.e., the guaranteed worst-case execution time of the root segment, is reached. Whenever the deadline occurs, the last version of the available result is delivered to the client. When scheduling an anytime algorithm, the completion of the root segment of the anytime algorithm must be guaranteed in order that a result of sufficient quality is available at this instant. The remaining time until the deadline is used to improve this result. The WCET problem of an anytime algorithm is thus reduced to finding a guaranteed upper bound for the WCET of the root segment. A loose upper bound of the WCET is of no serious concern, since the slack time between BCET and WCET is used to improve the result.

Most iterative algorithms are anytime algorithms. Anytime algorithms are used in pattern recognition, planning, and control.

An anytime algorithm should have the following properties [Zil96]:

(i) Measurable quality: It must be possible to measure the quality of a result.
(ii) Monotonic: The quality of the result must be a non-decreasing function of time and should improve with every iteration.
(iii) Diminishing returns: The improvement of the result should get smaller as the number of iterations increases.
(iv) Interruptability: After the completion of the root segment, the algorithm can be interrupted at any time and deliver a reasonable result.

10.2.4   State of Practice

The previous discussion shows that the analytic calculation of a reasonable upper WCET bound of an S-task which does not make use of operating system services is possible under restricting assumptions. There are a number of tools that support such an analysis [Wil08]. It requires an annotated source program that contains programmer-supplied application-specific information to ensure that the program terminates and a detailed model of the behavior of the hardware to achieve a reasonable upper WCET bound.

Bounds for the WCET of all time-critical tasks are *needed* in almost all hard real-time applications. This important problem is solved in practice by combining a number of diverse techniques:

  (i) Use of a restricted architecture that reduces the interactions among the tasks and facilitates the a priori analysis of the control structure. The number of explicit synchronization actions that require context switches and operating system services is minimized.
 (ii) The design of WCET models and the analytic analysis of sub-problems (e.g., the maximum execution time analysis of the source program) such that an effective set of test cases biased toward the worst-case execution time can be generated automatically.
(iii) The controlled measurement of sub-systems (tasks, operating system service times) to gather experimental WCET data for the calibration of the WCET models.
(iv) The implementation of an *anytime algorithm*, where only a bound for WCET of the root segment must be provided.
 (v) The extensive testing of the complete implementation to validate the assumptions and to measure the safety margin between the assumed WCET and the actual measured execution times.

The state of current practice is not satisfactory, because in many cases the minimal and maximum execution times that are observed during testing are taken for the BCET and WCET. Such an observed upper bound cannot be considered a guaranteed upper bound. It is to be hoped that in the future the WCET problem will get easier, provided simple processors with private scratchpad memory will form the components of multi-processor systems-on-chips (MPSoCs).

10.3   Static Scheduling

In static or pre-run time scheduling, a feasible schedule of a set of tasks is calculated off-line. The schedule must guarantee all deadlines, considering the resource, precedence, and synchronization requirements of all tasks. The construction of such a schedule can be considered as a constructive sufficient schedulability test. The precedence relations between the tasks executing in the different nodes can be depicted in the form of a *precedence graph* (Fig. 10.5).

Fig. 10.5 Example of a
precedence graph of a
distributed task set [Foh94]

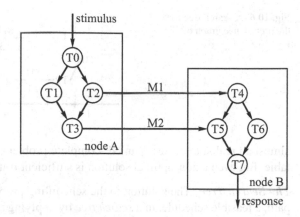

10.3.1 Static Scheduling Viewed as a Search

Static scheduling is based on strong regularity assumptions about the points in time when future service requests will be honored. Although the occurrence of external events that demand service is not under the control of the computer system, the recurring points in time when these events will be serviced can be established a priori by selecting an appropriate sampling rate for each class of events. During system design, it must be ascertained that the sum of the maximum delay times until a request is recognized by the system plus the maximum transaction response time is smaller than the specified service deadline.

The Role of Time A static schedule is a periodic time-triggered schedule. The timeline is partitioned into a sequence of basic granules, the *basic cycle time*. There is only one interrupt in the system: the periodic clock interrupt denoting the start of a new basic granule. In a distributed system, this clock interrupt must be globally synchronized to a precision that is much better than the duration of a basic granule. Every transaction is periodic, its period being a multiple of the basic granule. The least common multiple of all transaction periods is the *schedule period*. At compile time, the scheduling decision for every point of the schedule period must be determined and stored in a dispatcher table for the operating system for the full schedule period. At run time, the preplanned decision is executed by the dispatcher after every clock interrupt.

> **Example**: If the periods of all tasks are harmonic, e.g., either a positive or negative power of two of the full second, the schedule period is equal to the period of the task with the longest period.

Static scheduling can be applied to a single processor, to a multiple-processor, or to a distributed system. In addition to preplanning the resource usage in all nodes, the access to the communication medium must also be preplanned in distributed systems. It is known that finding an optimal schedule in a distributed system is in

Fig. 10.6 A search tree for
the precedence graph of
Fig. 10.5

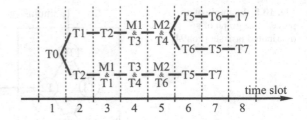

almost all realistic scenarios an NP-complete problem, i.e., computationally intractable. But even a non-optimal solution is sufficient if it meets all deadlines.

The Search Tree The solution to the scheduling problem can be seen as finding a
path, a feasible schedule, in a *search tree* by applying a search strategy. An example
of a simple search tree for the precedence graph of Fig. 10.5 is shown in Fig. 10.6.
Every level of the search tree corresponds to one unit of time. The depth of the
search tree corresponds to the period of the schedule. The search starts with an
empty schedule at the root node of this tree. The outward edges of a node point to
the possible alternatives that exist at this point of the search. A path from the root
node to a particular node at level n records the sequence of scheduling decisions that
have been made up to time-point n. Each path to a leaf node describes a complete
schedule. It is the goal of the search to find a complete schedule that observes all
precedence and mutual exclusion constraints, and which completes before the deadline. From Fig. 10.6, it can be seen that the lower branch of the search tree will lead
to a shorter overall execution time than the upper branches.

A Heuristic Function Guiding the Search To improve the efficiency of the search,
it is necessary to guide the search by some heuristic function. Such a heuristic function can be composed of two terms, the actual cost of the path encountered until the
present node in the search tree, i.e., the present point in the schedule, and the estimated cost until a goal node. Fohler [Foh94] proposes a heuristic function that estimates the time needed to complete the precedence graph, called TUR (time until
response). A lower bound of the TUR can be derived by summing up the maximum
execution times of all tasks and message exchanges between the current task and the
last task in the precedence graph, assuming true parallelism constrained by the com-
petition for CPU resources of tasks that reside at the same node. If this necessary
TUR is not short enough to complete the precedence graph on time, all the branches
from the current node can be pruned and the search must backtrack.

10.3.2   *Increasing the Flexibility in Static Schedules*

One of the weaknesses of static scheduling is the assumption of strictly periodic
tasks. Although the majority of tasks in hard real-time applications is periodic, there
are also sporadic service requests that have hard deadline requirements. An example

of such a request is an *emergency stop* of a machine. Hopefully, it will never be requested—the mean time between emergency stops can be very long. However, if an emergency stop is requested, it must be serviced within a small specified time interval.

The following three methods increase the flexibility of static scheduling:

(i)   The transformation of sporadic requests into periodic requests
(ii) The introduction of a sporadic server task
(iii) The execution of mode changes

Transformation of a Sporadic Request to a Periodic Request   While the future request times of a periodic task are known a priori, only the minimum inter-arrival time of a sporadic task is known in advance. The actual points in time when a sporadic task must be serviced are not known ahead of the request event. This limited information makes it difficult to schedule a sporadic request before run time. The most demanding sporadic requests are those that have a short response time, i.e., the corresponding service task has a low latency.

It is possible to find solutions to the scheduling problem if an independent sporadic task has a laxity l. One such solution, proposed by Mok [Mok83, p.44], is the replacement of a sporadic task T by a pseudo-periodic task T' as seen in Table 10.1.

This transformation guarantees that the sporadic task will always meet its deadline if the pseudo-periodic task can be scheduled. The pseudo-periodic task can be scheduled statically. A sporadic task with a short latency will continuously demand a substantial fraction of the processing resources to guarantee its deadline, although it might request service very infrequently.

Sporadic Server Task   To reduce the large resource requirements of a pseudo-periodic task with a long inter-arrival time (period) but a short latency, Sprunt et al. [Spr89] have proposed the introduction of a periodic server task for the service of sporadic requests. Whenever a sporadic request arrives during the period of the server task, it will be serviced with the high priority of the server task. The service of a sporadic request exhausts the execution time of the server. The execution time will be replenished after the period of the server. Thus, the server task preserves its execution time until it is needed by a sporadic request. The sporadic server task is scheduled dynamically in response to the sporadic request event.

Mode Changes   During the operation of most real-time applications, a number of different operating modes can be distinguished. Consider the example of a flight control system in an airplane. When a plane is taxiing on the ground, a different set of services is required than when the plane is flying. Better resource utilization can be realized if only those tasks that are needed in a particular operating mode must be scheduled. If the system leaves one operating mode and enters another, a corresponding change of schedules must take place.

During system design, one must identify all possible operating and emergency modes. For each mode, a static schedule that will meet all deadlines is calculated off-line. Mode changes are analyzed and the appropriate mode change schedules are developed. Whenever a mode change is requested at run time, the applicable mode change schedule will be activated immediately. We conclude this section with a comment by Xu and Parnas [Xu91, p.134].

> For satisfying timing constraints in hard real-time systems, predictability of the systems behavior is the most important concern; pre-run time scheduling is often the only practical means of providing predictability in a complex system.

10.4   Dynamic Scheduling

After the occurrence of a significant event, a dynamic scheduling algorithm determines *online* which task out of the ready task set must be serviced next. The algorithms differ in the assumptions about the complexity of the task model and the future task behavior [But04].

10.4.1   Scheduling Independent Tasks

The classic algorithm for scheduling a set of periodic independent hard real-time tasks in a system with a single CPU, the *rate monotonic algorithm*, was published in 1973 by [Liu73].

Rate Monotonic Algorithm   The rate monotonic algorithm is a dynamic preemptive algorithm based on static task priorities. It makes the following assumptions about the task set:

  (i) The requests for all tasks of the task set $\{T_i\}$ for which hard deadlines must be satisfied are periodic.
 (ii) All tasks are independent of each other. There exist no precedence constraints or mutual exclusion constraints between any pair of tasks.
(iii) The deadline interval of every task T_i is equal to its period p_i.
 (iv) The required maximum computation time of each task c_i is known a priori and is constant.
  (v) The time required for context switching can be ignored.

Table 10.1 Parameters of the pseudo-periodic task

Parameter	Sporadic task	Pseudo-periodic task
Computation time c	c	$c' = c$
Deadline interval d	d	$d' = c$
Period p	p	$p' = \min(1\text{-}1, p)$

(vi) The sum of the utilization factors μ of the n tasks with period p is given by

$$\mu = \Sigma c_i / p_i \le n\left(2^{1/n} - 1\right).$$

The term $n(2^{1/n} - 1)$ approaches $ln\ 2$, i.e., about 0.7, as n goes to infinity.

The rate monotonic algorithm assigns static priorities based on the task periods. The task with the shortest period gets the highest static priority, and the task with the longest period gets the lowest static priority. At run time, the dispatcher selects the task request with the highest static priority.

If all the assumptions are satisfied, the rate monotonic algorithm guarantees that all tasks will meet their deadline. The algorithm is optimal for single processor systems. The proof of this algorithm is based on the analysis of the behavior of the task set at the *critical instant*. A critical instant of a task is the moment at which the request of this task will have the longest response time. For the task system as a whole, the critical instant occurs when requests for all tasks are made simultaneously. Starting with the highest priority task, it can be shown that all tasks will meet their deadlines, even in the case of the critical instant. In a second phase of the proof, it must be shown that any scenario can be handled if the *critical-instant scenario* can be handled. For the details of the proof, refer to [Liu73].

It is also shown that assumption (vi) above can be relaxed in case the task periods are harmonic, i.e., they are multiples of the period of the highest priority task. In this case, the utilization factor μ of the n tasks,

$$\mu = \Sigma c_i / p_i \le 1,$$

can approach the theoretical maximum of unity in a single processor system.

In recent years, the rate monotonic theory has been extended to handle a set of tasks where the deadline interval can be different from the period [But04].

Earliest-Deadline-First (EDF) Algorithm This algorithm is an optimal dynamic preemptive algorithm in single processor systems which are based on dynamic priorities. The assumptions (i) to (v) of the rate monotonic algorithm must hold. The processor utilization μ can go up to 1, even when the task periods are not multiples of the smallest period. After any significant event, the task with the earliest deadline is assigned the highest dynamic priority. The dispatcher operates in the same way as the dispatcher for the rate monotonic algorithm.

Least-Laxity (LL) Algorithm In single processor systems, the least-laxity algorithm is another optimal algorithm. It makes the same assumptions as the EDF algorithm. At any scheduling decision instant, the task with the shortest laxity l, i.e., the difference between the deadline interval d and the computation time c

$$d - c = l$$

is assigned the highest dynamic priority.

In multiprocessor systems, neither the earliest-deadline-first nor the least-laxity algorithm is optimal, although the least-laxity algorithm can handle task scenarios, which the earliest-deadline-first algorithm cannot handle and vice versa.

10.4.2   Scheduling Dependent Tasks

From a practical point of view, results on how to schedule tasks with precedence and mutual exclusion constraints are much more important than the analysis of the independent task model. Normally, the concurrently executing tasks must exchange information and access common data resources to cooperate in the achievement of the overall system objectives. The observation of given precedence and mutual exclusion constraints is thus rather the norm than the exception in distributed real-time systems.

To solve this problem, the *priority ceiling protocol* was developed by [Sha90]. The *priority ceiling protocol* can be used to schedule a set of periodic tasks that have exclusive access to common resources protected by semaphores. These common resources, e.g., common data structures, can be utilized to realize an inter-task communication.

The priority ceiling of a semaphore is defined as the priority of the highest priority task that may lock this semaphore. A task T is allowed to enter a critical section only if its assigned priority is higher than the priority ceilings of all semaphores currently locked by tasks other than T. Task T runs at its assigned priority unless it is in a critical section and blocks higher priority tasks. In this case, it inherits the highest priority of the tasks it blocks. When it exits the critical section, it resumes the priority it had at the point of entry into the critical section.

The example of Fig. 10.7, taken from [Sha90], illustrates the operation of the priority ceiling protocol. A system of three tasks, *T1* (highest priority), *T2* (middle priority), and *T3* (lowest priority), compete for three critical regions protected by the three semaphores *S1*, *S2*, and *S3*.

Schedulability Test for the Priority Ceiling Protocol The following sufficient schedulability test for the priority ceiling protocol has been given by [Sha90, Theorem 16]. Assume a set of periodic tasks, *{T$_i$}* with periods p_i and computation times c_i. We denote the worst-case blocking time of a task t_i by lower priority tasks by B_i. The set of *n* periodic tasks *{T$_i$}* can be scheduled, if the following set of inequalities holds:

$\forall i, 1 \leq i \leq n$:

$$\left(c_1 / p_1 + c_2 / p_2 + \ldots + c_i / p_i + B_i / p_i\right) \leq i\left(2^{1/i} - 1\right)$$

In these inequalities, the effect of preemptions by higher priority tasks is considered in the first *i* terms (in analogy to the rate monotonic algorithm), whereas the

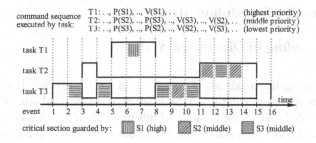

command sequence executed by task:
T1: .., P(S1), .., V(S1), ..   (highest priority)
T2: .., P(S2), .., P(S3), .., V(S3), .., V(S2), ..   (middle priority)
T3: .., P(S3), .., P(S2), .., V(S2), .., V(S3), ..   (lowest priority)

critical section guarded by: S1 (high)   S2 (middle)   S3 (middle)

Event	Action
1	T3 begins execution.
2	T3 locks S3.
3	T2 is started and preempts T3.
4	T2 becomes blocked when trying to access S2 since the priority of T2 is not higher than the priority ceiling of the locked S3. T3 resumes the execution of its critical section at the inherited priority of T2.
5	T1 is initiated and preempts T3.
6	T1 locks the semaphore S1. The priority of T1 is higher than the priority ceiling of all locked semaphores.
7	T1 unlocks semaphore S1.
8	T1 finishes its execution. T3 continues with the inherited priority of T2.
9	T3 locks semaphore S2.
10	T3 unlocks S2.
11	T3 unlocks S3 and returns to its lowest priority. At this point T2 can lock S2.
12	T2 locks S3.
13	T2 unlocks S3.
14	T2 unlocks S2.
15	T2 completes. T3 resumes its operation.
16	T3 completes.

Fig. 10.7 The priority ceiling protocol. (Example taken from Ref. [Sha90])

worst-case blocking time due to all lower priority tasks is represented in the term B_i/p_i. The blocking term B_i/p_i, which can become very significant if a task with a short period (i.e., small p_i) is blocked for a significant fraction of its time, effectively reduces the CPU utilization of the task system. In case this first sufficient schedulability test fails, more complex sufficient tests can be found in [Sha90]. The priority ceiling protocol is a good example of a predictable, but *non-deterministic* scheduling protocol.

Example: The NASA Pathfinder robot on MARS experienced a problem that was diagnosed as a classic case of priority inversion, due to a missing *priority ceiling protocol*. The full and most interesting story is contained in [Jon97]: *Very infrequently it was possible for an interrupt to occur that caused the (medium priority) communications task to be scheduled during the short interval while the (high priority) information bus thread was blocked waiting for the (low priority) meteorological data thread. In this case, the long-running communications task, having higher priority than the meteorological task, would prevent it from running, consequently preventing the blocked information bus task from running. After some time had passed, a watchdog timer would go off, notice that the data bus task had not been executed for some time, conclude that something had gone drastically wrong, and initiate a total system reset.*

10.5   Alternative Scheduling Strategies

10.5.1   Scheduling in Distributed Systems

In a control system, the maximum duration of an *RT transaction* is the critical parameter for the quality of control, since it contributes to the dead time of a control loop. In a distributed system, the duration of this transaction depends on the sum of the durations of all processing and communication actions that form the transaction. In such a system, it makes sense to develop a *holistic schedule* that considers all these actions together. In a time-triggered system, the processing actions and the communication actions can be *phase aligned* (see Sect. 3.3.4), such that a send slot in the communication system is available immediately after the WCET of a processing action.

It is already difficult to guarantee tight deadlines by dynamic scheduling techniques in a single processor event-triggered multi-tasking system if mutual exclusion and precedence constraints among the tasks must be considered. The situation is more complex in a distributed system, where the non-preemptive access to the communication medium must be considered. Tindell [Tin95] analyzes distributed systems that use the CAN bus as the communication channel and establishes analytical upper bounds to the communication delays that are encountered by a set of periodic messages. These results are then integrated with the results of the node-local task scheduling to arrive at the worst-case execution time of distributed real-time transactions. One difficult problem is the control of transaction jitter.

Since the worst-case duration of a RT transaction in an event-triggered distributed system can be exceedingly pessimistic, some researchers are looking at dynamic best-effort strategies and try to establish bounds based on a probabilistic analysis of the scheduling problem. This approach is not recommended in hard real-time systems, since the characterization of *rare events* is extremely difficult. Rare event occurrences in the environment, e.g., a lightning stroke into an electric power grid, will cause a highly correlated input load on the system (e.g., an alarm shower) that is very difficult to model adequately. Even an extended observation of a real-life system is not conclusive, because these rare events, by definition, cannot be observed frequently.

In soft real-time systems (such as multimedia systems) where the occasional miss of deadline is tolerable, probabilistic analysis is widely used. An excellent survey on the results of 25 years of research on real-time scheduling is contained in [Sha04].

10.5.2   Feedback Scheduling

The concept of feedback, well established in many fields of engineering, uses information about the *actual behavior* of a scheduling system to dynamically adapt the *scheduling algorithms* such that the intended behavior is achieved. Feedback scheduling starts with the establishment and observation of relevant performance parameters of the scheduling system. In a multimedia systems, the queue size that develops before a server process is an example of such a relevant performance parameter. These queue sizes are continuously monitored and the producer of information is controlled—either slowed down or speeded up—in order to keep the size of the queue between given levels, the *low* and *high watermark*.

By looking at the scheduling problem and control problem in an integrated fashion, better overall results can be achieved in many control scenarios. For example, the sample rate of a process can be dynamically adjusted based on the observed performance of the physical process.

Points to Remember

- A scheduler is called *dynamic* (or *online*) if it makes its scheduling decisions at run time, selecting one out of the current set of *ready* tasks. A scheduler is called static (or pre-run time) if it makes its scheduling decisions at compile time. It generates a dispatching table for the run time dispatcher off-line.
- A test that determines whether a set of ready tasks can be scheduled so that each task meets its deadline is called a *schedulability test*. We distinguish between *exact*, *necessary*, and *sufficient* schedulability tests. In nearly all cases of task dependency, even if there is only one common resource, the complexity of an exact schedulability test algorithm belongs to the class of NP-complete problems and is thus computationally intractable.
- While the future request times of a periodic task are known a priori, only the minimum interarrival time of a *sporadic* task is known in advance. The actual points in time when a sporadic task must be serviced are not known ahead of the request event.
- The *adversary argument* states that, in general, it is not possible to construct an optimal totally online dynamic scheduler if there are mutual exclusion constraints between a periodic and a sporadic task. The adversary argument accentuates the value of a priori information about the behavior in the future.
- In general, the problem of determining the worst-case execution time (WCET) of an arbitrary sequential program is unsolvable and is equivalent to the halting problem for Turing machines. The WCET problem can only be solved, if the programmer provides additional application-specific information at the source code level.
- In static or pre-run time scheduling, a feasible schedule of a set of tasks that guarantees all deadlines, considering the resource, precedence, and synchronization requirements of all tasks, is calculated off-line. The construction of such a schedule can be considered as a constructive sufficient schedulability test.

- The rate monotonic algorithm is a dynamic preemptive scheduling algorithm based on static task priorities. It assumes a set of periodic and independent tasks with deadlines equal to their periods.
- The Earliest-Deadline-First (EDF) algorithm is a dynamic preemptive scheduling algorithm based on dynamic task priorities. The task with the earliest deadline is assigned the highest dynamic priority.
- The Least-Laxity (LL) algorithm is a dynamic preemptive scheduling algorithm based on dynamic task priorities. The task with the shortest laxity is assigned the highest dynamic priority.
- The priority ceiling protocol is used to schedule a set of periodic tasks that have exclusive access to common resources protected by semaphores. The *priority ceiling* of a semaphore is defined as the priority of the highest priority task that may lock this semaphore.
- According to the priority ceiling protocol, a task T is allowed to enter a critical section only if its assigned priority is higher than the priority ceilings of all semaphores currently locked by tasks other than T. Task T runs at its assigned priority unless it is in a critical section and blocks higher priority tasks. In this case, it inherits the highest priority of the tasks it blocks. When it exits the critical section, it resumes the priority it had at the point of entry into the critical section.
- The critical issue in best-effort scheduling concerns the assumptions about the input distribution. Rare event occurrences in the environment will cause a highly correlated input load on the system that is difficult to model adequately. Even an extended observation of a real-life system is not conclusive, because these rare events, by definition, cannot be observed frequently.
- In soft real-time systems (such as multimedia systems) where the occasional miss of deadline is tolerable, probabilistic scheduling strategies are widely used.
- Anytime algorithms are algorithms that improve the quality of the result as more execution time is provided. They consist of a *root segment* that calculates a first approximation of the result of sufficient quality and a *periodic segment* that improves the quality of the previously calculated result.
- In *feedback scheduling*, information about the *actual behavior* of a scheduling system is used to dynamically adapt the *scheduling algorithms* such that the intended behavior is achieved.

Bibliographic Notes

Starting with the seminal work of Liu and Layland [Liu73] in 1973 on scheduling of independent tasks, hundreds of papers on scheduling are being published each year. In 2004 the *Real-Time System Journal* published a comprehensive survey *Real-Time Scheduling Theory: A Historical Perspective* [Sha04]. The book by Butazzo [But00] *Hard Real-Time Computer Systems* covers scheduling extensively. An excellent survey article on WCET analysis is contained in [Wil08]. Anytime algorithms are described in [Zil96].

Review Questions and Problems

10.1 Give taxonomy of scheduling algorithms.

10.2 Develop some necessary schedulability tests for scheduling a set of tasks on a single processor system.

10.3 What are the differences between *periodic* tasks, *sporadic* tasks, and *aperiodic* tasks?

10.4 Why is it hard to find the worst-case execution time (WCET) of a program?

10.5 What is the worst-case administrative overhead (WCAO)?

10.6 Given the following set of independent periodic tasks, where the deadline interval is equal to the period: {T1(5,8); T2(2,9); T3(4,13)}; (notation: task name(CPU time, period)).

    (a) Calculate the laxities of these tasks.

    (b) Determine, using a necessary schedulability test, if this task set is schedulable on a single processor system.

    (c) Schedule this task set on a two-processor system with the LL algorithm.

10.7 Given the following set of independent periodic tasks, where the deadline interval is equal to the period: {T1(5,8); T2(1,9); T3(1,5)}; (notation: task name(CPU time, period)).

    (a) Why is this task set not schedulable with the rate monotonic algorithm on a single processor system?

    (b) Schedule this task set on a single processor system with the EDF algorithm.

10.8 Why is it not possible to design, in general, an optimal dynamic scheduler?

10.9 Assume that the task set of Fig. 10.7 is executed without the priority ceiling protocol. At what moment will a deadlock occur? Can this deadlock be resolved by *priority inheritance*? Determine the point where the priority ceiling protocol prevents a task from entering a critical section.

10.10 Discuss the schedulability test of the priority ceiling protocol. What is the effect of blocking on the processor utilization?

10.11 What are the problems with dynamic scheduling in distributed systems?

10.12 Discuss the issue of temporal performance in best-effort distributed system.

10.13 What is the role of time in static scheduling?

10.14 List some properties of *anytime algorithms*!

10.15 What is *feedback scheduling*?

Chapter 11
System Design

Overview

This chapter on system design starts with a discussion on design in general. The designer must get a deep insight into all different aspects of the problem domain before she/he can design a proper architecture for an application. In computer system design, the most important goal is controlling the complexity of the evolving artifact. A thorough analysis of the purpose, the requirements, and constraints confine the design space and avoids the investigation of unrealistic design alternatives. The central step in the design of an architecture is the allocation of the identified system functions to nearly independent subsystems. The result of the architecture design phase is an *Interface Control Document* that precisely specifies the functional and temporal properties of the interfaces among the subsystems. A subsystem should have a high internal cohesion and simple external interfaces. In the following, different design styles such as *model-based design* and *component-based design* are discussed. The design of safety-critical systems starts with a *safety analysis* such as fault tree analysis and/or failure mode and effect analysis (FMEA) of the envisioned application and the development of a convincing *safety case*. Different standards that must be observed in the design of a safety-critical system are referred to, such as the IEC 61508 for electric and electronic equipment and the ARINC DO 178C standard for airborne equipment software. The elimination of all design errors, e.g., software errors or hardware errata of a large safety-critical system, is a major challenge. Design diversity can help to mitigate the problem of remaining design errors and support software maintainability. Maintainability of software is needed to correct design errors in the software and to adapt the software to the never-ending needs of an evolving application scenario. The final section of this chapter presents the principles that have been followed in the design of the time-triggered architecture.

© The Author(s), under exclusive license to Springer Nature Switzerland AG 2022   269
H. Kopetz, W. Steiner, *Real-Time Systems*,
https://doi.org/10.1007/978-3-031-11992-7_11

11.1   System Design

11.1.1   The Design Process

Design is an inherently creative activity, where both the *intuitive* and the *rational* problem-solving systems of the human mind are heavily involved. There is a common core to design activities in many diverse fields: building design, product design, and computer system design are all closely related. The designer must find a solution that *accommodates a variety of seemingly conflicting goals* to solve an often ill-specified design problem. In the end, what differentiates a good design from a bad design is often liable to subjective judgment.

> **Example:** Consider the design of an automobile [Wil21]. An automobile is a complex mass production product that is composed of a number of sophisticated subsystems (e.g., engine, transmission, chassis, etc.). Each of these subsystems itself contains hundreds of different components that must meet given constraints: functionality, efficiency, geometrical form, weight, dependability, and minimal cost. All these components must cooperate and interact smoothly to provide the *emergent* transportation service and the *look and feel* that the customer expects from the system *car*.

Let us, at the starting point, introduce a rather abstract definition of a system: A system, such as a car, is the result of the interactions of many related parts (subsystems) that forms a *whole* and provides a *purposeful service* to its environment. It is encapsulated by a *physical* or *virtual* skin that separates the system from its environment. In a computer system, the *interfaces* in the skin enable the exchange of messages between a system and its environment [Kop22, p. 10].

The first activity in the design of a system is the *purpose analysis*. During the *purpose analysis phase*, the goals and the economic and technical constraints of an envisioned solution are established. If the evaluation at the end of this phase results in a *go ahead* decision, a project team is formed to start the requirement analysis and the system design phase. There are three opposing views how to proceed in this first most critical phase when designing a large system:

(i)   A disciplined sequential approach, where every life cycle phase is thoroughly completed and validated before the next one is started (*grand design*)

(ii) A rapid-prototyping approach, where the implementation of a key part of the solution is started before the requirements analysis has been completed (*rapid prototyping*)

(iii) A hierarchical architecture design approach, where the gross system functions are allocated to top-level subsystems and the precise specification of the interactions among these subsystems is developed first

The rationale for the *grand design* is that a detailed and unbiased specification of the complete problem (the *What?*) must be available before a particular solution (the *How?*) is designed. The difficulty with *grand design* is that there are no clear *stopping rules*. The analysis and understanding of a large problem is never complete, and there are always good arguments for asking more questions concerning the requirements before starting with the *real* design work. Furthermore, the world evolves while the analysis is done, changing the original scenario. The phrase *paralysis by analysis* has been coined to point to this danger.

The rationale for the *rapid prototyping* approach assumes that, by investigating a particular solution of a key function at an early stage, a lot is learned about the problem space. The difficulties met during the search for a concrete solution guide the designer in asking the right questions about the requirements. The dilemma of rapid prototyping is that ad hoc implementations are developed with great expense. Since the first prototype does address limited aspects of the design problem only, it is often necessary to completely discard the first prototypes and to start all over again.

The rationale for the hierarchical architecture design approach is that it follows the well-proven principle of *design and conquer* to partition a complex design problem into nearly independent subsystems of lower complexity and establishes well-specified constraints (the precisely specified interfaces) for the development of the subsystems. It is the opinion of the authors that this third alternative, the hierarchical architecture design, is the most reasonable approach to the design of a large systems.

In the architecture design phase, a key designer should try to get a good understanding of the purpose and the gross architectural functions first, leaving detailed issues that affect only the internals of subsystems open. In his book [Bro10], Fred Brook states that *conceptual integrity of a design is the result of a single mind.* If it is not clear how to solve a particular problem, then a preliminary prototype of the most difficult part should be investigated with the explicit intent of discarding the solution if the looked-for insight has been gained.

Some years ago, Peters [Pet79] argued in a paper about design that design belongs to the set of *wicked problems*. Wicked problems are described by the following characteristics:

(i) A wicked problem cannot be stated in a definite way, abstracted from its environment. Whenever one tries to isolate a wicked problem from its surroundings, the problem loses its peculiarity. Every wicked problem is somehow unique and cannot be treated in the abstract.

(ii) A wicked problem cannot be specified without having a solution in mind. The distinction between specification (*what?*) and implementation (*how?*) is not as easy as is often proclaimed in academia.

(iii) Solutions to wicked problems have no stopping rule: for any given solution, there is always a better solution. There are always good arguments to learn more about the requirements to produce a better design.

(iv) Solutions to wicked problems cannot be right or wrong; they can only be *better* or *worse*.

(v) There is no definite test for the solution to a wicked problem: whenever a test is *successfully* passed, it is still possible that the solution will fail in some other way.

11.1.2   The Role of Constraints

Every design is embedded in a design space that is bounded by a set of known and unknown *constraints*. In some sense, constraints are *antonyms* to requirements. It is good practice to start a design by capturing the constraints and classifying them into *soft constraints*, *hard constraints*, and *limiting constraints*. A *soft constraint* is a desired but not obligatory constraint. A *hard constraint* is a given mandatory constraint that must not be neglected. A *limiting constraint* is a constraint that limits the utility of a design.

> **Example:** In building a house, the mandatory construction code of the area is a *hard constraint*, the orientation of the rooms and windows is a *soft constraint*, while the construction cost may be a *limiting constraint*.

Constraints limit the design space and help the designer to avoid the exploration of design alternatives that are unrealistic in the given environment. Constraints are thus our friends, not our adversaries. Special attention must be paid to the limiting constraints, since these constraints are instrumental for determining the value of a design for the client. It is good practice to precisely monitor the limiting constraints as a design proceeds. One important limiting constraint is the cost of the system.

11.1.3   System Design Versus Software Design

In the early days of *computer-application design*, the focus of design was on the *functional aspects* of software, with little regard for the *nonfunctional* properties of the computations that are generated by the software, such as *timing, energy efficiency*, or *fault tolerance*. This focus has led to *software design methods*—still prevalent today—that concentrate on the data transformation aspects of a program with little regard for non-functional requirements, such as timing or energy consumption.

> **Example:** A critical constraint in the design of a smart phone is the expected life of a battery load. This non-functional constraint is overlooked if the focus during the design is only on the functional properties of the design.

Software per se is a *plan* describing the operations of a real or virtual machine. A plan by itself (without a machine) does not have *any temporal dimension*, cannot have *state* (which depends on a precise notion of real time—see Sect. 4.2.1) and has no *behavior*. Only the *combination* of software and the targeted machine, the *platform*, produces behavior. This is one of the reasons why we consider the *component* and not the *job* (see Sect. 4.1.1) as the primitive construct at the level of architecture design of an embedded system.

The PIM defines a component's intended behavior, including the requirements on its non-functional properties. This intended behavior of a component can be realized by different implementation technologies or their combinations (i.e., there exist different ways to realize the PSM):

(i) By designing software for a programmable computer, resulting in a flexible component consisting of a local operating system with middleware and application software modules.
(ii) By developing software for a field-programmable gate array (FPGA) that implements the component's functionality by the proper interconnection of a set of highly concurrent logic elements.
(iii) By developing an application specific integrated circuit (ASIC) that implements the functionality of the component directly in hardware.

Multiple PIM components can be implemented in a single physical chip in all three implementation technologies if the component characteristic of a self-contained hardware/software unit and the interface specifications remain.

Example: Considering the implications of Pollack's rule (see Sect. 8.3.2), we conjecture that in the domain of embedded real-time systems predictable sequential processors combined with the appropriate system and application software will form the IP-cores, the components, of the envisioned multiprocessor systems-on-chips (MPSoC) of the embedded system of the future.

While it is rather simple to establish self-containment and interface equivalence in an MPSoC implementation, it is difficult in programmable computers, especially if they implement a multilayer software stack between the hardware and the application software (e.g., hypervisor, operating system, middleware). To be used for hard real-time systems, the behavior of the software stack on the target machine must be well understood. Such an understanding can only be established if the software layers are designed with real time in mind (e.g. see Sect. 14.4.1). As a result, components implementing common IT solutions will fail to ensure self-containment and interface equivalence. Similarly, the temporal hardware performance of many of today's sophisticated sequential processors with multiple levels of caching and speculative execution is difficult to specify.

The integration of multiple components into a single chip may have short- to mid-term economic benefits but comes with significant risks.

(i) Incomplete understanding of component interaction: even if different components are allocated to different cores in a multi-core SoC, the components may interact in nonobvious ways. For example, shared memory buses or the energy consumption from a shared battery may be easily overlooked or oversimplified in the component integration process.
(ii) Testability: testing without probe effect is a challenge (see Sect. 12.2.1) as the testing subsystem must also be integrated (partially) on the same chip.
(iii) Scalability: integration of multiple components may cause a specialization of the chip that limits its deployment possibilities. Furthermore, it is more difficult to exchange integrated components than components implemented on different chips.

(iv) Fault Containment: some fault classes (see Sect. 6.1.1) will cause multiple, if
not all, components on the chip to fail. All the components integrated on the
single chip belong to the same fault-containment unit (FCU) for these faults.

Example: The failure of the chip's power supply will cause all its integrated components
to fail.

Viewed from the outside, the services of a component must be *agnostic* of the
chosen implementation technology. Only then it is possible to change the implementation of a component without any effects at the system level. However, from
the point of view of some of the non-functional component characteristics such as
energy consumption, silicon real-estate requirements, flexibility to change, or non-
recurring development costs, different component implementations have vastly different characteristics. In a number of applications, it is desired to develop at first a
hardware-agnostic model of the services of a component at the architecture level
and to postpone the detailed decisions about the final implementation technology of
the component to a later stage.

Example: In a product for mass-market consumer appliance, it makes sense to first develop
a prototype of a component in software-on-a-CPU and to decide later, after the market
acceptance of the product has been established, to shift the implementation to an FPGA
or ASIC.

11.2   Design Phases

Design is a creative holistic human activity that cannot be reduced to following a set
of rules out of a design rule-book. Design is an art, supplemented by scientific principles. It is therefore in vain to try to establish a complete set of design rules and to
develop a fully automated design environment. Design tools can assist a designer in
handling and representing the design information and can help in the analysis of
design problems. As the capabilities of design and automation tools grow, so do the
designer's possibilities. However, tools cannot replace a creative designer.
Furthermore, tools themselves can become quite sophisticated and may require tool
qualification, i.e., rigorous evidence for their correct operation.

In theory, the design process should be structured into a set of distinct phases:
purpose analysis, requirements capture, architecture design, detailed component
design and implementation, component validation, component integration, system
validation, and finally system commissioning. In practice, such a strict sequential
decomposition of the design process is hardly possible, since the full scope of a new
design problem is not grasped until the design process is well under its way, requiring frequent iterations among the design phases.

The focus of this chapter is on the architecture design phase, while the validation
phases are covered in Chap. 12.

11.2.1 Purpose Analysis

Every rational design is driven by a given *purpose*. The purpose puts the design into the wider context of user expectations and economic justification and thus precedes the requirements. *Purpose analysis*, i.e., the analysis why a new system is needed and what is the ultimate goal of a design, must precede the requirements analysis, which already limits the scope of analysis and directs the design effort to a specific direction. Critical purpose analysis is needed in order to put the requirements into the proper perspective.

> **Example:** The purpose of acquiring a car is to provide a transportation service. There are other means of transportation, e.g., public transport, which should be considered in the purpose analysis phase.

In every project, there is an ongoing conflict between *what is desired* and *what can be done* within the given technical and economic constraints. A good under-standing and documentation of these technical and economic constraints reduces the design space and helps to avoid exploring unrealistic design alternatives.

11.2.2 Requirements Capture

The focus of the requirements phase is to get a good understanding and a concise documentation of the requirements and constraints of the essential system functions that provide the economic justification of the project. There is always the temptation to get sidetracked by irrelevant details about representational issues that obscure the picture of the whole. Many people find it easier to work on a well-specified detailed side problem than to keep focus on the critical system issues. It requires an experi-enced designer to decide between a *side problem* and a *critical system issue*.

Every requirement must be accompanied by an acceptance criterion that allows to measure, at the end of the project, whether the requirement has been met. If it is not possible to define a distinct acceptance test for a requirement, then the require-ment cannot be very important: it can never be decided whether the implementation is meeting this requirement or not. A critical designer will always be suspicious of postulated requirements that cannot be substantiated by a rational chain of argu-ments that, at the end, leads to a measurable contribution of the stated requirement to the purpose of the system.

In the domain of embedded systems, a number of *representation standards* and *tools* have been developed to support the system engineer (see Sect. 11.3.3). Standards for the uniform representation of requirements are of particular impor-tance, since they simplify the communication among designers and users.

11.2.3   Architecture Design

After the purpose of the system has been established and agreed upon among all involved parties and the main requirements have been captured and documented, the most crucial phase of the life cycle, the design of the system architecture, follows. In the context of distributed real-time systems, the architecture design establishes the decomposition of the overall system into nearly independent and stable *subsystems*, the purpose of each subsystem and the precise specification of the linking interfaces among the subsystems.

Systems with stable intermediate forms evolve much more rapidly than systems without them [Sim81]. Stable intermediate forms—such as the stable subsystems—are encapsulated by simple and firm interfaces. These interfaces provide the essential constraints for the later design of the subsystems.

Architecture design is a highly iterative design phase. Different decompositions should be sketched and each one of the proposed subsystems of a decomposition should be evaluated from a number of different viewpoints such as functionality, interface complexity, dependability, hardware cost, software cost, maintainability, and interface stability with respect to foreseen changes. After a most agreeable structure of the subsystems has been found, a set of *Interface Control Documents (ICDs)* [Cad22] of the linking interfaces of all subsystem must be produced.

The results of the architecture design phase are the gross specifications of the functions of the subsystems and the detailed specifications of the linking interfaces among the subsystems, recorded in the *Interface Control Documents* (ICDs). The detailed interface specification must first look at the required information flow among the subsystems. If, at the level of the architecture, the information flow is *unidirectional*, then this unidirectionality must be preserved in the detailed communication protocol design. Time-triggered protocols provide error detection at a receiver of unidirectional information flow without any involvement of the sender. The elimination of backward error propagation of a faulty receiver to a correct sender *by design* is one of the most important and beneficial characteristics of a time-triggered communication protocol. The information items, the maximum size of the messages among the subsystems, and the periodic instants when the time-triggered messages will be sent are part of the interface specification that is contained in the ICDs. The development and careful maintenance of the ICDs, probably the most important documents for the system development, should receive highest management attention. It is a good practice to make the organizational interfaces among project groups isomorphic to the technical interfaces contained in the ICDs. This isomorphism improves the stability and precision of the ICDs, since technical and organizational responsibility go hand in hand.

11.2.4 Design of Components

At the end of the architectural design phase, the requirements have been allocated to subsystems, and the *linking interfaces* (LIFs) of the subsystems are precisely specified in the value domain and in the temporal domain. The design effort can now be broken down into a set of concurrent design activities, each one focusing on the design, implementation, and testing of a single subsystem.

In case of a very large system, the decomposition of a system into subsystems at the top level of the architecture can be followed by a set of independent architecture design phases of each subsystems at the level below. At the end of this recursive decomposition, top level specifications for the components are available. Remember that a component is hardware/software unit that provides its services across the linking interfaces of the component.

The detailed design of the *local interface* (see Sect. 4.4.5) between a component and its local environment (e.g., the controlled object or other clusters) is not covered in the architectural design phase, since only the *semantics*, but not the *syntax*, of these local interfaces are needed for the cluster-LIF specification of the components. It is up to the detailed design of a component to specify and implement these local interfaces. In some cases, such as the design of the concrete man-machine interface for the operator, this can be a major activity.

The detailed steps that have to be taken in order to implement the design of a component depend on the chosen implementation technology. If the services of a component are implemented by a *software-on-a-CPU* design, then the necessary design steps will differ radically from a design that targets an ASIC as its final outcome. Since the focus of this book is on the topic of architecture design of embedded systems, we do not cover the detailed component implementation techniques for the different implementation technologies at any length.

11.3 Design Styles

11.3.1 Model-Based Design

In Chap. 2 of this book, we emphasize the role of model building for the understanding of any real-world scenario. Model-based design is a design method that establishes a useful framework for the development and integration of executable models of the controlled object and of the controlling computer system early in the design cycle.

After finishing the *purpose analysis* of a control system, in *model-based design*, executable high-level models of the *controlled object* (the plant) and the *controlling computer system* are developed in order that the dynamic interaction of these models can be studied at a high level of abstraction.

The first step in model-based design is concerned with the identification and mathematical modeling of the dynamics of the controlled object, i.e., the plant. Where possible, the results of the plant model are compared with experimental data from the operation of a real plant in order to validate the *faithfulness* of the plant model. In the second step, the mathematical analysis of the dynamic plant model is used for the synthesis of control algorithms that are tuned to the dynamics of the given plant model. In the third step, the executable plant model and the executable controlling computer model are integrated in a simulated environment, such that the correct interplay between the models can be validated and the quality of the considered control algorithms can be investigated in relation to the plant model. Although this simulation will often be operated in simulated time (*SIL—software in the loop*), it is important that the phase relationship between the messages exchanged between the plant model and the controlling computer model in the simulated environment and the (later) real-time control environment is exactly the same [Per10]. This exact phase relationship among the messages ensures that the *message order* in the simulations and in the target system will be alike. Finally, in the fourth phase the controlling computer system model is translated, possibly automatically, to the target execution environment of the control computers.

In *hardware-in-the-loop simulations (HIL)*, the simulation models must be executed in real time, since subsystems of the simulation are formed by the final target hardware.

Model-based design makes it possible to study the system performance not only under normal operating conditions, but also in *rare-event* situations, e.g., when a critical part of the system has failed. During the simulation, it is possible to tune the control algorithms such that a safe operation of the plant in a *rare-event scenario* can be maintained. Furthermore, a model-based design environment can be used for automated testing and the training of plant operators.

Example: It is a standard procedure to train pilots on a simulator in order to get them acquainted with the necessary control actions in case of a rare-event incident that cannot be reproduced easily during the flight of a real airplane.

A key issue in model-based design focuses on the specification of the linking interface (LIF) between the plant model and the controlling computer system model. As already discussed in Sect. 4.4.5, this interface specification must cover the value dimension and the temporal dimension of the messages that cross the LIF. The semantic interface models must be presented in an executable form, such that the simulation of the complete control system can be performed on a high level of abstraction and the automatic generation of the code for the target control system is supported. A widely used tool environment for model-based design is the MATLAB design environment [Att09].

11.3.2   Component-Based Design

In many engineering disciplines, large systems are built from prefabricated components with known and validated properties. Components are connected via stable, understandable, and standardized interfaces. The system engineer has knowledge about the global properties of the components—as they relate to the system functions—and of the detailed specification of the component interfaces. Knowledge about the internal design and implementation of the components is neither needed nor available in many cases. A prerequisite for such a constructive approach to system building is that the validated properties of the components are not affected by the system integration. This composability requirement is an important constraint for the selection of a platform for the component-based design of large distributed real-time systems.

Component-based design is a *meet-in-the middle* design method. On the side the *functional and temporal requirements* of the components are derived top-down from the desired application functions. On the other side, the *functional and temporal capabilities* of the available components are provided by the component specifications (bottom-up). During the design process, a proper match between component requirements and component capabilities must be established. If there is no component available that meets the requirements, a new component must be developed.

A prerequisite of any component-based design is a crystal clear component concept that supports the precise specification of the services that are delivered and acquired across the component interfaces. The notion of component as a hardware-software unit, introduced in Sect. 4.1.1, provides for such a component concept. In many non-real-time applications, a software unit is considered to form a component (i.e., *a software-only component*). In real-time systems, where the temporal properties of components are as important as the value properties, the notion of a software-only component is of questionable utility, since no temporal capabilities can be assigned to software without associating the software with a concrete machine. The specification of the temporal properties of the API (application programming interface) between the application software and the execution environment (middleware, operating system) of the concrete machine is so involved that a simple specification of the temporal properties of the API is hardly possible. If the mental effort needed to understand the specification of the component interfaces is in the same order of magnitude as the effort needed to understand the internals of the component operation, the abstraction of a component does not make sense anymore. Thus, in real-time systems, a component must always be specified in terms of software and hardware: the PIM specifies *requirements* on the non-functional properties which the PSM must implement (see Sect. 11.1.3).

The temporal capabilities of a (hardware-software) component are determined by the frequency of the oscillator that drives the component hardware. According to Sect. 8.2.3, this frequency can be lowered if the voltage is lowered (*voltage-frequency scaling*), resulting in substantial savings of the energy required to perform the computation of the component at the expense of extending the real time needed for the execution of the computation. A holistic resource scheduler that is aware of the temporal needs and the energy requirements can match the temporal capabilities of a component to the temporal requirements of the application, thus saving energy. Energy saving is very important in mobile battery-operated devices, an important sector of the embedded systems market.

11.3.3   Architecture Design Languages

The representation of the platform independent model, the PIM (the design at the architectural level—see Sect. 4.4), e.g., in the form of components and messages, requires a notation that is apt for this purpose.

In 2007, the Object Management Group (OMG) extended the Unified Modeling Language (UML) by a profile called MARTE (Modeling and Analysis of Real-Time and Embedded system) that extends UML to support the specification, design, and analysis of embedded real-time systems at the architectural level [OMG98]. UML-MARTE targets the modeling of both the software part and of the hardware part of an embedded system. The core concepts of UML-MARTE are expressed in two packages, the *foundation package* that is concerned with structural models and the *causality package* that focuses on behavioral modeling and timing aspects. In UML-MARTE, the fundamental unit of behavior is called an *action* that transforms a set of inputs into a set of outputs, taking a specified duration of real time. Behaviors are composed out of *actions* and are initiated by *triggers*. The UML-MARTE specification contains a special section on the modeling of time. It distinguishes between three different kinds of *time abstractions*: (i) *logical time* (called *causal/temporal*) that is only concerned with temporal order without any notion of a temporal metric between events; (ii) *discrete time* (called *clocked-synchronous*) where the continuum of time is partitioned by a clock into a set of ordered *granules* and where actions may be performed within a granule; and (iii) *real time* (called *physical/real time*) where the progression of real time is precisely modeled. For a detailed description of UML-MARTE refer to [OMG08]. SysML [Hau06] is another UML profile (addressing systems engineering) that is receiving more and more attention from the real-time industry.

Another example of an architecture design language is the AADL (Architecture Analysis and Design Language) developed at the Carnegie Mellon University Software Engineering Institute and standardized in 2004 by the Society of Automotive Engineers (SAE). AADL has been designed to specify and analyze the architecture of large embedded real-time systems. The core concept of AADL is the notion of a *component* that interacts with other components across interfaces. An AADL component is a *software unit* enhanced by attributes that capture the characteristics of the machine that is bound to the software unit, such that timing requirements and the worst-case execution time (WCET) of a computation can be expressed. AADL components interact exclusively though defined interfaces that are bound to each other by *declared connections*. AADL supports a graphical user interface and

contains language constructs that are concerned with the implementation of components and the grouping of components into more abstract units called *packages*.
There are tools available to analyze an AADL design from the point of view of timing and reliability. For a detailed description of AADL, refer to [Fei06]. The AADL
modeling of Integrated-Modular Avionics (IMA) using TTEthernet (see Sect. 7.5.2)
is given in [Rob16].

A detailed comparison of UML, SysML, AADL, and MARTE is given in
[Eve10]. Further architecture languages are GIOTTO and System C:

GIOTTO [Hen03] is a language for representing the design of a time-triggered
embedded system at the architectural level. GIOTTO provides for intermediate
abstractions that allow the design engineer to annotate the functional programming
modules with temporal attributes that are derived from the high-level stability analysis of the control loops. In the final development step, the assignment of the software modules to the target architecture, these annotations are constraints that must
be considered by the GIOTTO compiler.

System C is an extension of C++ that enables the seamless hardware/software
co-simulation of a design at the architectural level and provides for a step-by-step
refinement of a design down to the register transfer level of a hardware implementation or to a C program [Bla09]. System C is well suited to represent the functionality
of a design at the PIM level.

11.3.4   Test of a Decomposition

We do not know how to measure the quality of the result of the architecture design
phase on an absolute scale. The best we can hope to achieve is to establish a set of
guidelines and checklists that facilitate the comparison of two design alternatives
relative to each other. It is good practice to develop a project-specific checklist for
the comparison of design alternatives at the beginning of a project. The guidelines
and checklists presented in this section can serve as a starting point for such a
project-specific checklist.

Functional Coherence   A component should implement a self-contained function
with high internal coherence and low external interface complexity. If the component is a gateway, i.e., it processes input/output signals from its environment, only
the abstract message interface, the LIF, to the cluster and not the local interface to
the environment (see Sect. 4.3.1) is of concern at the level of architecture design.
The following list of questions is intended to help determine the functional coherence and the interface complexity of a component:

(i) Does the component implement a self-contained function?
(ii) Is the g-state (ground-state) of the component well defined?
(iii) Is it sufficient to provide a single level of error recovery after any failure, i.e.,
    a restart of the whole component? A need for a multilevel error recovery is
    always an indication of a weak functional coherence.

(iv) Are there any control signals crossing the message interface or is the interface of a component to its environment a strict data-sharing interface? A strict data-sharing interface is simpler and should therefore be preferred. Whenever possible, try to keep the temporal control within the subsystem that you are designing (e.g., on input, *information pull* is preferable over *information push*; see Sect. 4.4.1)!

 (v) How many different data elements are passed across the message interface? Are these data elements part of the interface model of the component? What are the timing requirements?

(vi) Are there any phase-sensitive data elements passed across the message interface?

Testability Since a component implements a single function, it must be possible to test the component in isolation. The following questions should help to evaluate the testability of a component:

 (i) Are the temporal as well as the value properties of the message interface precisely specified such that they can be simulated in a test environment?

 (ii) Is it possible to observe all input/output messages and the g-state of a component without the probe effect?

(iii) Is it possible to set the g-state of a component from the outside to reduce the number of test sequences?

(iv) Is the component software deterministic, such that the same input cases, will always lead to the same results?

 (v) What are the procedures to test the fault-tolerance mechanisms of the component?

(vi) Is it possible to implement an effective *built-in self-test* into the component?

Dependability The following checklist of questions refers to the dependability aspects of a design:

 (i) What is the effect of the worst malicious failure of the component to the rest of the cluster? How is it detected? How does this failure affect the minimum performance criterion?

 (ii) How is the rest of the cluster protected from a faulty component?

(iii) In case the communication system fails completely, what is the local control strategy of a component to maintain a safe state?

(iv) How long does it take other components of the cluster to detect a component failure? A short error detection latency simplifies the error handling drastically.

 (v) How long does it take to restart a component after a failure? Focus on the fast recovery from any kind of a single fault—a single Byzantine fault [Dri03]. The zero fault case takes care of itself and the two or more independent Byzantine fault case is expensive, unlikely to occur, and unlikely to succeed. How complex is the recovery?

(vi) Are the normal operating functions and the safety functions implemented in different components, such that they are in different FCUs?

(vii) How stable is the message interface with respect to anticipated change requirements? What is the probability and impact of changes of a component on the
rest of the cluster?

Energy and Power   Energy consumption is a critical non-functional parameter of
a mobile device. Power control helps to reduce the silicon die temperature and consequently the failure rate of devices:

(i) What is the energy budget of each component?
(ii) What is the peak power dissipation? How will peak power affect the temperature and the reliability of the device?
(iii) Do different components of an FCU have different power sources to reduce the
possibility of common mode failures induced by the power supply? Is there a
possibility of a common mode failure via the grounding system (e.g., lightning
stroke)? Are the FCUs of an FTU electrically isolated?

Physical Characteristics   There are many possibilities to introduce common-mode
failures by a careless physical installation. The following list of questions should
help to check for these:

(i) Are mechanical interfaces of the replaceable units specified, and do these
mechanical boundaries of replaceable units coincide with the diagnostic
boundaries?
(ii) Are the FCUs of an FTU (see Sect. 6.4.2) mounted at different physical loca-
tions, such that spatial proximity faults (e.g., a common mode external fault
such as water, EMI, and mechanical damage in case of an accident) will not
destroy more than one FCU?
(iii) What are the cabling requirements? What are the consequences of transient
faults caused by EMI interference via the cabling or by bad contacts?
(iv) What are the environmental conditions (temperature, shock, and dust) of the
component? Are they in agreement with the component specifications?

11.4   Design of Safety-Critical Systems

The economic and technological success of embedded systems in many applications leads to an increased deployment of computer systems in domains where a
computer failure can have severe consequences. A computer system becomes
safety-critical (or a *hard real-time system*) when a failure of the computer system
can have catastrophic consequences, such as the loss of life, extensive property
damage, or a disastrous damage to the environment.

> **Example:** Some examples of safety-critical embedded systems are as follows: a flight-
> control system in an airplane, an electronic stability program in an automobile, a train-
> control system, a nuclear reactor control system, medical devices such as heart pacemakers,
> the control of the electric power grid, or a control system of a robot that interacts
> with humans.

11.4.1   What Is Safety?

Safety can be defined as *the probability that a system will survive a given time-span without the occurrence of a critical failure mode that can lead to catastrophic consequences*. In the literature [Lal94], the magical number 10^9 hours, i.e., 115,000 years, is the MTTF (mean time to failure) that is associated with safety-critical operations. Since the hardware reliability of a VLSI component is less than 10^9 hours, a safety-aware design must be based on hardware-fault masking by redundancy. It is impossible to achieve confidence in the correctness of the design to the level of the required MTTF in safety-critical applications by testing only—extensive testing can establish confidence in a MTTF in the order of 10^4 to 10^5 hours [Lit93]. A formal reliability model must be developed in order to establish the required level of safety, considering the experimental failure rates of the subsystems and the redundant structure of the system.

Mixed-Criticality Architectures   Safety is a system property—the overall system design determines which subsystems are safety-relevant and which subsystems can fail without any serious consequences on the remaining safety margin. In the past, many safety-critical functions have been implemented on dedicated hardware, physically separated from the rest of the system. Under these circumstances, it is relatively easy to convince a certification authority that any unintended interference of safety-critical and non-safety-critical system functions is barred by design. However, as the number of interacting safety-critical functions grows, a sharing of communication and computational resources becomes inevitable. This results in a need of *mixed-criticality architectures*, where applications of different criticality can coexist in a single integrated architecture and the probability of any unintended interference, both in the domains of value and time, among these different-criticality applications must be excluded by architectural mechanisms. If mixed-criticality partitions are established by software on a single CPU, the *partitioning system software*, e.g., a hypervisor, is assigned the highest criticality level of any application software module that is executed on this system.

Fail-Safe Versus Fail-Operational   In Sect. 1.5.2, a *fail-safe system* has been defined as a system, where the application can be put into a *safe state* in case of a failure. At present, the majority of industrial systems that are safety-relevant fall into this category.

> **Example:** In most scenarios, a robot is in a safe state when it ceases to move. A robot control system is safe if it either produces correct results (both in the domain of value and time) or no results at all, i.e., the robot comes to a standstill. The safety requirement of a robot control system is thus a *high error detection coverage* (see Sect. 6.1.2).

In a number of applications, there exists a basic mechanical or hydraulic control system that keeps the application in a safe state in case of a failure of the computer control system that optimizes the performance. In this case, it is sufficient if the computer system is guaranteed to fail *cleanly* (see Sect. 6.1.3), i.e., inhibits its outputs when a failure is detected.

> **Example:** The ABS system in a car optimizes the braking action, depending on the surface condition of the road. If the ABS system fails cleanly, the conventional hydraulic brake system is still available to bring a car to a safe stop.

There exist safety-relevant embedded applications where the physical system requires the continuous computer control in order to maintain a safe state. A total loss of computer control may cause a catastrophic failure of the physical system. In such an application, which we call *fail-operational*, the computer must continue to provide an acceptable level of service, if failures occur within the computer system.

> **Example:** In a modern airplane, there is no mechanical or hydraulic backup to the computer-based flight control system. Therefore, the flight control system must be *fail-operational*.

Fail-operational systems require the implementation of active redundancy (as discussed in Sect. 6.4) to mask component failures.

In the future, it is expected that the number of *fail-operational systems* will increase for the following reasons:

(i) The cost of providing two subsystems based on different technologies—a basic mechanical or hydraulic backup subsystem for basic safety functions and an elaborate computer-based control system to optimize the process—will become prohibitive. The aerospace industry has already demonstrated that it is possible to provide fault-tolerant computer systems that meet challenging safety requirements.

(ii) If the difference between the functional capabilities of the computer-based control system and the basic mechanical safety system increases further and the computer system is available most of the time, then the operator may not have any experience in controlling the process safely with the basic mechanical safety system anymore.

(iii) In some advanced processes, computer-based nonlinear control strategies are essential for the safe operation of a process. They cannot be implemented in a simple safety system anymore.

(iv) The decreasing hardware costs make fail-operational (fault-tolerant) systems that require no expensive on-call maintenance competitive in an increasing number of applications.

(v) Autonomous systems, like self-driving cars, cannot rely on a human backup to exercise a mechanic or hydraulic backup.

11.4.2   Safety Analysis

The architecture of a safety-critical system must be carefully analyzed before it is put into operation in order to reduce the probability that an *accident* caused by a computer failure will occur.

Damage is a pecuniary measure for the loss in an accident, e.g., death, illness, injury, loss of property, or environmental harm. Undesirable conditions that have the potential to cause or contribute to an accident are called *hazards*. A hazard is thus a *dangerous state* that can lead to an accident, given certain environmental triggering conditions. Hazards have a *severity* and a *probability*. The severity is related to the worst potential damage that can result from the accident associated with the hazard. The severity of hazards is often classified in a severity class. The product of *hazard severity* and *hazard probability* is called *risk*. The goal of safety analysis and safety engineering is to identify hazards and to propose measures that eliminate or at least reduce the hazard or reduce the probability of a hazard turning into a catastrophe, i.e., to minimize the risk [Lev95]. A risk originating from a particular hazard should be reduced to a level that is *as low as reasonably practical (ALARP)*. This is a rather imprecise statement that must be interpreted with good engineering judgment. An action that is provided to reduce the risk associated with a hazard to a tolerable level is called a *safety function*. *Functional safety* encompasses the analysis, design, and implementation of safety functions. There exists an international standard, IEC 61508 on *functional safety*.

> **Example:** A risk minimization technique is the implementation of an independent safety monitor that detects a hazardous state of the controlled object and forces the controlled object into a safe state.

In the following, we discuss two safety analysis techniques, *fault tree analysis* and *failure mode and effect analysis*.

Fault Tree Analysis A *fault tree* provides graphical insight into the possible combinations of component failures that can lead to a particular system failure, i.e., an accident. Fault tree analysis is an accepted methodology to identify hazards and to increase the safety of complex systems [Xin08]. The fault tree analysis begins at the system level with the identification of the undesirable failure event (the *top event* of the fault tree). It then investigates the subsystem failure conditions that can lead to this top event and proceeds down the tree until the analysis stops at a basic failure, usually a component failure mode (events in ellipses). The parts of a fault tree that are still undeveloped are identified by the diamond symbol. The failure conditions can be connected by the AND or the OR symbol. AND connectors typically model redundancy or safety mechanisms.

> **Example:** Figure 11.1 depicts the fault tree of an electric iron. The undesirable top event occurs if the user of the electric iron receives an electric shock. Two conditions must be satisfied for this event to happen: the metal parts of the iron must be under high voltage (hazardous state) and the user must be in direct or indirect contact with the metal parts, i.e., the user either touches the metal directly or touches a wet piece of cloth that conducts the electricity. The metal parts of the iron will be under high voltage if the insulation of a wire that touches the metal inside the iron is defective and the *ground-current monitor* that is supposed to detect the hazardous state (the metal parts are under high voltage) is defective.

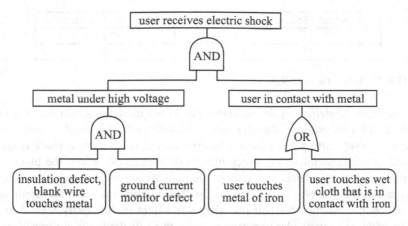

Fig. 11.1 Fault tree for an electric iron

Fault trees can be formally analyzed with mathematical techniques. Given the probability of basic component failures, the probability of the top event of a static fault tree can be calculated by standard combinatorial approaches.

Warm and cold spares, shared pools of resources, and sequence dependencies in which the order of the failure occurrence determines the state of the system require more elaborate modeling techniques. A fault tree that cannot by analyzed by combinatorial approaches is called a *dynamic fault tree*. A dynamic fault tree is transformed into a Markov chain that can be solved by numerical techniques. There are excellent computer tools available that assist the design engineer in evaluating the reliability and safety of a given design, e.g., Mobius [Dea02].

Failure Mode and Effect Analysis (FMEA) Failure mode and effect analysis (FMEA) is a bottom-up technique for systematically analyzing the effects of possible failure modes of components within a system to detect weak spots of the design and to prevent system failures from occurring. FMEA requires a team of experienced engineers to identify all possible failure modes of each component and to investigate the consequences of every failure on the service of the system at the system/user interface. The failure modes are entered into a standardized work sheet as sketched in Fig. 11.2.

A number of software tools have been developed to support the FMEA. The first efforts attempted to reduce the bookkeeping burden by introducing customized spreadsheet programs. Further efforts have been directed toward assisting the reasoning process and to provide a system-wide FMEA [Sta03].

FMEA is complementary to the fault tree analysis. While the fault tree analysis starts from the undesirable top event and proceeds down to the component failures that are the cause of this system failure, the FMEA starts with the components and investigates the effects of the component failure on the system functions.

Component	Failure mode	Failure effect	Probability	Criticality

Fig. 11.2 Worksheet for an FMEA

Dependability Modeling A *dependability model* is a model of a distributed system constructed for the purpose of analyzing the *reliability of behavior* of the envisioned system. A good starting point for a reliability model is a structure block diagram derived from the architectural representation of the design, where the blocks are components and the connection among components are the dependencies among the components. The blocks are annotated with the failure rates and the repair rates of components, where the repair rate after a transient fault, closely related to the g-state cycle, is of particular importance, since most of the faults are transients. If there is any dependency among the failure rates of components, e.g., caused by the co-location of components on the same hardware unit, these dependencies must be carefully evaluated, since the correlated failures of components have a strong impact on the overall reliability. The correlation of failures among replicated components in a fault-tolerant design is of particular concern. There are a number of software tools to evaluate the reliability and availability of a design, such as the Mobius tool [Dea02].

The dependability analysis establishes the *criticality* of each function for the analyzed mission. The criticality determines the level of attention that must be given to the component that implements the function in the overall design of the system.

An example for the criticality level assignment of functions with respect to the airworthiness of a computer system onboard an aircraft is given in Table 11.1.

11.4.3   Safety Case

A *safety case* is a combination of a set of sound and well-documented arguments supported by analytical and experimental evidence concerning the safety of a given design. The safety case must convince an independent certification authority that the system under consideration is safe to deploy. What exactly constitutes a proper safety case of a safety-critical computer system is a subject of intense debate.

Outline of the Safety Case The safety case must argue why it is extremely unlikely that faults will cause a catastrophic failure. The arguments that are included in the safety case will have a major influence on design decisions at later stages of the project. Hence, the outline of the safety case should be planned during the early stages of a project.

At the core of the safety case is a rigorous analysis of the envisioned hazards and faults that could arise during the operation of the system and could cause

Table 11.1 Criticality level

Criticality	Failure of function
Level A	Results in catastrophic failure condition for the aircraft
Level B	Results in hazardous/severe major failure condition for the aircraft
Level C	Results in major failure condition for the aircraft
Level D	Results in minor failure condition for the aircraft
Level E	Has no effect on aircraft operational capability or pilot workload

Adapted from [ARI92]

catastrophic effects, such as harm to humans, economic loss, or severe environmental damage. The safety case must demonstrate that sufficient provisions (engineering and procedural) have been taken to reduce the risk to a level that is acceptable to society and why some other possible measures have been excluded (maybe due to economic or procedural reasons). The evidence is accumulated as the project proceeds. It consists of management evidence (ensuring that all prescribed procedures have been followed), design evidence (demonstrating that an established process model has been followed), and testing and operational evidence that is collected during the test phases and the operational phases of the target system or similar systems. The safety case is thus a *living document*.

A safety case will combine evidence from independent sources to convince the certification authority that the system is safe to deploy. Concerning the type of evidence presented in a safety case, it is commonly agreed that:

(i) Deterministic evidence is preferred over probabilistic evidence (see Sect. 5.6).
(ii) Quantitative evidence is preferred over qualitative evidence.
(iii) Direct evidence is preferred over indirect evidence.
(iv) Product evidence is preferred over process evidence.

Computer systems can fail for external and internal reasons (refer to Sect. 6.1). External reasons are related to the operational environment (e.g., mechanical stress, external electromagnetic fields, temperature, wrong input) and to the system specification. The two main internal reasons for failure are the following:

(i) The computer hardware fails because of a random physical fault. Section 6.4 presented a number of techniques how to detect and handle random hardware faults by redundancy. The effectiveness of these fault-tolerance mechanisms must be demonstrated as part of the safety case, e.g., by fault injection (Sect. 12.5).
(ii) The design, which consists of the software and hardware, contains residual design faults. The elimination of the design faults and the validation that a design (software and hardware) is *fit for purpose* is one of the great challenges of the scientific and engineering community. No single validation technology can provide the required evidence that a computer system will meet ultrahigh dependability requirements.

Whereas standard fault-tolerance techniques, such as the replication of components for the implementation of triple-modular redundancy, are well established to mask the consequences of random hardware failures, there is no such standard technique known for the mitigation of errors in the design of software or hardware.

Properties of the Architecture   It is a common requirement of a safety-critical application that no single fault, which is capable of causing a catastrophic failure, may exist in the whole system. This implies that for a *fail-safe application*, every critical error of the computer must be detected within such a short latency that the application can be forced into the safe state *before* the consequences of the error affect the system behavior. In a *fail-operational application*, a safe system service must be provided even *after* a single fault in any one of the components has occurred.

Fault-Containment Unit (FCU)   At the architectural level, it must be demonstrated that *every* single fault can only affect a defined FCU and that it will be detected at the boundaries of this FCU. The partitioning of the system into independent FCUs is thus of utmost concern.

Experience has shown that there are a number of sensitive points in a design that can lead to a common-mode failure of all components within a distributed system:

 (i) A single source of time, such as a central clock.
 (ii) A babbling component that disrupts the communication among the correct components in a communication system with shared resources (e.g., a bus system).
(iii) A single fault in the power supply or in the grounding system.
(iv) A single design error that is replicated when the same hardware or system software is used in all components.

Design Faults   A disciplined software development process with inspections and design reviews reduces the number of design faults that are introduced into the software during initial development. Experimental evidence from testing, which in itself is infeasible to demonstrate the safety of the software in the ultra-dependable region, must be combined with structural arguments about the partitioning of the system into autonomous fault-containment units. The credibility can be further augmented by presenting results from formal analysis of critical properties and the experienced dependability of previous generations of similar systems. Experimental data about field-failure rates of critical components form the input to reliability models of the architecture to demonstrate that the system will mask random component failures with the required high probability. Finally, *diverse* mechanisms play an important role in reducing the probability of common-mode design failures.

Composable Safety Argument   Composability is another important architectural property and helps in designing a convincing safety case (see also Sect. 4.7.1). Assume that the components of a distributed system can be partitioned into two groups: one group of components that is involved in the implementation of safety-critical functions and another group of components that is not involved in safety-critical functions. If it can be shown at the architectural level that no error in any one of the not-involved components can affect the proper operation of the components that implement the safety-critical function, it is possible to exclude the not-involved components from further consideration during the safety case analysis.

11.4.4 Safety Standards

The increasing use of embedded computers in diverse safety-critical applications has prompted the appearance of many domain-specific safety standards for the design of embedded systems. This is a topic of concern, since differing safety standards are roadblocks to the deployment of a cross-domain architecture and tools. A standardized unified approach to the design and certification of safety-critical computer system would alleviate this concern.

In the following, we discuss two safety standards that have achieved wide attention in the community and have been practically used in the design of safety-relevant embedded systems.

IEC 61508 In 1998, the International Electrotechnical Commission (IEC) has developed a standard for the design of Electric/Electronic and Programmable Electronic (E/E/PE) safety-related systems, known as *IEC 61508 standard* on *functional safety*. The standard is applicable to any safety-related control or protection system that uses computer technology. It covers all aspects in the software/hardware design and operation of safety systems that operate on demand, also called *protection systems*, and safety-relevant control systems that operate in continuous mode.

> **Example:** An example for a *safety system that operates on demand* (a protection system) is an *emergency shutdown system* in a nuclear power plant.

> **Example:** An example of a *safety-relevant control system* is a control system in a chemical plant that keeps a continuous chemical process within safe process parameters.

The corner stone of *IEC 61508* is the accurate specification and design of the safety functions that are needed to reduce the risk to a level *as low as reasonably practical (ALARP)* [Bro00]. The safety functions should be implemented in an independent safety channel. Within defined system boundaries, the safety functions are assigned to safety integrity levels (SIL), depending on the tolerated *probability for a failure on demand* for protection systems and a *probability of failure per hour* for safety-relevant control systems (Table 11.2).

The *IEC 61508* standard addresses random physical faults in the hardware, design faults in hardware and software, and failures of communication in a distributed system. IEC 61508–2 deals with the contribution of fault tolerance to the dependability of the safety function. In order to reduce the probability of design faults of hardware and software, the standard recommends the adherence to a disciplined software development process and the provision of mechanisms that mitigate the consequences of remaining design faults during the operation of a system. It is

Table 11.2 Safety integrity level (SIL) of safety functions

Safety integrity level	Average tolerated probability for a failure per demand	Average tolerated probability for a failure per hour
SIL 4	$\geq 10^{-5}$ to $<10^{-4}$	$\geq 10^{-9}$ to $<10^{-8}$
SIL 3	$\geq 10^{-4}$ to $<10^{-3}$	$\geq 10^{-8}$ to $<10^{-7}$
SIL 2	$\geq 10^{-3}$ to $<10^{-2}$	$\geq 10^{-7}$ to $<10^{-6}$
SIL 1	$\geq 10^{-2}$ to $<10^{-1}$	$\geq 10^{-6}$ to $<10^{-5}$

interesting to note that dynamic reconfiguration mechanisms are not recommended in systems above SIL 1. IEC 61508 is the foundation for a number of domain-specific safety standards, such as the ISO 26262 standard for automotive applications, EN ISO 13849 for the machinery and off-highway industry, and IEC 60601 and IEC 62304 for medical devices.

> **Example:** [Lie10] gives an example for the assignment of the *automotive safety integrity level* (ASIL) according to ISO 26262 to the two tasks, the *functional task* and the *monitoring task*, of an electronic actuator pedal (EGAS) implementation. If a *certified monitoring task* that detects an unsafe state and is guaranteed to bring the system into a safe state is *independent* of the *functional task*, then the *functional task* does not have to be certified.

RTCA/DO-178C and DO-254   Over the past decades, safety-relevant computer systems have been deployed widely in the aircraft industry. This is the reason why the aircraft industry has extended experience in the design and operation of safety-relevant computer systems. The document *RTCA/DO-178C: Software Considerations in Airborne Systems and Equipment Certification* [ARI11] and the related document *RTCA/DO-254: Design Assurance Guidance for airborne electronic hardware* [ARI05] contain standards and recommendations for the design and validation of the software and hardware for airborne safety-relevant computer systems. These documents have been developed by a committee consisting of representatives of the major aerospace companies, airlines, and regulatory bodies and thus represent an international consensus view on a reasonable and practical approach that produces safe systems. Experience with the use of this standard has been gained within a number of major projects, such as the application of *RTCA/ DO-178B* (preceding RTCA/DO-178C) in the design of the Boeing 777 aircraft and follow-on aircrafts.

The basic idea of *RTCA/DO-178C* is a two-phase approach: in the first phase, the *planning phase*, the structure of the safety case, the procedures that must be followed in the execution of the project, and the produced documentation are defined. In the second phase, the *execution phase*, it is checked that all procedures that are established in the first phase are precisely adhered to in the execution of the project. The criticality of the software is derived from the criticality of the software-related function that has been identified during safety analysis and is classified according to Table 11.1. The rigor of the software development process increases with an increase in the criticality level of the software. The standard contains tables and checklists that suggest the design, validation, documentation, and project management methods that must be followed when developing software for a given criticality level. At higher criticality levels, the inspection procedures must be performed by personal that is independent from the development group. For the highest criticality level, *level A*, the application of formal methods is recommended, but not demanded.

When it comes to the elimination of design faults, both standards, *IEC 61508* and *RTCA/DO-178C*, demand a rigorous software development process, hoping that software is developed according to such a process will be free of design faults. From a certification point of view, an evaluation of the software product would be more appealing than an evaluation of the development process, but we must recognize there are fundamental limitations concerning the validation of a software product by testing [Lit95].

The *RTCA/DO-297 Integrated Modular Avionics (IMA) Development Guidance and Certification Considerations* standard addresses the role of design methodologies, architectures, and partitioning methods in the certification of modern integrated avionics systems in commercial aircraft. This standard also considers the contribution of time-triggered partitioning mechanisms in the design of safety-relevant distributed systems.

11.5   Design Diversity

Field data on the observed reliability of many large computer systems indicate that a significant and increasing number of computer system failures are caused by design errors in the software and not by physical faults of the hardware. While the problems of random physical hardware faults can be solved by applying redundancy (see Sect. 6.4), no generally accepted procedure to deal with the problem of design (software) errors has emerged. The techniques that have been developed for handling hardware faults are not directly applicable to the field of software, because there is no physical process that causes the aging of the software.

Software errors are design errors that have their root in the unmanaged complexity of a design. In [Boe01], the most common software errors are analyzed. Because many hardware functions of a complex VLSI chip are implemented in microcode that is stored in a ROM, the possibility of a design error in the hardware must be considered in a safety-critical system. The issue of a single design error that is replicated in the software of all nodes of a distributed system warrants further consideration. It is conceivable that an FTU built from nodes based on the same hardware and using the same system software exhibits common-mode failures caused by design errors in the software or in the hardware (micro-programs).

11.5.1   Diverse Software Versions

The three major strategies to attack the problem of unreliable software are the following:

(i) To improve the understandability of a software system by introducing a structure of conceptual integrity and by simplifying programming paradigms. This is, by far, the most important strategy that has been widely supported throughout this book.

(ii) To apply formal methods in the software development process so that the specification can be expressed in a rigorous form. It is then possible to verify formally—within the limits of today's technology—the consistency between a high-level specification expressed in a formal specification language and the implementation.

(iii) To design and implement diverse versions of the software such that a safe level of service can be provided even in the presence of design faults.

In our opinion, these three strategies are not contradictory, but complementary. An understandable and well-structured software system is a prerequisite for the application of any of the other two techniques, i.e., program verification and software diversity. In safety-critical real-time systems, all three strategies should be followed to reduce the number of design errors to a level that is commensurate with the requirement of ultrahigh dependability.

Design diversity is based on the hypothesis that different programmers using different programming languages and different development tools don't make the same programming errors. This hypothesis has been tested in a number of controlled experiments with a result that it is only partially encouraging [Avi85]. Design diversity increases the overall reliability of a system. It is, however, not justified to assume that the errors in the diverse software versions that are developed from the same specification are not correlated [Kni86].

The detailed analysis of field data of large software systems reveals that a significant number of system failures can be traced to flaws in the system specification. To be more effective, the diverse software versions should be based on different specifications. This complicates the design of the voting algorithm. Practical experience with non-exact voting schemes has not been encouraging [Lal94].

What place does software diversity have in safety-critical real-time systems? The following case study of a fault-tolerant railway signaling system that is installed in a number of European train stations to increase the safety and reliability of the train service is a good example of the practical utility of software diversity.

11.5.2   An Example of a Fail-Safe System

The VOTRICS train signaling system that has been developed by Alcatel [Kan95] is an industrial example of the application of design diversity in a safety-critical real-time environment. The objective of a train signaling system is to collect data about the state of the tracks in train stations, i.e., the current positions and movements of the trains and the positions of the switches, and to set the signals and shift the switches such that the trains can move safely through the station according to the given timetable entered by the operator. The safe operation of the train system is of utmost concern.

The VOTRICS system is partitioned into two independent subsystems. The first subsystem accepts the commands from the station operators, collects the data from the tracks, and calculates the intended position of the switches and signals so that the train can move through the station according to the desired plan. This subsystem uses a TMR architecture to tolerate a single hardware fault.

The second subsystem, called the *safety bag*, monitors the safety of the state of the station. It has access to the real-time database and the intended output commands of the first subsystem. It dynamically evaluates safety predicates that are derived from the traditional "rule book" of the railway authority. In case it cannot dynamically verify the safety of an intended output state, it has the authority to block the outputs to the switching signals, or to even activate an emergency shutdown of the complete station, setting all signals to red and stopping all trains. The safety bag is also implemented on a TMR hardware architecture.

The interesting aspect about this architecture is the substantial independence of the two diverse software versions. The versions are derived from completely different specifications. Subsystem one takes the operational requirements as the starting point for the software specification, while subsystem two takes the established safety rules as its starting point. Common mode specification errors can thus be ruled out. The implementation is also substantially different. Subsystem one is built according to a standard programming paradigm, while subsystem two is based on expert system technology. If the rule-based expert system does not come up with a positive answer within a pre-specified time interval, a violation of a safety condition is assumed. It is thus not necessary to analytically establish a WCET for the expert system (which would be very difficult).

The system has been operational in different railway stations over a number of years. No case has been reported where an unsafe state remained undetected. The independent safety verification by the safety bag also has a positive effect during the commission phase, because failures in subsystem one are immediately detected by subsystem two.

From this and other experiences, we can derive the general principle that in a safety-critical system, the execution of every safety-critical function must be monitored by a second independent channel based on a diverse design. *There should not be any safety-critical function on a single channel system.*

11.5.3   Multilevel System

The technique described above can also be applied to fail-operational applications that are controlled by a two-level computer system (Fig. 11.3). The higher-level computer system provides full functionality and has a high-error detection coverage. If the high-level computer system fails, an independent and differently designed

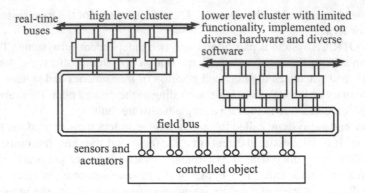

Fig. 11.3 Multilevel computer system with diverse software

lower-level computer system with reduced functionality takes over. The reduced functionality must be sufficient to guarantee safety.

Such an architecture has been deployed in the computer system for the space shuttle [Lee90, p. 297]. Along with a TMR system that uses identical software, a fourth computer with diverse software is provided in case of a design error that causes the correlated failure of the complete TMR system. Diversity is deployed in a number of existing safety-critical real-time systems, as in the Airbus fly-by-wire system [Tra88], and in railway signaling [Kan95].

11.6   Design for Maintainability

The total cost of ownership of a product is not only the cost of the initial acquisition of the product, but the sum of the acquisition cost, the cost of operation, the expected maintenance cost over the product life, and finally, at the end of the product lifetime, the cost of product disposal. *Design for maintainability* tries to reduce the expected maintenance cost over the product lifetime. The cost of maintenance, which can be higher than the cost of the initial acquisition of the product, is strongly influenced by the product design and the maintenance strategy.

11.6.1   Cost of Maintenance

In order to be able to analyze the cost structure of a maintenance action, it is necessary to distinguish between two types of maintenance actions: *preventive mainte-nance* and *on-call maintenance*.

Preventive maintenance (sometimes also called *scheduled* or *routine mainte-nance*) refers to a maintenance action that is scheduled to take place periodically at planned intervals, when the plant or machine is intentionally shut down for maintenance. Based on knowledge about the increasing failure rate of components and the results of the analysis of the anomaly detection database (see Sect. 6.3), components that are *expected to fail* in the near future are identified and replaced during preventive maintenance. An effective scheduled maintenance strategy needs extensive component instrumentation to be able to continually observe component parameters and learn about the imminent wear-out of components by statistical techniques.

On-call maintenance (sometimes also called *reactive maintenance*) refers to a maintenance action that is started after a product has failed to provide its service. By its nature, it is unplanned. In addition to the *direct repair cost*, the on-call maintenance costs comprise the cost of *maintenance readiness* (to ensure the immediate availability of a repair team in case a failure occurs) and the cost of *unavailability of service* during the interval starting with the failure occurrence until the repair action has been completed. If an assembly line must be stopped during the *unavailability-of-service interval*, the cost of *unavailability of service* can be substantially higher than the initial product acquisition cost and the repair cost of the failed component.

> **Example:** It is a goal of plant managers to reduce the probability for the need of on-called maintenance action as far as possible, ideally to zero.

> **Example:** In the airline industry, unscheduled maintenance of an airplane means lost connections and extra cost for the lodging of passengers.

Another aspect that influences the cost of maintenance relates to the question whether *permanent hardware faults* or *software errors* are considered. The repair of a permanent hardware fault requires the physical replacement of the broken component, i.e., the spare part must be available at the site of failure and must be installed by a physical maintenance action. Given an appropriate infrastructure has been set up, the repair of a software fault can be performed remotely by downloading a new version of the software via the Internet with minimal or without any human intervention.

11.6.2 Maintenance Strategy

The design for maintenance starts with the specification of a maintenance strategy for a product. The maintenance strategy will depend on the *classification of components*, on the *maintainability/reliability/cost* tradeoff of the product, and the *expected use* of the product.

Component Classification   Two classes of components must be distinguished from the point of view of maintenance: components that exhibit *wear-out failures* and components that exhibit *spontaneous failures*. For components that exhibit wear-out failures, physical parameters must be identified that indicate the degree of wear-out. These parameters must be continually monitored in order to periodically establish the degree of wear-out and to determine whether a replacement of the component must be considered during the next scheduled maintenance interval.

> **Example:** Monitoring the temperature or the vibration of a *bearing* can produce valuable information about the degree of wear-out of the bearing before it actually breaks down. In some manufacturing plants, more than one hundred thousand sensors are installed to monitor wear-out parameters of diverse physical components.

If it is not possible to identify a measurable wear-out parameter of a component or to measure such a parameter, another conservative technique of maintenance is the *derating of components* (i.e., operating the components in a domain where there is minimal stress on the components) and the systematic replacement of components after a given interval of deployment during a scheduled maintenance interval. This technique is, however, quite expensive.

For components with a spontaneous failure characteristic, such as many electronic components, it is not possible to estimate the interval that contains the instant of failure ahead of time. For these components, the implementation of *fault tolerance*, as discussed in Sect. 6.4, is the *technique of choice* to shift on-call maintenance to preventive maintenance.

Maintainability/Reliability Tradeoff This tradeoff determines the design of the field-replaceable units (FRU) of a product. An FRU is a unit that can be replaced in the field in case of failure. Ideally, an FRU consists of one or more FCUs (see Sect. 6.1.1) in order that effective diagnosis of an FRU failure can be performed. The size (and cost) of an FRU (a *spare part*) is determined by a cost analysis of a maintenance action on one side the and the impact of the FCU structure on the reliability of the product on the other side. In order to reduce the time (and cost) of a repair action, the mechanical interfaces around an FRU should be easy to connect and disconnect. Mechanical interfaces that are easy to connect or disconnect (e.g., a plug) have a substantially higher failure rate than interfaces that are firmly connected (e.g., a solder connection). Thus, the introduction of FRU structure will normally *decrease* the product reliability. The most reliable product is one that cannot be maintained. Many consumer products fall into this category, since they are designed for optimal reliability—if the product is broken, it must be replaced as whole by a new product.

Expected Use The expected use of a product determines whether a failure of the product will have serious consequences—such as the *downtime* of a large assembly line. In such a scenario, it makes economic sense to implement a fault-tolerant electronic system that masks a spontaneous permanent failure of an electronic device. At the next scheduled maintenance interval, the broken device can be replaced, thus restoring the fault-tolerance capability. Hardware fault tolerance thus transforms the expensive on-call maintenance action to a lower-cost scheduled maintenance action. The decreasing cost of electronic devices on the one side and the increasing labor cost and the cost of production loss during on-call maintenance on the other side shift the break-even point for many electronic control systems toward fault-tolerant systems.

In an ambient intelligence environment, where smart Internet-enabled devices are placed in many homes, the maintenance strategy must ensure that non-experts can replace broken parts. This requires an elaborate diagnostic subsystem that diagnoses a fault to an FRU and orders the spare part autonomously via the Internet. If the spare part is delivered to the user, the inexperienced user must be capable to replace the spare part with minimal effort in order to restore the fault tolerance of the system with minimum mental and physical effort.

Example The maintenance strategy of the *Apple iPhone* relies on the complete replacement of a broken hardware device, eliminating the need for setting up an elaborate hardware maintenance organization. Software errors are corrected semi-automatically by downloading a new version of the software from the *Apple iTunes store.*

11.6.3   Software Maintenance

The term *software maintenance* refers to all needed software activities to provide a useful service in a changing and evolving environment. These activities include:

- *Correction of software errors*: It is difficult to deliver error-free software. If dormant software errors are detected during operation in the field, the error must be corrected and a new software version must be delivered to the customer.
- *Elimination of vulnerabilities:* If a system is connected to the Internet, there is a high probability that any existing vulnerability will be detected by an intruder and used to attack and damage a system that would otherwise provide a reliable service.
- *Adaptation to evolving specifications:* A successful system changes its environment. The changed environment puts new demands on the system that must be fulfilled in order to keep the system relevant for its users.
- *Addition of new functions:* Over time, new useful system functions will be identified that should be included in a new version of the software.

The connection of an embedded system to the Internet is a mixed blessing. On the one side, it makes it possible to provide Internet-related services and to download a new version of the software remotely, but on the other side it enables an adversary to exploit vulnerabilities of a system that would be irrelevant if no Internet connection were provided.

Any embedded system that is connected to the Internet must support a *secure download service* [Obe09]. This service is absolutely essential for the continued remote maintenance of the software. The secure download must use strong cryptographic methods to ensure that an adversary cannot get control of the connected hardware device and download a software of its liking.

Example: A producer of modems sold ten thousands of modems all over the world before hackers found out that the modems contained a *vulnerability*. The producer did not consider to provide the infrastructure for a secure download service for installing a new corrected version of the software remotely.

11.7   The Time-Triggered Architecture

In this final section of the chapter on System Design, we present a short overview of the design principles of the time-triggered architecture that has been developed, first at the Technical University of Berlin and then at the Technical University of Vienna over a period of 20 years. The TTA integrates many of the ideas presented in this book. After building a number of industrial prototypes at the Technical University of Vienna, the company TTTech (time-triggered technology) has been formed to further develop and market the time-triggered technology to the industrial community. Today, the TTA is deployed in a number of challenging aerospace, automotive, and industrial projects.

In the following paragraphs, we discuss the key architectural principles of the TTA. A *principle* is an accepted statement about some fundamental insight in a domain of discourse. Principles establish the basis of the formulation of operational rules, i.e., the *services* of an architecture.

11.7.1   Principle of a Consistent Global Time

Of distinct importance for the TTA is the *principle of a consistent global time* (see Sect. 2.5). Embedded computer systems must interact with the physical environment that is ruled by the progression of physical time. The progression of physical time is thus a first-order citizen and not an add-on to the cyber-model that is the basis of the computer control of the physical environment. The availability of a fault-tolerant sparse global time base in every node of a large embedded system is at the foundation of the TTA. This global time base helps to simplify a design. In the TTA, this global time base is used

- To establish a *consistent temporal order* of all relevant events and to solve the problem of simultaneity in a distributed computer system. The consistent temporal order is a prerequisite for introducing the notion of a consistent state of a distributed system. A consistent notion of state is needed when a new (repaired) component must be reintegrated into a running system.
- To precisely specify the temporal properties of interfaces in *Interface Control Documents*, such that a component-based design-style can be followed.
- To establish *conflict-free time-controlled communication channels* for the transport of time-triggered (TT) messages. TT messages have a short and predictable latency and thus help to reduce the dead time in distributed phase-aligned control loops.

- To construct unidirectional multicast communication channels with error detection at the receiver, where a failure of a receiver cannot propagate back to the sender.
- To build systems with *deterministic behavior* that are easy to understand, avoid the occurrence of *Heisenbugs*, and support the straightforward implementation of fault-masking by redundancy.
- To monitor and extend (by state estimation) the *temporal accuracy of real-time* data and to ensure that a control action at the interface to the physical environment is based on data that is temporally accurate.

11.7.2   Principle of Component Orientation

The notion of a component introduced in Chap. 4 is the *primitive structure element* of the time-triggered architecture. A TTA component is a self-contained hardware/ software unit that interacts with its environment solely by the exchange of messages. A component is a unit of design and a unit of fault containment (FCU). The message-based linking interfaces (LIFs) of a component are precisely specified in the domains of time and value and are agnostic about the concrete implementation technology of a component. A component can be used on the basis of their *interface specification*, contained in the *Interface Control Documents*, without knowing the internals of the component's implementation. The time-triggered integration framework of the TTA ensures that real-time transactions that span over more than one component have defined end-to-end temporal properties. In a time-critical RT transaction, the computations of the components and the message transport by the time-triggered communication system can be *phase-aligned*.

A component can be expanded to a new cluster without changing the specification of the LIF of the original component that becomes the *external LIF* of the new cluster. After such an expansion, the external LIF is provided by a gateway component that supports on the other side a second LIF to the new (expanded) cluster (the *cluster LIF*). We thus have a *recursive component concept* in the TTA. Depending on the point of view taken, a set of components can be viewed as a cluster (focus on the *cluster LIF* of the gateway component) or as a single component (focus on the *external LIF* of the gateway component). This recursive component concept makes it possible to build well-structured systems of arbitrary size within the TTA.

Components can be integrated to form hierarchical structures or network structures. In a hierarchical structure, a designated gateway component links different levels of the hierarchy. We now take the view from the lower level of the hierarchy. The designated gateway component has two LIFs, one to the lower level of the hierarchy (the *cluster LIF*) and another one to the higher level of the hierarchy (the *external LIF*). Since the different hierarchical levels can obey different architectural styles, the designated gateway component must resolve the ensuing property mismatches. The *external LIF* of the gateway component, viewed from the lower level, is a *local unspecified interface* of the cluster. Vice versa, the *cluster LIF* of the gateway component becomes a *local unspecified interface* when viewed from above.

11.7.3   Principle of Coherent Communication

The basic communication mechanism of the TTA is the unidirectional BMTS (basic message transport service) that follows, whenever possible, the fate-sharing model of the Internet. The fate-sharing model was formulated by David Clark, an architect of the DARPA net, as follows: *The fate-sharing model suggests that it is acceptable to lose the state information associated with an entity if, at the same time, the entity itself is lost* [Cla88, p.3]. The fate- sharing principle demands that all state information that is associated with a message transfer must be stored in the end points of the communication. Even in the design of the time-triggered network-on-chip (TTNoC), the fate-sharing principle was considered. The fate-sharing principle can be applied in a safety-critical configuration if the end systems are guaranteed to be fail-silent. Otherwise, information about the intended temporal behavior of the end systems has to be also stored in an independent FCU, a *guardian* or the *network*, to contain the faulty behavior of a babbling node (see also Sect. 7.1.2).

As long as different subsystems of the TTA are connected by time-triggered communication systems, such as the TTNoC, TTP, or TTEthernet, the BMTS is characterized by a *constant* transport delay and a minimal jitter of one granule of the global time. If a message is transported via an event-triggered protocol (e.g., in the Internet), no such temporal guarantee can be given.

This single coherent communication mechanism makes it possible to move a component (which can be an IP-core of a system-on-chip) to another physical location without changing the basic communication mechanism among the components.

11.7.4   Principle of Fault Tolerance

The principle of fault tolerance deals with the occurrence of faults in safety-critical systems. It states that in any complex safety-critical application, the occasional failure of a subsystem during the lifetime of the subsystem cannot be avoided and therefore fault-tolerance mechanisms must be put into place to ensure that such a failure will not lead to an accident.

Most dependability engineers that are working on the design and validation of safety-critical systems would agree that there is strong experimental evidence that it is impossible to overcome the constraints that are summed up in the following *four impossibility results*.

- It is impossible to find all design faults in a large and complex monolithic software system. Experience with large systems (more than a million lines of code) has shown that it is practically impossible to eliminate all design faults. Although most design faults that occur during normal operation can be eliminated during operational testing, some design faults in the code that deal with the extremely rare *edge cases* often remain undetected.

- It is impossible to avoid single event upsets in non-redundant hardware during the lifetime of an ultra-dependable system. A single-event upset (SEU) is a hardware failure that can be caused by cosmic radiation. Normally, an SEU does not permanently damage the hardware. Li et al. [Li17] have shown that a single bit-flip in a middle layer of a machine learning convolution network can cause the misclassification of a *truck* as a *bird*.
- It is impossible to establish the ultrahigh dependability of a large monolithic system by testing and simulation. In an archival paper on *Validation of Ultra-High Dependability for Software-Based Systems*, Littlewood and Strigini [Lit93] argue that no solution exists for the validation of ultrahigh dependability by testing of a monolithic systems relying on complex software.
- It is impossible to precisely specify all edge cases that can be encountered in real-life situations. Uncovered edge cases, such as a non-thought-of situation in a driving scenario, can be the cause of an accident.

Bibliographic Notes

Many books have been written about design, most of them emanating from the field of architecture design. *Design Methods, Seeds of Human Futures* by Jones [Jon78] takes an interdisciplinary look at design and makes an enjoyable reading for a computer scientist, as well as the book *A Pattern Language* [Ale97] by Christopher Alexander. The excellent books *Systems Architecting, Creating and Building Complex Systems* [Rec91] and *The Art of Systems Architecting* [Rec02] by Eberhardt Rechtin presents many empirically observed design guidelines that have been an important input for writing this chapter. The problem of software design for embedded systems is discussed in [Lee02]. The book *Embedded System Design* by Gajski [Gaj09] covers topics of hardware synthesis and verification.

Points to Remember

- In his recent book [Bro10], Fred Brook states that *conceptual integrity of a design is the result of a single mind.*
- Constraints limit the design space and help the designer to avoid the exploration of design alternatives that are unrealistic in the given environment. Constraints are thus our friends, not our adversaries.
- Software per se is an *action plan* describing the operations of a real or virtual machine. A plan by itself (without a machine) does not have *any temporal dimen-*

sion, cannot have *state*, and has no *behavior*. This is one of the reasons why we consider the *component* and not the *job* as the primitive construct at the level of architecture design of an embedded system.

- *Purpose analysis*, i.e., the analysis why a new system is needed and what is the ultimate goal of a design must precede the requirements analysis.
- The analysis and understanding of a large problem is never complete, and there are always good arguments for asking more questions concerning the requirements before starting with the *real* design work. The paraphrase *paralysis by analysis* has been coined to point out this danger.
- Model-based design is a design method that establishes a useful framework for the development and integration of executable models of the controlled object and of the controlling computer system.
- Component-based design is a *meet-in-the middle* design method. On the one side, the *functional and temporal requirements* on the components are derived top-down from the desired application functions. On the other side, the *functional and temporal capabilities* of the components are contained in the specifications of the available components.
- Safety can be defined as *the probability that a system will survive a given time-span without the occurrence of a critical failure mode that can lead to catastrophic consequences*.
- *Damage* is a pecuniary measure for the loss in an accident, e.g., death, illness, injury, loss of property, or environmental harm. Undesirable conditions that have the potential to cause or contribute to an accident are called *hazards*. A hazard is thus a *dangerous state* that can lead to an accident, given certain environmental triggering conditions.
- Hazards have a *severity* and a *probability*. The severity is related to the worst potential damage that can result from the accident associated with the hazard. The severity of hazards is often classified in a severity class. The product of *hazard severity* and *hazard probability* is called *risk*.
- The goal of safety analysis and safety engineering is to identify hazards and to propose measures that eliminate or at least reduce the hazard or reduce the probability of a hazard turning into a catastrophe, i.e., to minimize the risk.
- A *safety case* is a combination of a sound set of well-documented arguments supported by analytical and experimental evidence concerning the safety of a given design. The safety case must convince an independent certification authority that the system under consideration is safe to deploy.
- It is a goal of plant managers to reduce the probability for the need of on-called maintenance action as far as possible, ideally to zero.
- The connection of an embedded system to the Internet is a mixed blessing. On the one side, it makes it possible to provide Internet-related services and to download a new version of the software remotely, but on the other side, it enables an adversary to exploit vulnerabilities of a system that would be irrelevant if no Internet connection were provided.

Review Questions and Problems

11.1 Discuss the advantages and disadvantages of *grand design* versus *incremental development*.

11.2 Which are the characteristics of a "wicked" problem?

11.3 Why is the notion of a component, a hardware/software unit, introduced as the basic building block of a system? What are the problems with the notion of a software component in the context of real-time system design?

11.4 Discuss the different types of constraints that restrict a design. Why is it important to explore these constraints before starting a design project?

11.5 *Model-based design* and *component-based design* are two different design strategies. What are the differences?

11.6 What are the concepts behind UML-MARTE and AADL?

11.7 Which are the results of the architecture design phase?

11.8 Establish a checklist for the evaluation of a design from the point of view of *functional coherence, testability, dependability, energy and power*, and *physical installation*.

11.9 What is a safety case? Which is the preferred evidence in a safety case?

11.10 Explain the safety analysis techniques of fault tree analysis and failure mode and effect analysis!

11.11 What is the key idea behind the safety standard IEC 6105?

11.12 What is a SIL?

11.13 What are the advantages and disadvantages of design diversity?

11.14 Discuss the reliability/maintainability tradeoff!

11.15 Why do we need to *maintain* software?

11.16 Why is a secure download service essential if an embedded system is connected to the Internet?

Chapter 12
Validation

Overview

This chapter deals with assessment technologies. These technologies must convince a designer, user, or a certification authority that the developed computer system is safe to deploy and will fulfill its intended function in the planned real-world environment. In the first section of this chapter, we elaborate on the differences between *validation* and *verification*. Validation deals with the consistency between the informal *model of the user's intention* and the behavior of the *system-under-test (SUT)*, while verification deals with the consistency between a given *(formal) specification* and the *SUT*. The missing link between validation and verification are errors in the specification. The following section deals with the challenges of testing and the preferred validation technique. At the core of testing are the interference-free observability of results and the controllability of the inputs. The design for testability provides a framework that supports these characteristics. In most cases, only a tiny fraction of the input space can be examined by test cases. The proper selection of test cases should justify the assumption that, given the results of the test cases are correct, the system will operate correctly all over the input domain. In digital systems, the validity of such an induction is doubtful, since digital inputs are not continuous but discrete—a single bit-flip can make a correct result erroneous. The decision whether the result of a test input is correct is delegated to a *test oracle*. The automation of test oracles is another challenge in the domain of testing. Model-based design, where a model of the plant and a model of the computer controller are interconnected to study the performance of closed-loop control systems is a promising route toward the automation of the test oracle. Given that a complete formal model of a design is available, formal methods can be deployed to check whether selected properties hold in all possible states of the model. In the last few years, the technique of model checking has matured such that it can handle systems of industrial size. The correct operation of the fault-masking mechanisms of a fault-tolerant system can only be assessed if the input space is extended to include the faults the system is supposed to tolerate. In the last section, the topics of physical fault

© The Author(s), under exclusive license to Springer Nature Switzerland AG 2022   307
H. Kopetz, W. Steiner, *Real-Time Systems*,
https://doi.org/10.1007/978-3-031-11992-7_12

injection and software-based fault injection are covered. Since any physical sensor or actuator will eventually fail, fault-injection campaigns must establish the safe operation of a system even in the case that any particular sensor or actuator has failed.

12.1   Validation Versus Verification

An essential fraction—up to 50%—of the development costs of a real-time computer system is devoted to ensure that the system is *fit-for-purpose*. In safety-critical applications that must be certified, this fraction is even higher.

When developing an embedded computer system, it is useful to distinguish between three different types of system representations [Gau05]:

(i) The informal *model of the user's intention* that determines the role of the embedded computer system in the given real-world application context. In the embedded world, this model deals with the relationships of the computer inputs and outputs to the effects in the physical environment. This model is usually not fully documented and informal, since in most cases it is not possible to think about and formalize all system aspects that are relevant in a real-world scenario.

(ii) The model of the *system specification* which captures and documents, either in natural language or in some formal notation, the intentions of the client and the obligations of the system developers *as understood by the person or the group of persons who develop the (formal) specification.*

(iii) The *system under test (SUT)* (the result of the system development) that should perform the system functions according to the *model of the user's intention*.

Verification establishes the consistency between a *(formal) system specification* and the *SUT*, while *validation* is concerned with the consistency between the *model of the user's intention* and the *SUT*. The missing link between verification and validation is the relation between the *(informal) model of the user's intention* and the *(formal) specification* of the system. We call errors that occur in this phase of development *specification errors*, while we call errors that occur during the transformation of a given specification to the SUT *implementation errors*. While *verification* can, in theory, be reduced to a formal process, *validation* must examine the system's behavior in the real world. If properties of a system have been formally verified, it still has not been established whether the existing formal specification captures all aspects of the intended behavior in the user's environment, i.e., if it is free of *specification errors*. Sometimes the term *specification testing* is used to find out whether the specification is consistent with the model of the user's intentions [Gau05].

Validation, specification testing, and verification are thus three complementary means to support quality assurance. The prime validation method is *testing*, while the ideal verification method is *formal analysis*.

During *testing*, the behavior of a real-time computer system is exercised at carefully selected points of the input domain, and the corresponding results in the domains of value and time are classified as *correct* or *erroneous*. It is assumed, given that the test cases have been properly selected and correctly executed, that the *induction* that the program will operate correctly at all points of the enormous input space is justified. In a digital system, where the change of a single bit can have drastic consequences on the behavior, this induction is *fragile*. If we take a purely probabilistic point of view, an estimate that the *mean time to failure* (MTTF) of the SUT will be larger than a given *number of hours* can only be made if operational tests have been executed for a duration that corresponds to this *number of hours* [Lit93]. In practice, this means that it is not possible to establish an MTTF of more than 10^3–10^5 hours by operational testing. This is orders of magnitude lower than the desired MTTF of safety-critical systems, which is in the order of 10^9 hours.

The main shortcoming of formal methods is the missing link between the informal *model of the user's intention* and the *formal specification* that is the reference for assessing the correctness of the system. This gap is systematically explored in [Kop22]. In particular, the *specification dilemma* is discussed [Kop22, Section 11.1]:

> The proposal to use tokens of a formal symbolic language is only a partial solution of this specification dilemma. Although the relations among the tokens of a formal language are precisely defined, the token themselves are not grounded in the real world. (A token is a *meaningless placeholder* in a formal symbolic language, e.g., a variable. The process of assigning a real-world meaning to a token, typically through natural language, is called *grounding*.)

Furthermore, only a subset of the properties relevant for the system operation can be captured in *formal properties* that are examined during the formal analysis.

12.2   Testing Challenges

Observability of the outputs of the *SUT* and *controllability* of the test inputs are the core of any testing activity.

In non-real-time systems, the *observability* and *controllability* are provided by test and debug monitors that halt the program flow at a test point and give the tester the opportunity to monitor and change program variables. In distributed real-time systems, such a procedure is not suitable for the following two reasons:

(i) The temporal delay introduced at the test points modifies the temporal behavior of the system in such a manner that existing errors can be hidden and new errors can be introduced. This phenomenon is called the *probe effect*. Probe effects have to be avoided when testing real-time systems.

(ii) In a distributed system, there are many *loci* of control. The halting of one control path introduces a temporal distortion in the coordinated control flow that can lead to new errors.

12.2.1 Design for Testability

By *design for testability*, we mean the design of a framework and the provision of mechanisms that facilitate the testing of a system. The following techniques improve the testability:

 (i) Partitioning the system into composable subsystems with observable and in the domains of value and time well-specified interfaces. It is then possible to test each subsystem in isolation and to limit the integration effects to the testing of the *emerging* behavior. Probe effect-free observability of subsystem outputs at any level of the architecture was one of the motivations to provide *multicasting* in the basic message transport service (see Sect. 4.1.1) of the time-triggered architecture.

 (ii) Establishment of a static temporal control structure such that the temporal control structure is *independent* of the input data. It is then possible to test the temporal control structure in isolation.

(iii) Reducing the temporal dimension of the input space by introducing a sparse time base of proper granularity. The granularity of this time base should be sufficient for the application at hand but should not be any smaller. The smaller the granularity of the sparse time base, the larger the potential input space in the temporal domain. By decreasing the size of the input space or by increasing the number of non-redundant test cases the *test coverage*, i.e., the fraction of the total input space that is covered by the tests, can be improved.

(iv) Publication of the ground state of a node in a g-state message at the periodic reintegration point. The ground state can then be observed by an independent component without probe effect.

 (v) Provision of determinism in the software such that the same output messages will be produced if the same input messages are applied to a component.

Because of their deterministic properties and their static control structure, time-triggered systems are easier to test than event-triggered systems.

12.2.2 Test Data Selection

During the test phase, only a tiny fraction of the potential input space of a computer system can be exercised. The challenge for the tester is to find an effective and representative *test data set* that will give the designer confidence that the system will work correctly for *all* inputs. In this section, we present some methods for test data selection:

Random Test Data Selection: Test data are selected randomly without any consideration of the program structure or the operational profile of use.

Requirements Coverage: In this method, the requirements specification is the starting point for selecting the test data. For each one of the given requirements, a set of test cases is designed to check whether the requirement is satisfied. The hidden assumption in this criterion is that the set of requirements is complete.

White Box Testing: The internal structure of a system is examined to derive a set of test data such that some kind of coverage criterion is satisfied, e.g., that all statements have been executed or that all branches of a program have been tested. This test data selection criterion is most effective for unit testing, where the internals of the component implementation are available.

Model-Based Test Data Selection: The test data is derived from a model of the system under test and a model of the physical plant. Model-based test data selection can be automated, since the correctness of test results can be related to a performance criterion of the physical process.

> **Example**: Consider the case where a *controller of an automotive engine* is tested versus a *model of this engine*. The model of the engine has been extensively validated with respect to the operation of the real engine and is assumed to be correct. The control algorithms that are implemented in the controller determine the performance parameters of the engine such as energy efficiency, torque, pollution, etc. By observing the performance parameters of the engine, we can detect anomalies that are caused by a misbehavior of the controller software.

Operational profile: The basis of the test data selection is the operational profile of the *system under test* in the given application context. This test data selection criterion misses rare events.

Peak load: A hard real-time system must provide the specified timely service under all conditions covered by the load and fault hypothesis, i.e., also under peak loads that are caused by *rare events*. The peak load scenario puts extreme stress on the system and should be tested extensively. The behavior of the system in above peak load situations must also be tested. *If peak load activity is handled correctly, the normal load case will take care of itself.* In most cases, it is not possible to generate rare events and peak load in the real-world operational environment. Therefore peak load testing is best performed in a model-based test environment.

Worst-Case Execution Time (WCET): To determine the WCET of a task experimentally, the task source code can be analyzed to generate a test data set that is biased toward the worst-case execution time.

Fault-Tolerance Mechanisms: Testing the correctness of the fault-tolerance mechanism is difficult, because faults are not part of the *normal* input domain. Mechanisms must be provided that can activate the faults during the test phase. For example, software- or hardware-implemented fault injection can be used to test the correctness of the fault-tolerance mechanisms (see the following Sect. 12.5).

Cyclic systems: If a system has a cyclic behavior (many control systems are cyclic), the crossing of a particular phase of the cycle is a repetitive event in the temporal domain. In many cyclic systems, it is sufficient to test all events that occur in a single cycle.

The above list of test data selection criteria is not complete. A survey study on the effectiveness of different test data selection criteria is contained in [Jur04].

Selecting the test data by using a combination of the above criteria seems to be more effective than relying on a single criterion in isolation.

In order to be able to judge the quality of a test data set, different *coverage measures* have been introduced. A *coverage measure* describes the degree to which a test data set exercises the SUT. Common coverage criteria are:

  (i) *Function coverage*—has every function been exercised?
 (ii) *Statement coverage*—has every statement of the source code been executed?
(iii) *Branch coverage*—has every branch instruction been executed in all directions?
(iv) *Condition coverage*—has every Boolean condition been fully exercised?
 (v) *Fault coverage*—have the fault-tolerant mechanisms been tested for every fault
      that is contained in the fault hypothesis?

12.2.3   Test Oracle

Given that a set of test cases has been selected, a method must be provided to determine whether the result of a test case produced by the SUT is acceptable or not. In the literature, the term *test oracle* is used to refer to such a method. The design of an *algorithmic test oracle* is a prerequisite of any test automation, which is needed for reducing the cost of testing.

In practice, the judgment whether the result of a test case is in conformance with a natural language representation of the *model of the user's intention* is often delegated to a human. Model-based design and model-based testing can help to partially solve the problem.

The structured design process, discussed in Sect. 4.4 and Chap. 11, distinguishes between the PIM (platform-independent model) and the PSM (platform-specific model) of a component. An executable representation of the complete interface behavior (in the domains of value and time) at the PIM level of a design can act as the reference for the adjudication of a test result at the PSM level and thus help to detect *implementation errors*. The *oracle challenge* is thus shifted from the PSM level to the PIM level. Since the PIM is developed in an early phase of the design, errors can be captured early in the life cycle, which reduces the cost associated with correction of the errors.

The LIF specification of a component (see Sect. 4.4.2) should contain *input assertions* and *output assertions*. Input assertions limit the input space of the component and exclude input data that the component is not designed to handle. Output assertions help to immediately detect errors that occur inside a component. Both input assertions and output assertions can be considered to act as a *test oracle light* [Bar01]. Since the PIM is not resource constrained, the wide use of input assertions and output assertions at the PIM level can help to debug the PIM specification. In the second phase, when the PIM is transformed to the PSM, some of these assertions can be removed to arrive at an efficient code for the target machine.

In a time-triggered system, where the temporal control structure is static, the detection of temporal errors in the execution of a program at the PSM level is straightforward and can be automated.

12.2.4   System Evolution and Technology Readiness Levels (TRLs)

Most successful systems evolve over time. Existing deficiencies are corrected, and new functions are introduced in new versions of the system. The validation of these new versions must be concerned with two issues:

(i) Regression *testing:* checking that the functionality of the previous version (that must be supported in the new version) has not been modified and is still correct.
(ii) *New-function testing*: checking that the functionality that is new to the latest version is implemented correctly.

Regression testing can be automated by executing the test data set of the previous version on the new version. The anomalies that have been detected in the old version (see Sect. 6.3 on anomaly detection) can be the source of new test cases that put extraordinary stress on the system. Bertolino [Ber07] lists six questions that have to be addressed when designing a specific test campaign:

(i) *WHY* do we perform the test? Is the objective of the test campaign to find residual design errors, to establish the reliability of a product before it can be released, or to find out whether the system has a usable man-machine interface?
(ii) *HOW* to choose the test cases? There are different options to choose the test cases: random, guided by the operational profile, looking at a specific demand on the system (e.g., a shutdown scenario of a nuclear reactor), or based on knowledge about the internal structure of the program (see Sect. 12.2.2).
(iii) *HOW MUCH* testing is sufficient? The decision about how many test cases must be executed can be derived from *coverage analysis* or from *reliability considerations.*
(iv) *WHAT* is it that we execute? What is the system under test—a module, the system in a simulated environment or the system in its real-world target environment.
(v) *WHERE* do we perform the observation? This depends on the structure of the system and the possibility to observe the behavior of subsystems without disturbing the system operation.
(vi) *WHEN* is it in the product life cycle that we perform the test? In the early phases of the life cycle, testing can only be performed in an artificial laboratory environment. The real test comes when the system is exercised in its target environment.

Technology readiness levels (TRLs) systematically define the maturity of a technical solution during its development. Originally introduced by NASA in the 1970s and 1980s, TRLs are now widely adopted, including in EU-funded research programs starting with the Horizon 2020 framework program. The TRLs differ slightly with different organizations. For example, the European TRLs [EAR14] distinguish nine levels. They range from basic research (TRL 1), where basic principles are observed, to lab validation (TRL 4), to the system proven in the operational environment (TRL 9).

12.3   Testing of Component-Based Systems

The component-based design of embedded applications, which is the focus of this book, requires appropriate strategies for the validation of component-based systems. A component is a hardware/software unit that encapsulates and hides its design and makes its services available at the message-based LIF (linking interface) of the component. While the component provider has knowledge about the internals of a component and can use this knowledge to arrive at effective test cases, the component user sees a component as a black box and must use and test the component on the basis of the given interface specification.

12.3.1   Component Provider

A component provider sees a component independent from the context of use. The provider must be concerned with the correct operation of the component in all possible user scenarios. In our component model, the user scenarios can be parameterized via the TII (Technology-Independent Interface, see Sect. 4.4.3). The component provider must test the proper functioning of the component in the full parameter space supported by the TII. The component provider has access to the source code and can monitor the internal execution within the component across the TDI (Technology-Dependent Interface, see Sect. 4.4.4), which is normally not utilized by the component user.

12.3.2   Component User

The *component user* is concerned with the performance of the component in its concrete *context of use* that is defined by a concrete parameter setting of the component for the given application. The component user can assume that the provider has tested the functions of the component as specified by the interface model and will put the focus of testing on the effects of component integration and the emerging behavior of a set of components, which is outside the scope of testing of the component provider.

In the first step, a component user must validate that the *prior properties of the component*, i.e., the properties that the component supplier has tested in isolation, are not refuted by the integration of the component. The component integration framework plays an important role in this phase.

In an event-triggered system, queues must be provided at the entry and exit of a component to align the component performance with the pending user requests, with the user's capability to absorb the results in a timely manner and with the transport capabilities of the communication system. Since every queue has a potential for overflow, flow-control mechanisms are needed across component boundaries. In the test phase, the reaction of the component to queue overflow must be examined.

The integration of components can give rise to planned or unanticipated emergent behavior that is caused by the component interactions. *Emergent behavior is that which cannot be predicted through analysis at any level simpler than that of the system as a whole* [Dys98, p. 9]. This definition of emergent behavior makes it clear that the detection and handling of emergent behavior is in the realm of a component user and not of the component supplier. Mogul [Mog06] lists many examples of emergent behavior in computer systems that are caused by the interaction of components. The appearance of emergent behavior is not well-understood and a subject of current research (see also Sect. 2.4).

12.3.3 Communicating Components

During system integration, commercial-off-the-shelf (COTS) components or application-specific components are connected by their corresponding linking interfaces (LIFs). The message exchange across these linking interfaces must be carefully tested. In Sect. 4.6, we have introduced three levels of a LIF specification: the *transport level*, the *operational level*, and the *semantic level*. The LIF tests can follow along these three levels. The test at the transport level and the operational level, which must be precisely specified, can be performed mechanically, while the test of the meta-level (the semantics) will normally need human intervention. The multicast capability of the BMTS (basic message transport service—see Sect. 4.1.1) enables the probe effect-free observation of the information exchanged among communicating components.

In *model-based design* (Sect. 11.3.1), executable models of the behavior of the *physical plant* and of the *control algorithms for the computer system* are developed in parallel. At the PIM level, these models, embodied in components, can be linked in a simulation environment such that the interaction of these components can be observed and studied. In a control system, the performance (quality of control) of the closed loop system can be monitored and used to find the optimal control parameter setting. The simulation will normally operate on a different timescale than the target system. In order to improve the faithfulness of the simulation with respect to the target system, the phase relationships of the messages exchanged between the PIM components of the plant and of the controller should be the same as the phase relationships in the final implementation, the PSM. This constant phase relationship will avoid many subtle design errors caused by an uncontrolled and unintended phase relationship among messages [Per10].

12.4   Formal Methods

By the term *formal methods*, we mean the use of mathematical and logical techniques to express, investigate, and analyze the specification, design, documentation, and behavior of computer hardware and software. In highly ambitious projects, formal methods are applied to *prove formally* that a piece of software implements the specification correctly. John Rushby [Rus93, p.87] summarizes the benefits of formal methods as follows:

Formal methods can provide important evidence for consideration in certification, but they can no more "prove" that an artifact of significant logical complexity is fit for its purpose than a finite-element calculation can "prove" that a wing span will do its job. Certification must consider multiple sources of evidence, and ultimately rests on informed engineering judgment and experience.

12.4.1   Formal Methods in the Real World

Any formal investigation of a real-world phenomenon requires the following steps to be taken:

 (i) *Conceptual model building:* This important *informal* first step leads to a precise natural language representation of the real-world phenomenon that is the subject of investigation.
 (ii) *Model formalization:* In this second step, the natural language representation of the problem is transformed and expressed in a formal specification language with precise syntax and semantics. All assumptions, omissions, or misconceptions that are introduced in this step will remain in the model and limit the validity of the conclusions derived from the model. Different degrees of rigor can be distinguished.
(iii) *Analysis of the formal model*: In the third step, the problem is formally analyzed. In computer systems, the analysis methods are based on discrete mathematics and logic. In other engineering disciplines, the analysis methods are based on different branches of mathematics, e.g., the use of differential equations to analyze a control problem.
(iv) *Interpretation of the results*: In the final step, the results of the analysis must be interpreted and applied to the real world.

Only step (iii) out of these four steps can be fully mechanized. Steps (i), (ii), and (iv) will always require human involvement and human intuition and are thus as fallible as any other human activity.

An *ideal and complete* verification environment takes the *specification*, expressed in a formally defined specification language, the *implementation*, written in a formally defined implementation language, and the *parameters of the execution environment* as inputs and establishes mechanically the *consistency* between specification and implementation. In the second step, it must be ensured that all assumptions and architectural mechanisms of the target machine (e.g., the properties and timing of the instruction set of the hardware) are consistent with the model of computation that is defined by the implementation language. Finally, the correctness of the verification environment itself must be established.

12.4.2 Classification of Formal Methods

Rushby [Rus93] classifies the use of formal methods in computer science according to the increasing rigor into the following three levels:

 (i) *Use of concepts and notation of discrete mathematics.* At this level, the sometimes ambiguous natural language statements about requirements and specification of a system are replaced by the symbols and conventions of discrete mathematics and logic, e.g., set theory, relations, and functions. The reasoning about the completeness and consistency of the specification follows a semiformal manual style, as it is performed in many branches of mathematics.
 (ii) *Use of formalized specification languages with some mechanical support tools.* At this level, a formal specification language with a fixed syntax is introduced that allows the mechanical analysis of some properties of the problems expressed in the specification language. At level (ii), it is not possible to generate complete proofs mechanically.
 (iii) *Use of fully formalized specification languages with comprehensive support environments, including mechanized theorem proving or proof/model checking.* At this level, a precisely defined specification language with a direct interpretation in logic is supplied, and a set of support tools is provided to allow the mechanical or automated analysis of specifications expressed in the formal specification language.

12.4.3 Benefits of Formal Methods

Level (i) methods: The compact mathematical notation introduced at this level forces the designer to clearly state the requirements and assumptions without the ambiguity of natural language. Since familiarity with the basic notions of set theory and logic is part of an engineering education, the disciplined use of level (i) methods will improve the communication within a project team and within an engineering organization and enrich the quality of documentation. Since most of the serious faults are introduced early in the life cycle, the benefits of the level (i) methods are most pronounced at the early phases of requirements capture and architecture design. Rushby [Rus93, p.39] sees the following benefits in using level (i) methods early in the life cycle:

- The need for effective and precise communication between the software engineer and the engineers from other disciplines is greatest at an early stage, when the interdependencies between the mechanical control system and the computer system are specified.
- The familiar concepts of discrete mathematics (e.g., set, relation) provide a repertoire of mental building blocks that are precise, yet abstract. The use of a precise notation at the early stages of the project helps to avoid ambiguities and misunderstandings.
- Some simple mechanical analysis of the specification can lead to the detection of inconsistencies and omission faults, e.g., that symbols have not been defined or variables have not been initialized.
- The reviews at the early stages of the life cycle are more effective if the requirements are expressed in a precise notation than if ambiguous natural language is used.
- The difficulty to express vague ideas and immature concepts in a semiformal notation helps to reveal problem domains that need further investigation.

Level (ii) methods: Level (ii) methods are a mixed blessing. They introduce a rigid formalism that is cumbersome to use, without offering the benefit of mechanical proof generation. Many of the specification languages that focus on the formal reasoning about the temporal properties of real-time programs are based at this level. Level (ii) formal methods are an important intermediate step on the way to provide a fully automated verification environment. They are interesting from the point of view of research.

Level (iii) methods: The full benefits of formal methods are only realized at this level. However, the available systems for verification are not complete in the sense that they cover the entire system from the high level specification to the hardware architecture. They introduce an intermediate level of abstraction that is above the functionality of the hardware. Nevertheless, the use of such a system for the rigorous analysis of some critical functions of a distributed real-time system, e.g., the correctness of the clock synchronization, can uncover subtle design faults and lead to valuable insights. The NASA technical memorandum [Sim11] defines a *Modeling and Verification Evaluation Score (MVES), a metric that is intended to estimate the amount of trust that can be placed on the evidence that is obtained* (by formal verification activities). It studies the formal verification process and artifacts of the TTEthernet protocol (see Sect. 7.5.2) as a use case.

12.4.4   Model Checking

In the last few years, the verification technique of *model checking*, a level (iii) method, has matured to the point that it can be used for the analysis and the support of the certification of safety-critical designs [Cla03]. Given a formal behavioral model of the specification and a formal property that must be satisfied, the *model checker* checks automatically whether the property holds in all states of the system model. In case the model checker finds a violation, it generates a concrete counter example. The main problem in model checking is the state explosion. In the last few years, clever formal analysis techniques have been developed to get a handle on the state explosion problem such that systems of industrial size can be verified by model checking.

12.5   Fault Injection

Fault injection is the intentional introduction of faults by software or hardware in order to validate the system behavior under fault conditions. During a fault-injection experiment, the target system is exposed to two types of inputs: the *injected faults* and the *input data*. The faults can be seen as *another type of input* that activates the fault-management mechanisms. Careful testing and debugging of the fault-management mechanisms are necessary because a notable number of system fail ures are caused by errors in the fault-management mechanisms.

Fault injection serves two purposes during the evaluation of a dependable system:

(i) *Testing and Debugging:* During normal operation, faults are *rare events* that occur only infrequently. Because a fault-tolerance mechanism requires the occurrence of a fault for its activation, it is very cumbersome to test and debug the operation of the fault-tolerance mechanisms without artificial fault injection.

(ii) *Dependability Forecasting:* This is used to get experimental data about the likely dependability of a fault-tolerant system. For this second purpose, the types and distribution of the expected faults in the envisioned operational envi-ronment must be known.

Table 12.1 comparison of the two different purposes of fault injection.

It is possible to inject faults into the state of the computation (*software-implemented fault injection*) or at the physical level of the hardware (*physical fault injection*).

Table 12.1 Fault injection for testing and debugging versus dependability forecasting [Avr92]

	Testing and debugging	Dependability forecasting
Injected faults	Faults derived from the specified fault hypothesis	Faults expected in the operational environment
Input data	Targeted input data to activate the injected faults	Input data taken from the operational environment
Results	Information about the operation and effectiveness of the fault-tolerance mechanisms	Information about the envisioned dependability of the fault-tolerant system

12.5.1   Software-Implemented Fault Injection

In *software-implemented fault injection*, errors are seeded into the memory of the computer by a fault-injection software tool. These seeded errors mimic the effects of hardware faults or design faults in the software. The errors can be seeded either randomly or according to some preset strategy to activate specific fault-handling tasks.

Software-implemented fault injection has a number of potential advantages over physical fault injection:

(i) Predictability: The space (memory cell) *where* and the instant *when* a fault is injected is fixed by the fault-injection tool. It is possible to reproduce every injected fault in the value domain and in the temporal domain.

(ii) Reachability: It is possible to reach the inner registers of large VLSI chips. Pin-level fault injection is limited to the external pins of a chip.

(iii) Less effort than physical fault injection: The experiments can be carried out with software tools without any need to modify the hardware.

12.5.2   Physical Fault Injection

During physical fault injection, the target hardware is subjected to adverse physical phenomena that interfere with the correct operation of the computer hardware. In the following section, we describe a set of hardware fault-injection experiments that have been carried out on the MARS (maintainable real-time system) architecture in the context of the ESPRIT Research Project *Predictably Dependable Computing Systems (PDCS)* [Kar95, Arl03].

The objective of the MARS fault-injection experiments was to determine the *error detection coverage* of the MARS nodes experimentally. Two replica-determinate nodes receive identical inputs and should produce the same result. One of the nodes is subjected to fault injections (the *FI-node*); the other node serves as a reference node (a *golden node*). As long as the consequences of the faults are detected within the FI-node, and the FI-node turns itself off, or the FI-node produces a detectably incorrect result message, the error has been classified as detected. If the FI-node produces a result message different from the result message of the golden node without any error indication, a fail-silence violation has been observed.

Table 12.2 Characteristics of different physical fault-injection techniques

Fault-injection technique	Heavy ion	Pin-level	EMI
Controllability, space	Low	High	Low
Controllability, time	None	High/medium	Low
Flexibility	Low	Medium	High
Reproducibility	Medium	High	Low
Physical reachability	High	Medium	Medium
Timing measurement	Medium	High	Low

Three different fault-injection techniques were chosen at three different sites (see Table 12.2). At Chalmers University in Goeteborg, the CPU chip was bombarded with α particles until the system failed. At LAAS in Toulouse, the system was subjected to pin-level fault injection, forcing an equi-potential line on the board into a defined state at a precise moment of time. At the Technische Universität Wien, the whole board was subjected to electromagnetic interference (EMI) radiation according to the IEC standard IEC 801-4.

Many different test runs, each one consisting of 2000–10,000 experiments, were carried out with differing combinations of error detection techniques enabled. The results of the experiments can be summarized as follows:

(i) With all error detection mechanisms enabled, no fail-silence violation was observed in any of the experiments.

(ii) The end-to-end error detection mechanisms and the double execution of tasks were needed in experiments with every one of the three fault-injection methods if error detection coverage of >99% must be achieved.

(iii) In the experiment that used heavy ion radiation, a triple execution was needed to eliminate all coverage violations. The triple execution consisted of a test run with known outputs between the two replicated executions of the application task. This intermediate test run was not needed in the EMI experiments and the pin-level fault injection.

(iv) The bus guardian unit was needed in all three experiments if a coverage of >99% must be achieved. It eliminated the most critical failure of a node, the *babbling idiots*.

A detailed description of the MARS fault-injection experiments and a comparison with software-fault injection carried out on the same system is contained in [Arl03].

12.5.3 Sensor and Actuator Failures

The sensors and actuators, placed at the interface between the *physical world* and *cyberspace*, are physical devices that will eventually fail, just like any other physical device. The failures of sensors and actuators are normally not spontaneous *crash failures* but manifest themselves either as transient malfunctions or a gradual drift away from the correct operation, often correlated with extreme physical conditions (e.g., temperature, vibration). An undetected sensor failure produces erroneous inputs to the computational tasks that, as a consequence, will lead to erroneous outputs that can be safety-relevant. Therefore it is state-of-the-art that any industrial-strength embedded system must have the capability to detect or mask the failure of any one of its sensors and actuators. This capability must be tested by fault-injection experiments, either software-based or physical.

An actuator is intended to transform a digital signal, generated in cyberspace, to some physical action in the environment. The incorrect operation of an actuator can only be observed and detected if one or more sensors observe the intended effect in the physical environment. This error detection capability with respect to actuator failures must also be tested by fault-injection experiments (see also the example in Sect. 7.2.5.1).

In safety-critical applications, these fault-injection tests must be carefully documented, since they form a part of the safety case.

Points to Remember

- An essential fraction—up to 50%—of the development costs of a real-time computer system is devoted to ensure that the system is *fit-for-purpose*. In safety-critical applications that must be certified, this fraction is even higher.
- *Verification* establishes the consistency between a *(formal) specification* with the system under test *(SUT)*, while *validation* is concerned with the consistency between the *model of the user's intention* with the *SUT*. The missing link between verification and validation is the relation between the *model of the user's intention* and the *(formal) specification* of the system.
- If a purely probabilistic point of view is taken, then an estimate that the *mean time to failure* (MTTF) of the SUT will be larger than a given *number of hours* can only be made if system tests have been executed for a duration that is larger than this *number of hours*.
- The modification of the behavior of the object under test by introducing a test probe is called the *probe effect*.
- *Design for testability* establishes a framework where test-outputs can be observed without a probe effect and where test inputs can be controlled at any level of the system architecture.
- It is a challenge for the tester to find an effective and representative set of test data that will give the designer confidence that the system will work correctly for all inputs. A further challenge relates to finding an effective automatable test oracle.
- In the last few years, clever formal techniques have been developed to get a handle on the state explosion problem such that systems of industrial size can be verified by model checking.

- *Fault injection* is the intentional activation of faults by hardware or software means to be able to observe the system operation under fault conditions. During a fault-injection experiment, the target system is exposed to two types of inputs: the injected faults and the input data.
- The sensors and actuators, placed at the interface between the *physical world* and *cyberspace*, are physical devices that will eventually fail, just like any other physical device. Therefore it is state-of-the-art that any industrial-strength embedded system must have the capability to detect or mask the failure of any one of its sensors and actuators.

Bibliographic Notes

In the survey article *Software Testing Research: Achievements, Challenges, Dreams*, Bertoloni [Ber07] gives an excellent overview of the state of the art in software testing and some of the open research challenges. The research report "Formal Methods and the Certification of Critical Systems" [Rus93] by John Rushby is a seminal work on the role of formal methods in the certification of safety-critical systems.

Review Questions and Problems

12.1. What is the difference between validation and verification?

12.2. Describe the different methods for test data selection!

12.3. What is a test oracle?

12.4. How does a component provider and component user test a component-based system?

12.5. Discuss the different steps that must be taken to investigate a real-world phenomenon by a formal method. Which one of these steps can be formalized, which cannot?

12.6. In Sect. 12.4.2, three different levels of formal methods have been introduced. Explain each one of these levels and discuss the costs and benefits of applying formal methods at each one of these levels.

12.7. What is model checking?

12.8. What is the "probe effect"?

12.9. How can the "testability" of a design be improved?

12.10. What is the role of testing during the certification of an ultra-dependable system?

12.11. Which are the purposes of fault-injection experiments?

12.12. Compare the characteristics of hardware and software fault-injection methods.

Chapter 13
Internet of Things

Overview

The connection of *physical things* to the Internet makes it possible to access remote sensor data and to control the physical world from a distance. The mash-up of captured data with data retrieved from other sources, e.g., with data that is contained in the Web, gives rise to new synergistic services that go beyond the services that can be provided by an isolated embedded system. The *Internet of Things* is based on this vision. A *smart object*, which is the building block of the Internet of Things, is just another name for an embedded system that is connected to the Internet. There is another technology that points in the same direction—the *RFID technology*. The RFID technology, an extension of the ubiquitous optical bar codes that are found on many everyday products, requires the attachment of a smart low-cost electronic ID-tag to a product such that the identity of a product can be decoded from a distance. By putting more intelligence into the ID tag, the *tagged thing* becomes a *smart object*. The novelty of the Internet of Things (IoT) is not in any new disruptive technology, but in the pervasive deployment of *smart objects*.

At the beginning of this chapter, the vision of the IoT is introduced. The next section elaborates on the forces that drive the development of the IoT. We distinguish between *technology push* and *technology pull* forces. The *technology push forces* see in the IoT the possibility of vast new markets for novel ICT products and services, while the *technology pull forces* see the potential of the IoT to increase the productivity in many sectors of the economy, to provide new services, e.g., for an aging society, and to promote a new lifestyle. Section 13.3 focuses on the technology issues that have to be addressed in order to bring the IoT to a mass market. Section 13.4 discusses the RFID technology, which can be seen as a forerunner of the IoT. The topic of wireless sensor networks, where *self-organizing smart objects* build ad hoc networks and collect data from the environment, is covered in Sect. 13.5. The pervasive deployment of smart objects that collect data and control the physical environment from a distance poses a severe challenge to the security and safety of the world and the privacy of our lives.

H. Kopetz, W. Steiner, *Real-Time Systems*,
https://doi.org/10.1007/978-3-031-11992-7_13

13.1   The Vision of the Internet of Things (IoT)

Over the past 50 years, the Internet has exponentially grown form a small research network, comprising only a few nodes, to a worldwide pervasive network that services more than a billion users. The further miniaturization and cost reduction of electronic devices make it possible to expand the Internet into a new dimension: to *smart objects*, i.e., everyday physical things that are enhanced by a small electronic device to provide local intelligence and connectivity to the cyberspace established by the Internet. The small electronic device, a *computational component* that is attached to a *physical thing*, bridges the gap between the physical world and the information world. A *smart object* is thus a *cyber-physical system* or an *embedded system*, consisting of a *thing* (the physical entity) and a *component* (the computer) that processes the sensor data and supports a wireless communication link to the Internet.

> **Example**: Consider a *smart refrigerator* that keeps track of the availability and expiry date of food items and autonomously places an order to the next grocery shop if the supply of a food item is below a given limit.

The novelty of the IoT is not in the functional capability of a *smart object*—already today many embedded systems are connected to the Internet—but in the expected size of billions or even trillions of *smart objects* that bring about novel technical and societal issues that are related to size. Some examples of these issues are as follows: authentic identification of a *smart object*, autonomic management and self-organization of networks of smart objects, diagnostics and maintenance, context awareness and goal-oriented behavior, and intrusion of the privacy. Special attention must be given to *smart objects* that can act—more or less autonomously—in the physical world and can *physically* endanger people and their environment.

The advent of low-power wireless communication enables the communication with a *smart object* without the need of a physical connection. Mobile *smart objects* can move around in the physical space while maintaining their identity. The wide availability of signals from the global positioning system (GPS) makes it possible to make a smart object *location and time-aware* and offer services that are tuned to the current context of use.

We can envision an *autonomic smart* object that has access to a domain-specific knowledge base—similar to the *conceptual landscape* introduced in Sect. 2.2—and is empowered with reasoning capabilities to orient itself in the selected application domain. Based on the capability level of a smart object, [Kor10] distinguish between *activity-aware*, *policy-aware*, *and process*-aware smart objects.

> **Example**: A *pay-per-use smart tool* is an activity-aware smart object that collects data about the time and intensity of its use and transmits the data autonomously to the billing department. A *policy-aware smart tool* will know about its use cases and will ensure that it is not used outside the contracted use cases. A *process-aware smart tool* will reason about its environment and guide the user how to optimally apply the tool in the given scenario.

According to the IoT vision, a *smart planet* will evolve, where many of the everyday things around us have an identity in cyberspace, acquire intelligence, and mash-up information from diverse sources. On the *smart planet*, the world economy and support systems will operate more smoothly and efficiently. But the life of the average citizen will also be affected by changing the relation of power between those that have access to the acquired information and can control the information and those that do not.

13.2   Drivers for an IoT

Which are the forces that drive the development of the Internet of Things? They are on both sides of the technology landscape: *technology push forces* and *technology pull forces*. The technology push forces see in the IoT a vast new market for the deployment of current and future information and communication technologies (ICT). The IoT will help to utilize existing and new factories, provide new employment opportunities in the ICT sector, and contribute to the further development of the ICT technologies in general.

In this section, the focus is mainly on *technology pull forces*. Which areas of our economy, society, and life in general will benefit from the wide deployment of the IoT? The following analysis is not exhaustive—we are only highlighting some sectors where, according to our present understanding, the wide deployment of the IoT technology will have a major impact.

13.2.1   Uniformity of Access

The Internet has achieved the worldwide interoperability of heterogeneous end systems over a wide variety of communication channels. The IoT should extend this interoperability to the universe of heterogeneous *smart objects*. From the point of view of reduction of the cognitive complexity (see Chap. 2), the IoT can make a very significant contribution: the establishment of a *uniform access pattern* to things in the physical world.

13.2.2   Logistics

The first commercial application of a forerunner of the IoT, the RFID (radio frequency identification—see Sect. 13.4) technology, is in the area of logistics. With the decision of some major corporations to base their supply chain management on RFID technology, the development of low-cost RFID tags and RFID readers has moved forward significantly. There are many quantitative advantages in using RFID technology in supply chain management: the movement of goods can be tracked in real time, shelf space can be managed more effectively, inventory control is improved, and above all, the amount of human involvement in the supply chain management is reduced considerably.

13.2.3   Energy Savings

Already today, embedded systems contribute to energy savings in many different sectors of our economy and our life. The increased fuel efficiency of automotive engines, the improved energy-efficiency of household appliances, and the reduced loss in energy conversion are just some examples of the impact of this technology on energy savings. The low cost and wide distribution of IoT devices opens many new opportunities for energy savings: individual climate and lighting control in residential buildings, reduced energy loss in transmission by the installation of *smart grids*, and better coordination of energy supply and energy demand by the installation of smart meters. The *dematerialization of parts of our life* such as the replacement of physical meetings by virtual meetings and the delivery of information goods such as the daily paper, music, and videos by the Internet lead to substantial energy savings.

13.2.4   Physical Security and Safety

A significant technology pull for the IoT technology comes from the domains of physical security and safety. Automated IoT-based access control systems to buildings and homes and IoT-based surveillance of public places will enhance the overall physical security. Smart passports and IoT-based identifications (e.g., a smart key to access a hotel room or a smart ski lift ticket) simplify admission controls checks and increase the physical security, while reducing human involvement in these checks. Car-to-car and car-to-infrastructure communication will alert the driver of dangerous traffic scenarios, such as an icy road or an accident, and reduce the number of accidents.

IoT technology can help to detect counterfeit goods and suspected unapproved spare parts that are becoming a safety risk in some application domains such as the airline or automotive industry.

On the other side, safety and security applications intrude into the privacy of our lives. It is up to policy makers to draw the fine line between the rights of a person to individual privacy and the desire of the public to live in a safe environment. It is up to the scientists and engineers to provide the technology so that the political decisions can be flawlessly implemented.

13.2.5 Industrial

In addition to streamlining the supply chain and the administration of goods by the application of RFID technology, the IoT can play a significant role in reducing maintenance and diagnostic cost. The computerized observation and monitoring of industrial equipment does not only reduce maintenance cost because an abnormality can be detected before it leads to a failure, but also improves the safety in the plant (see also Sect. 11.6).

A smart object can also monitor its own operation and call for preventive or spontaneous maintenance in case a part wears out or a physical fault is diagnosed. Automated fault diagnosis and simple maintenance are absolutely essential prerequisites for the wide deployment of the IoT technology in the domain of ambient intelligence.

13.2.6 Medical

The wide deployment of IoT technology in the medical domain is anticipated. Health monitoring (heart rate, blood pressure, etc.) or precise control of drug delivery by a smart implant is just two potential applications. Body area networks that are part of the clothing can monitor the behavior of impaired persons and send out alarm messages if an emergency is developing. Smart labels on drugs can help a patient to take the right medication at the right time and enforce drug compliance.

> **Example:** A heart pacemaker can transmit important data via a Bluetooth link to a mobile phone that is carried in the shirt pocket. The mobile phone can analyze the data and call a doctor in case an emergency develops.

13.2.7 Lifestyle

The IoT can lead to a change in lifestyle. A smart phone can function as a browser for smart objects and augment the reality we see with background information retrieved from a diversity of context-dependent databases.

13.3 Technical Issues of the IoT

13.3.1 Internet Integration

Depending on the computational capabilities and the available energy, a smart object can be integrated into the Internet either directly or indirectly via a *base station* that is connected to the Internet. The indirect integration will be chosen when the smart object has a very limited power budget. Application-specific power-optimized protocols are used to connect the smart object to a nearby *base station*. The base station that is not power constrained can act as a standard Web server that provides gateway access to the reachable smart objects.

> **Example**: In an RFID system, the RFID reader acts as a base station that can read the local RFID tags. The reader can be connected to the Internet. In *Sensor Networks*, one or more *base stations* collect data from the sensor nodes and forward the data to the Internet.

A number of major companies have formed an alliance to develop technical solutions and standards to enable the direct integration of low-power smart objects into the Internet. The Internet Engineering Task Force (IETF) working group *6LoWPAN* (*IPv6 over Low-Power Wireless Personal Area Networks*) has standardized energy-efficient solutions for the integration of the IPv6 standard with the IEEE 802.15.4 wireless near field communication standard.

Guaranteeing the safety and information security of IoT-based systems is considered to be a difficult task. Many smart objects will be protected from general Internet access by a tight firewall to avoid that an adversary can acquire control of a smart object. The important topic of privacy and security in the IoT is further addressed in Sect. 13.4.5.

13.3.2   Naming and Identification

The vision of the IoT (that all of the billions of smart objects can communicate via the Internet) requires a well-thought-out *naming architecture* in order to be able to identify a smart object and to establish an access path to the object.

Every name requires a *named context* where the name can be resolved. The recursive specification of naming context leads to a *hierarchical name structure*—the naming convention adhered to in the Internet. If we want a name to be universally interpretable without reference to a specific naming context, we need a single context with a universally accepted name space. This is the approach taken by the RFID community, which intends to assign an *Electronic Product Code (EPC)* to every physical smart object (see also Sect. 13.4.2). This is more ambitious than the forerunner, the optical bar code, which assigns a unique identifier only to a class of objects.

Isolated Objects The following three different object names have to be distinguished when we refer to the simple case of an isolated object:

*   *Unique object identifier (UID)*: refers to the physical identity of a specific object. The Electronic Product Code (EPC) of the RFID community is such a UID.
*   *Object type name*: refers to a class of objects that ideally have the same properties. It is the name that is encoded in the well-established *optical bar code*.
*   *Object role name*: In a given use context, an object plays a specific role that is denoted by the *object role name*. At different times, the same object can play dif-

ferent roles. An object can play a number or roles and a role can be played by a number of objects.

Example: The assumption that all objects that have the same *object type name* are identical does not always hold. Consider the case of an *unapproved spare part* that has the same visible properties and is of the same type as an *approved spare part* but is a *cheaper copy* of the approved part.

Example: An *office key* is an *object role name* for a physical object type that unlocks the door of an office. Any instance of the *object type* is an *office key*. When the lock in the office door is changed, a different object type assumes the role of the *office key*. A particular office key can also unlock the laboratory. It then plays two roles, the role of an *office key* and the role of a *laboratory key*. A *master key* can open any office—there are thus two different keys that play the same role.

Composite Objects   Whenever a number of objects are integrated to form a composite object, a *new whole*, i.e., new object, is created that has an emerging identity that goes beyond the identities of the constituent objects. The composite object resembles a *new concept* (see Sect. 2.2.1) that requires a *new name*.

Example: *George Washington's axe* is the subject of a story of unknown origin in which the famous artifact is *still George Washington's axe* (a composite object) despite having had both its head replaced twice and its handle replaced three times [Wik10].

A composite object that provides an emergent service requires its own UID that is hardly related to the UIDs of its constituent parts. The different names, *UID*, *object type name*, and *object role name* must be introduced at the level of composite objects as well. Since a composite object can be an *atomic unit* at the next level of integration, the name space must be built up recursively.

Which one of the object names, introduced in the above paragraphs, should be the access points for the communication with the Internet? It will be difficult to manage the *communication complexity* if all objects that are contained in *multilevel composite objects* can be accessed *anytime* at *anyplace*.

It is evident that the introduction of a flat name space for all smart objects of the universe, as stipulated by the EPC, is only a starting point. More research on the proper design of name space architectures in the IoT is needed.

13.3.3   Near-Field Communication

The IoT requires, in addition to the established WLANs (wireless local area networks), short-range energy-efficient WPANs (wireless personal area networks) in order to enable the energy-efficient wireless access to *smart objects* over a small distance. The IEEE 802.15 standard working group develops standards for WPAN networks. Among the networks that are conforming to the 802.15 standards are the *Bluetooth* network and the *ZigBee* network.

Originally, *Bluetooth* has been introduced as a wireless alternative to the RS232 wire-bound communication channel [Bar07]. Bluetooth, standardized in IEEE

802.15.1, defines a complete WPAN architecture, including a security layer. At the physical level, it achieves a data rate of up to 3 Mbit/second over a distance of 1 m (Class 3—maximum transmission power of 1 mW) to 100 m (Class 1—maximum transmission power 100 mW) using the transmission technology of frequency *hopping*. Bluetooth allows multiple devices to communicate over a single adapter.

The *ZigBee* alliance is a group of companies that develops a secure WPAN that is intended to be simpler, more energy efficient, and less expensive than Bluetooth [Bar07]. *ZigBee* uses high-level communication protocols based on the IEEE 802.15.4 standard for low-power digital radios. ZigBee devices are requested to have a battery life of more than a year.

The NFC (near-field communication) standard [Fin03], an extension of the ISO/ IEC 14443 proximity card standard, is a short-range high-frequency wireless communication technology which enables the exchange of data between devices over a distance of less than 20 cm. The technology is compatible with both existing smartcards and readers, as well as with other NFC devices, and is thereby compatible with the existing contactless infrastructure already in use for public transportation and payment. NFC is primarily aimed for use in mobile phones.

13.3.4   *IoT Device Capabilities Versus Cloud Computing*

Smart objects that have access to the Internet can take advantage of services that are offered by the *cloud* (large data centers that provide their services through the Internet). The division of work between a smart object and the cloud will be determined, to a considerable degree, by privacy and energy considerations [Kum10]. If the energy required to execute a task locally is larger than the energy required to send the task parameters to a server in the cloud, then the task is a candidate for remote processing. However, there are other aspects that influence the decision about work distribution: autonomy of the smart object, response time, reliability, and security.

13.3.5   *Autonomic Components*

The large number of smart objects that are expected to populate our environment requires an autonomic system management without the need of frequent human interactions. This autonomic management must cover *network service discovery*, *system configuration and optimization*, *diagnosis of failures and recovery after failures*, and *system adaptation and evolution*. There is a need for a *multilevel autonomic management*, starting with the fine-grained management of components up to the coarse-grained management of massive assemblies of components or large systems.

Fig. 13.1 Model of an
autonomic component.
(Adapted from [Hue08])

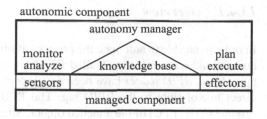

Figure 13.1 shows the generic MAPE-K (Monitoring, Analyzing, Planning, Execution with a Knowledgebase) architecture of an *autonomic component* [Hue08]. An *autonomic component* consists of two *independent fault-containment units* (FCU), a *managed component* and an *autonomy manager*. The *managed component* can be a single component, a cluster of components, or a part of a larger system. The *autonomy manager* consists of a monitor that observes and analyzes information about the behavior of the managed component, a planning module that develops and evaluates alternative plans to achieve the stated goals, and finally an interface to the managed object that allows the autonomy manager to influence the behavior of the *managed component*. The autonomy manager maintains a *knowledge base* with static and dynamic entries. The *static entries* are provided a priori, i.e., at design time. They set up the goals, beliefs, and generic structure of the knowledge base, while the dynamic entries are filled in during operation to capture the acquired information about concrete situational parameters. The *multicast communication primitive* makes it possible for the autonomy manager to observe the behavior of the managed component and its interactions with the environment without any *probe effect*.

In its simplest form, the autonomy manager recognizes objects and object changes (events) and assigns them to known concepts (see Sect. 2.2.1). It then selects an action based on *event-condition-action* rules. If more than one action is applicable, it uses utility functions to select the action with the highest utility value for achieving the desired goals. In a more advanced form, the autonomy manager is based on a cognitive architecture that supports some form of advanced reasoning and can improve its decision by evaluating past decisions and by the incorporation of learning [Lan09].

13.4   RFID Technology

The easy and fast identification of *things* is required in many situations, e.g., in stores, warehouses, supermarkets, etc. For this purpose, an optical barcode is attached to an object. Reading an optical barcode requires the careful positioning of the object by a human such that a direct line-of-sight between the barcode and the barcode reader is established.

13.4.1   Overview

In order to be able to automate the process of object identification and eliminate the human link, *electronic tags* (called RFID tags) that can be read from a small distance by an *RFID reader* have been developed. An RFID reader does not require a direct line-of-sight to the RFID tag. The RFID tag stores the unique Electronic Product Code (EPC) of the attached object. Since an RFID tag has to be attached to every object, the cost of an RFID tag is a major issue. Due to the standardization of the RFID technology by the International Standard Organization (ISO) and the massive deployment of RFID technology, the cost of an RFID tag has been reduced significantly over the last few years.

The RFID reader can act as a gateway to the Internet and transmit the object identity, together with the read-time and the object location (i.e., the location of the reader) to a remote computer system that manages a large database. It is thus possible to track objects in real time.

> **Example**: An *electronic ski pass* is an RFID tag that is queried by the reader that is built into the admission gate to a ski lift. Based on the object identifier, a picture of the person that owns the ski pass is displayed to the operator and the gate is opened automatically if the operator does not intervene.

13.4.2   The Electronic Product Code (EPC)

Whereas an optical barcode denotes a product class (all boxes of the same product have the same barcode), the EPC of an RFID tag denotes an object instance (every box has a unique identifier). It is the intent of the EPC to assign a *unique identifier (UID)* to every identifiable *thing* on the globe, i.e., a unique name to each *smart object* of the IoT.

The EPC is managed by the international organization *EPC global*. In order to cope with the huge number of things the EPC must identify, the EPC contains a number of fields. A small header field determines the structure of the remaining fields. A typical EPC has a length of 96 bits and contains the following fields:

- Header (8 bits): defines the type and the length of all subsequent fields.
- EPC Manager (28 bits): specifies the entity (most often the manufacturer) that assigns the object class and serial number in the remaining two fields.
- Object Class (24 bits): specifies a class of objects (similar to the optical bar code).
- Object Identification Number (36 bits): contains the serial number within the object class.

The EPC is unique product identification but does not reveal anything about the properties of the product. Two *things* that have the same properties, but are designed by two different manufacturers, will have completely different EPCs. Normally, the unique EPC is used as a key to find the product record in a *product database*. The product record contains all required information about the attributes of the product.

13.4.3 RFID Tags

A RFID tag contains as its most important data element the EPC of the associated *physical thing*. A number of different RFID tags have been developed and standardized. Basically, they fall into two main categories: *passive RFID tags* and *active RFID tags*.

Passive RFID Tags Passive tags do not have their own power supply. They get the power needed for their operation from energy harvested out of the electric field that is beamed on them by the RFID reader. The energy required to operate a passive tag of the latest generation is below 30 µW and the cost of such a tag is below 5 ¢. Passive tags contain in addition to a standardized EPC (Electronic Product Code) as a unique identification number selected other information items about product attributes. Due to the low level of the available power and the cost pressure on the production of RFID tags, the communication protocols of passive RFID tags do not conform to the standard Internet protocols. Specially designed communication protocols between the RFID tag and the RFID reader that consider the constraints of passive RFID tags have been standardized by the ISO (e.g., ISO 18000-6C also known as the EPC global Gen 2) and are supported by a number of manufacturers. The parameters of a typical low-cost passive RFID tag are given in Table 13.1.

Active RFID Tags Active tags have their own on-board power supply, e.g., a battery that gives them the capability to support many more services than passive tags. The lifetime of an active tag is limited by the lifetime of the battery, typically in the order of a year. Active tags can transmit and receive over a longer distance than passive tags, typically in the order of hundreds of meters, can have sensors to monitor their environment (e.g., temperature, pressure), and sometimes support standard

Table 13.1 Parameters of a typical low-cost passive RFID tag

Storage	128–512 bits of read-only storage
Memory	32–128 bits of volatile read-write memory
Gate count	1000–10,000 gates
Operating frequency	868–956 MHz (UHF)
Clock cycles per read	10,000 clock cycles
Scanning range	3 meters
Performance	100 read operations per second
Tag power source	Passively powered by reader via RF signal
Power consumption	10 microwatts

Adapted from [Jue05]

Internet communication protocols. In some sense, an active RFID tag resembles a small embedded system. The ISO standard 18000-7 specifies the protocol and the parameters for the communication with an active tag in the 433 MHz range. The reduction of the power consumption of an active RFID tag in *sleep* mode is a topic of current research.

13.4.4   RFID Readers

The RFID reader is a gateway component between the world of RFID tags and the Internet. These two worlds are characterized by different architectural styles, naming conventions, and communication protocols. On the Internet side, an RFID reader looks like a standard Web server that adheres to all Internet standards. On the RFID side, the RFID reader respects the RFID communication protocol standards. The RFID reader has to resolve all property mismatches.

13.4.5   RFID Security

Whenever we connect a computer to the Internet, sensitive security issues arise [Lan97] that must be addressed. Standard security techniques are based on the deployment of cryptographic methods, like *encryption, random number generation*, and *hashing* as outlined in Sect. 6.2. The execution of cryptographic methods requires energy and silicon real estate, which are not sufficiently available in all smart objects, such as low-cost RFID tags. The often-heard argument that computationally constrained RFID tagged objects will disappear in the near future as the microelectronic devices become cheaper overlooks the price pressure on simple RFID tags. If low-cost RFID tags are placed on billions of retail products, even a one-cent increase in the cost of a tag for the provision of cryptographic capabilities will be shunned.

The information security threats in the IoT can be classified into three groups: (I) the threats that compromise the *authenticity* of information, (ii) the threats to *privacy* caused by a pervasive deployment of IoT products, and (iii) *denial of service* threats. We assume that the vast majority of IoT devices are connected to the cyberspace by a wireless connection. A wireless connection always presents a serious vulnerability since it opens the door to an unnoticed observation of the traffic by an adversary.

Authentication   It is a basic assumption in the IoT that the *electronic device*, e.g., a RFID tag, represents a unique *physical thing* in cyberspace and that this link between the electronic device and the physical thing which has been established by a *trusted authority* can be relied upon. This belief in *tag authenticity* can be shaken easily.

Scanning and replicating an unprotected tag is relatively easy, since a tag is nothing else than a string of bits that can be copied without difficulty.

Attaching another *physical thing*—e.g., a *faked product*—to an authentic tag can break the link between the *physical thing* and the *tag*, the representative of the physical thing in cyberspace. This kind of attack has to be addressed in the level of *physical design of a smart object* and cannot be dealt with by cyberspace security methods.

The known techniques to ensure the authenticity of the *thing* behind a low-cost RFID tag are quite limited. A tag is a bit-string that can be read by any commodity reader and can be copied to produce a *cloned tag*. Even a digital signature could not prevent *cloning of tags*. *Men in the middle attacks*, where an attacker mimics a correct tag, might break the established link between the *reader* and the *tag*. Accessing the product database can detect the existence of *cloned tags* by discovering that the uniqueness property of the EPC has been violated, but it cannot eliminate cloning.

> **Example**: Accessing the product database can identify a counterfeit piece of art that carries a cloned tag and finding out that the genuine object is at a location that is different from the tag reader.

Tamper-proof tags that physically break when they are detached from the *thing* they have been attached to by the trusted authority are one solution to the problem of physical tag stealing. In order to be able to ascertain the authenticity of valuable things, *physical one-way functions* (Poufs) have been proposed [Pap02]. An example for a POWF is a transparent optical device with a random three-dimensional microstructure that is attached to the thing in a tamper-proof manner. Since the randomness of the structure cannot be controlled during the manufacturing process, it is impossible to produce two POWF that are alike. When read by a laser under a specific angle, a POWF response is a bit stream that is characteristic for this unique POWF. Depending on the reading angle, different characteristic bit streams can be retrieved. These bit streams can be stored in the product database. It is difficult for an adversary to clone a POWF, because it is a *thing with random characteristic physical properties* that cannot be represented by a mathematical function. A POWF is not a *construct of cyberspace* that can be copied or reconstructed.

Privacy The main privacy concern in the RFID world is the *clandestine reading* of a tag by an unauthorized reader. Since low-cost RFID tags are unprotected and can be read by commodity readers, clandestine tag tracking by unauthorized readers discloses valuable information to an adversary. If the adversary uses a sensitive reader with a high-power antenna output (*rogue reading*), he can significantly extend the nominal read range. The information about EPC codes and other attributes that are contained in the tag can be linked with the identity of the person carrying the tag in order to construct a personal profile. Since a low-cost tag does not have the cryptographic capability to authenticate the reader, it will disclose its information whenever it is queried. Clandestine tag reading can be prevented by permanently killing the tag as soon as the tag enters the consumer domain, i.e., at the point-of-sale. Tag killing enforces consumer privacy effectively. However, if tags support the functionality of tag killing, a *vulnerability* with respect to availability is established.

Example: By analyzing the tagged medication a person is carrying, an adversary could infer information about the health condition of the person.

Example: If—as has been proposed—money bills contain an RFID tag, an adversary with a hidden reader could determine unnoticeably the amount of money a person is carrying in her/his briefcase.

Example: In a commercial setting, an adversary with a hidden reader could periodically monitor the inventory of goods in a competing supermarket.

Another privacy enforcement technique does not *prevent* but *detects* clandestine reading. A consumer can carry a special *monitoring tag* that alerts the carrier whenever a clandestine reading attack is detected. The monitoring tag transforms a *clandestine reading action* to an *open reading action* and thus exposes the hidden adversary.

Denial of Service   A *denial of service attack* tries to make a computer system unavailable to its users. In any wireless communication scenario, such as an RFID system or a sensor network, an adversary can jam the ether with high-power signals of the appropriate frequency in order to interfere with the communication of the targeted devices. In the Internet, an adversary can send a coordinated burst of service requests to a site to overload the site such that legitimate service requests cannot be handled any more (see also Sect. 6.2.2 on botnets).

Some RFID tags support—as a privacy enhancement mechanism—the functionality to put a tag into a sleep mode or to permanently kill a tag. An adversary can use this functionality to interfere with the proper operation of the service.

Example: At an automated supermarket checkout, an RFID reader determines the purchased goods by reading the RFID tags of the items in the shopping cart. If an adversary disables some tags, the respective items will not be recognized and don't appear on the bill.

13.5   Wireless Sensor Networks (WSN)

Recent progress in the field of Micro-Electro-Mechanical Systems (MEMS), low-power microelectronics, and low-power communication has made it possible to build small integrated *smart objects*, called *sensor nodes*, that contain a sensor, a microcontroller, and a wireless communication controller. A sensor node can acquire a variety of physical, chemical, or biological signals to measure properties of its environment. Sensor nodes are resource constrained. They are powered either by a small battery or by energy harvested from its environment and have limited computational power, a small memory, and constrained communication capabilities.

In order to monitor and observe a phenomenon, a number (from few tens to thousands) of sensor nodes are deployed, either systematically or randomly, in a *sensor field* to form an ad hoc self-organizing network—a wireless sensor network (WSN).

The WSN collects data about the targeted phenomenon and transmits the data via an ad hoc multi-hop communication channel to one or more base stations that can be connected to the Internet.

After a *sensor node* is deployed in a *sensor field*, it is left on its own and relies on its self-organizing capabilities. At first, it must detect its neighbors and establish communication. In the second phase, it must learn about the arrangement in which the nodes are connected to each other, *the topology of nodes*, and build up ad hoc multi-hop communication channels to a base station. In case of the failure of an active node, it must reconfigure the network.

Wireless sensor networks can be used in many different applications such as remote environment monitoring, surveillance, medical applications, ambient intelligence, and in military applications. The utility of a wireless sensor network is in the *collective emergent intelligence* of all active sensor nodes, not the contribution of any particular node.

A sensor network is operational as long as a minimum number of nodes is active and the connectivity of the active nodes to one of the base stations is maintained. In battery-powered sensor networks, the lifetime of the network depends on the energy capacity of the batteries and the power consumption of a node. When a sensor node has depleted its energy supply, it will cease to function and cannot forward messages to its neighbors any more. Energy conservation is thus of utmost importance in sensor networks. The design of the nodes, the communication protocols, and the design of the system and application software for sensor networks are primarily determined by this quest for energy efficiency and low cost.

Attempts were made to use the RFID infrastructure for the interconnection of autonomous low-cost RFID-based sensor nodes [Bha10]. These sensor nodes operate without a battery and harvest the energy either from the environment or the electromagnetic radiation emitted by the RFID reader. This technology has the potential to produce long-lasting, low-cost ubiquitous sensor nodes that may revolutionize many embedded applications.

Points to Remember

- According to the IoT vision, a *smart planet* will evolve, where many of the everyday things around us have an identity in cyberspace, acquire intelligence, and mash-up information from diverse sources.
- The *Electronic Product Code (EPC)* is a unique identifier for the naming of every physical smart object on the planet. This is more ambitious than the forerunner, the optical bar code, which assigns a unique identifier only to a class of objects. The EPC is managed by the international organization *EPC global*.
- A composite object requires its own UID that is only loosely related to the UIDs of its constituent parts. The different names, *UID*, *object type name*, and *object role name* must be introduced at the level of composite objects as well. Since a

composite object can be an *atomic unit* at the next level of integration, the name space must be built up recursively.

- The division of work between a smart object and the cloud will be determined, to a considerable degree, by energy considerations. If the energy required to execute a task locally is larger than the energy required to send the task parameters to a server in the cloud, the task is a candidate for remote processing.
- The autonomic management of smart objects must cover *network service discovery, system configuration and optimization, diagnosis of failures and recovery after failures*, and *system adaptation and evolution*.
- An RFID reader can act as a gateway to the Internet and transmit the object identity, together with the read-time and the object location (i.e., the location of the reader) to a remote computer system that manages a large database.
- The information security threats in the IoT can be classified into three groups: (i) the threats that compromise the *authenticity* of information, (ii) the threats to *privacy* caused by a pervasive deployment of IoT products, and (iii) *denial of service* threats.
- In order to avoid clandestine reading, a tag must authenticate the reader.
- It is difficult for an adversary to clone *physical one-way functions* (POWF), because it is a *thing with random characteristic physical properties* that cannot be represented by a mathematical function. A POWF is not a *construct of cyberspace* that can be copied or reconstructed.
- After a *sensor node* is deployed in a *sensor field*, it is left on its own and relies on its self-organizing capabilities. At first, it must detect its neighbors and establish communication. In the second phase, it must learn about the arrangement in which the nodes are connected to each other, *the topology of nodes*, and build up ad hoc multi-hop communication channels to a base station. In case of the failure of an active node, it must reconfigure the network.

Bibliographic Notes

In 2009, the European Union has published a Strategic Research Roadmap for the Internet of Things [Ver09] that discusses the vision of the IoT and relevant research issues up to the year 2020 and beyond. The excellent *RFID handbook* [Fin03] is a valuable reference for the RFID technology. The September 2010 special issue of the Proceedings of the IEEE [Gad10] is devoted to RFID technology. The January 2022 special issue of the Proceedings of the IEEE addresses *Future Networks with Wireless Power Transfer and Energy Harvesting* technologies [Cle22] which we expect to accelerate the IoT widespread adoption further.

Review Questions and Problems

13.1. What is the vision of the *Internet of Things* and which are the most pressing technical issues that must be resolved?

13.2. What are the drivers for the *Internet of Things*?

13.3. What is a *smart object*?

13.4. Discuss the naming of smart objects! What is a *UID*, a *type name*, a *role name*, or a *name of a composite object*?

13.5. Discuss the different standards for near-field communication!

13.6. What is the relation between the IoT and *cloud computing*?

13.7. Describe the MAPE-K model of an autonomic component!

13.8. What are the functions of an RFID reader?

13.9. What are typical parameters for low-cost RFID tags?

13.10. What is the Electronic Product Code (EPC) and what is its relation to the ubiquitous optical bar code?

13.11. What is a *physical one-way function* (POWF)? Where is it needed?

13.12. What are the three main security threats in the RFID field?

13.13. How is a sensor node deployed in a sensor field?

13.14. Describe the self-organizing capabilities of a sensor node!

Review Questions and Problems

Chapter 14
Cloud and Fog Computing

Overview

Cloud computing has evolved from commercializing excess capacity in data centers in the early 2000s. Operators of large data centers then realized that their resource utilization was, on average, so low that it became profitable to rent out resources like computation time and storage to paying customers. As a result, the cloud computing paradigm has been quickly adopted worldwide with soaring success. By the end of 2020, about six hundred hyper-scale data centers offering cloud computing have been in operation. Cloud computing is frequently used to realize soft real-time systems, like video-streaming, E-commerce, or office and collaboration applications. However, this chapter is rather interested in exploring how cloud computing fits in as a design principle for distributed embedded applications that form hard real-time systems.

We start this chapter by outlining significant differences between the *world of cloud* and the world of *RT systems*. We argue that the world of cloud on its own is unfit for hard real-time systems because of its inability to provide timeliness guarantees and introduce *fog computing* (also called *edge computing*) as a means to leverage cloud benefits. Fog computing is an architectural style that differentiates between the *embedded*, the *fog*, and the *cloud layers*. In contrast to the cloud, the fog layer is close to the controlled object and consists of *fog nodes* characterized by their ability to provide *resource pooling, northbound and southbound connectivity*, and *configuration of the fog and embedded layer*. We detail the main benefits and risks of fog computing and continue with key fog and cloud technologies along said characteristics. We discuss *virtualization* as a method for resource pooling, introduce *time-triggered virtual machines* (TTVM), and give a prospect of *containers* and *serverless computing*. *Northbound connectivity* connects the fog layer to the cloud, while *southbound connectivity* connects the fog layer to the embedded layer. We discuss the main differences between northbound and southbound connectivity. We also give examples for the *configuration characteristic*. We conclude this chapter with a discussion of use cases of cloud computing and fog computing for hard

H. Kopetz, W. Steiner, *Real-Time Systems*,
https://doi.org/10.1007/978-3-031-11992-7_14

real-time systems and explain how the *Nerve* software platform for industrial fog computing satisfies the fog node characteristics.

14.1 Introduction

In Chap. 1, we define the main functional requirements of an RT system as data collection, direct digital control, and man-machine interaction. In the past, the collected data has been transported to an enterprise computer for archival and further analysis. However, with the movement of many enterprise computing functions into the cloud and the requirement to remotely monitor, diagnose, and update RT systems via the Internet, the need for interfacing a real-time control system with the cloud arose. In addition, once cloud connectivity is established, novel types of RT systems emerge.

> **Example:** Self-driving cars are among the most complex RT systems. They typically use diverse sensors and *high-definition maps* to safely identify their position and maneuver. In contrast to traditional digital maps stored on car-local storage, digital maps as a *cloud service* enable the map data to be treated as *RT image* with a *temporal accuracy* determined by the map update frequency. This RT image may also include information on weather, traffic situations, or accidents. Cars can collectively and on the fly update the RT image, for example, using the *Sensor Ingestion Interface Specification* published by HERE [HER15] or the ETSI Technical Report on *Analysis of the Collective Perception Service (CPS)* [ETS19].

Although it would be economically attractive to off-load more and more RT system functions to the cloud, significant limitations apply as the connectivity to the cloud and computations in the cloud generally do not meet the temporal requirements for RT system functions. Thus, there exists a boundary between the *world of the cloud* and the *world of real-time systems*, and this boundary must be in close spatial proximity to the controlled object. *Fog nodes* are an implementation of this boundary. They interface the RT system to the cloud and provide centralized computing capabilities.

With the advent of the cloud computing paradigm, its use for executing real-time tasks has been studied. For soft real-time tasks, it quickly became a success. We call the respective services *soft real-time services*, and services like video-on-demand or collaborative office tools are nowadays in widespread use. However, the execution of hard real-time tasks in the cloud, which we call *hard real-time services*, has shown to be a significant challenge, and to the best knowledge of the authors, there does not exist a single cloud provider that offers the capabilities to execute hard real-time services in a pure cloud computing environment today.

The said challenge originates from the complexity of the distributed computer that forms the cloud and its connectivity to the real-time entities, which causes high temporal variation (i.e., jitter) both in the communication to the cloud and the task execution times in the cloud itself. Thus, the action delay of messages to and from the cloud and their processing will be of low quality and, in practice, may not even be reliably calculable.

Example: Tools to measure Internet latency are commonly available. For example, the *ping* command can measure the round-trip delay from a pinging entity to a cloud server and back again. Pinging www.google.com from central Europe results in round-trip delays of about 20 ms on average, but occasionally, the delays significantly exceed 100 ms. Another standard tool is *traceroute*, which measures the number of hops between a sender and a receiver. A brief experiment shows that common domains like www.google.com or www.aws.com are reachable in between ten and twenty hops (again from central Europe).

Hard real-time services require calculable bounds on the worst-case communication time and worst-case execution times. *Neither is possible using today's cloud technologies, thus violating the core cloud computing assumption that the cloud's underlying HW/SW system will provide adequate performance (in this case guaranteed timeliness) automatically.* However, emerging technologies like 5G or the IETF *detnet* standards for communication and the use of specialized real-time hardware in the cloud may gradually address these open problems.

As the response time of a cloud service to the consumer is also an essential property for soft real-time services, cloud providers have quickly realized that the geographic co-location of their data centers with the users significantly reduces communication latency. As we continue to distribute the resources and co-locate them even further toward the consumers, a new architectural style emerges, *fog computing*.

14.2   Characteristics of the Cloud

The world of the cloud is fundamentally different from the world of real-time systems. The dynamic resource sharing in the cloud increases dependability and performance at the cost of real-time guarantees. We summarize the main differences between the two worlds in Table 14.1.

Let us look in some detail on the following *essential characteristics* of cloud computing:

Resource Pooling. In simplified terms, a cloud is a data center that pools resources and provides them to many consumers. Such resources are, for example, storage, processing, memory, and network bandwidth, and the cloud will ensure that each

Table 14.1 Comparison *world of cloud* to *world of RT systems*

Characteristic	World of cloud	World of RT systems
Timeliness	Designed for average-case	Designed for worst-case with guarantees
Dependability	Availability	Reliability and safety
Flexibility	Consumer shall perceive infinite resources with fast scale-up/down	Infrequent updates within well-defined bounds
Physical structure	Data centers	Embedded systems
Accessibility	Broad and global	Restricted

consumer gets access to its share of the resources. However, a consumer will typically be unaware of the precise location of the provided resources and will not be able to request the resource location precisely. This characteristic, multiple consumers sharing a pool of resources, is called *resource pooling*.

Broad Network Access. The resources in the cloud must be easily accessible through standardized communication means. Furthermore, as users may operate various devices with different communication capabilities (e.g., from smart phones to workstations), the cloud also needs to support multiple communication standards.

Elastic Compute. Cloud computing inherently supports even massive and rapid changes in consumer resource demands. On the one hand, the *on-demand self-service* characteristic says that cloud computing enables the consumer to increase its share of the resources in an automated manner without the human intervention of the cloud service provider. On the other hand, the *rapid elasticity* characteristic further says that the cloud will quickly handle such changes in consumer resource demands. As the resource usage per consumer may vary vastly over time, cloud computing must also have a *measured service* characteristic that essentially implements metering methods for bookkeeping of how much resources have been allocated to or consumed by the consumer.

The outstanding economic benefit of cloud computing in various use cases and potential risks and challenges have been analyzed early on, as discussed by Armbrust et al. [Arm10]. Not surprisingly, the economic benefit is highest in use cases that effectively leverage the cloud computing characteristics discussed above. Such use cases are the ones in which the consumer demand varies highly, but there are also use cases in which the resource demand is not predictable, such as a new Web startup offering a novel SaaS. Armbrust et al. also give a third class of use cases— computational problems that the customer prepares as highly parallelizable can be solved faster by increasing the consumer resource provision for specific periods. We continue the discussion of cloud use cases in the context of real-time systems in Sect. 14.5.1.

In terms of risks, some of the typical ones of a data center remain, for example, the failure of individual resources or large-scale damages. However, cloud providers can mitigate these risks like traditional data center operators and may have even better options in terms of mitigation strategies as they often operate multiple cloud sites that may serve as backup for each other. Nevertheless, some new risks arise inherently with cloud computing or significantly increase compared to traditional data centers. First, as multiple consumers share the same resource pool, these consumers now pose security threats to each other (Armbrust et al. call this *internal-facing security*). Indeed, large cloud providers invest heavily to ensure no information leakage between consumers. They do so, for example, by using formal methods. Secondly, as multiple consumers share the benefits of the cloud infrastructure, they also are commonly affected by its failure. For example, the complexity of the protocols and algorithms that allow scaling the resources quickly and automatically with the consumer demand increases the risk of common software failures.

NIST defines the cloud computing paradigm in a special publication [Mel11], which is accepted as state of the art by the cloud industry. In addition to the *essential characteristics* we discussed in this section, it also specifies the following *three service models*.

The service models are:

- *Software as a Service (SaaS)*
- *Infrastructure as a Service (IaaS)*
- *Platform-as-a-Service (PaaS)*

The service model most commonly used by the general public is *Software as a Service (SaaS)*, in which a consumer uses an application hosted in the cloud, for example, a webmail application. The only interface to such applications may be the Web browser. However, the consumer of cloud services often is not the end customer but rather another business requiring different service models. One is *Infrastructure as a Service (IaaS)*, in which the consumer (e.g., a business) has control over the applications deployed in the cloud, the operating system, and the storage. Finally, *Platform-as-a-Service (PaaS)* is another service model in-between SaaS and IaaS, in which the consumer develops and deploys its applications in the cloud but is limited to platform capabilities that the cloud provides.

> **Example:** In the previous digital maps example, the end customer may use the map service as a SaaS, while the digital map provider may not operate an infrastructure of its own but be a consumer of IaaS of a cloud provider.

Service-Level Agreement The cloud computing provider and the consumer enter a legal contract, called the service-level agreement (SLA), that defines the cloud service delivered by the provider and the costs thereby incurred by the consumer. Some cloud computing providers indicate average response times (of the service executed in the cloud) in their SLAs. However, as to the authors' knowledge, SLAs do not guarantee worst-case response times.

14.3 The Advent of Fog Computing

In contrast to cloud computing, fog computing does not have a broadly accepted definition. Although NIST defines a *Fog Computing Conceptual Model* [Ior18], its uptake is far behind the NIST cloud computing definition. Thus, in this chapter, we define fog computing in light of its usage as a design principle for distributed embedded systems and give our interpretation of how it relates to similar concepts like edge computing.

14.3.1 Fog Computing for Distributed Embedded Systems

Fog computing links the cloud and embedded *architectural styles* (see Sect. 2.2.4) and introduces new design principles, rules, and conventions. Fog computing for distributed embedded systems is, thus, an architectural style by itself in which the distributed computer that controls the physical process consists of embedded nodes, at least one *fog node*, and cloud services. We call the one or multiple fog nodes the *fog layer*, which has the following three characteristics:

(i) Resource Pooling: the fog layer is provisioned to execute real-time services for different, potentially independent applications. In case a fog node allows the execution of hard real-time services, it is equipped with adequate computational resources.

(ii) Connectivity:

   (a) The fog layer is connected to the cloud through a *northbound* interface.

   (b) The fog layer is connected to embedded nodes through a *southbound* interface that communicates via a real-time network with the embedded nodes and may connect directly to sensors, actuators, or both as well.

   (c) The fog layer implements a *gateway component* (see Sect. 4.5) that links the cloud and embedded *architectural styles*.

(iii) Configuration:

   (a) The fog layer can configure at least some communication and computation resources accessible via its southbound interface.

   (b) The fog layer itself is configurable via its northbound interface.

We call the cloud the *cloud layer* and the embedded nodes the *embedded layer*. Figure 14.1 depicts the three layers in a fog computing architecture.

Example: Automated Drilling Machine—a fog computing architecture implementing an automated drilling machine may be realized as follows. The controlled object is an automated drilling machine consisting of a camera, a positioning system for the drill, and the drill itself. The machine is automatically supplied with workpieces, and it is the task of the distributed computer to determine the target location, size, and depth for the drill holes and perform the actual drilling. The camera captures the visual appearance of the workpiece (before, during, and after the drilling). The camera then sends this image via the real-time network to the fog node, which produces the concrete set points for the location of the drill, drill speed, and drill force. The fog node sends these set points to the embedded nodes,

Fig. 14.1 Example of fog computing architecture

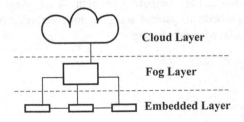

which apply them to the positioning system and the drill. Additionally, the fog node sends pictures of the workpiece before and after the drilling to the cloud for quality control.

If multiple fog nodes form the fog layer, they may operate as an ensemble by showing a resource pooling characteristic, similar to cloud computing.

Example: Automated Drilling Factory Floor—a factory floor may implement a hundred drilling machines serviced by ten fog nodes. If one of the fog nodes fails, the remaining non-faulty fog nodes will automatically service the affected drilling machines. The fog nodes are sufficiently provisioned for this use case.

Fog nodes can be stationary or mobile. For example, fog nodes can be located in an industrial automation setting at the factory floor to control individual production cells or industrial robots. Fog nodes can also be mobile, for example, integrated inside an automobile.

14.3.2 Fog Computing Benefits and Risks

It is well known since many years that the management of large and complex real-time systems can be simplified by architectures that embrace abstraction and divide-and-conquer principles [Kah79]. In particular, a fog architecture has the following benefits compared with a traditional embedded-only architecture.

(i) Partitioning. Modern real-time systems become more and more complex, and we must use simplification strategies to keep them comprehensible. Partitioning the system into well-defined subsystems is a simplification strategy (see Scct. 2.1.4). Different subsystems often have different computational characteristics and, thus, a fog computing architecture can optimize the overall real-time system by appropriately mapping subsystems to the appropriate cloud, fog, or embedded layer. Furthermore, the layered structure of fog computing can also guide the partitioning process and provide basic mechanisms to control subsystem interfaces on which richer control mechanisms can build.

(ii) System-of-Systems. The fog layer can also abstract details of the embedded layer. On the one hand, abstraction is another complexity simplification strategy. On the other hand, abstraction enables fog computing as evolutionary architecture for system-of-systems formation (see Sect. 4.7.3).

Example: Collaborating self-driving cars are constituent systems. They will be produced by different car manufacturers and can evolve uncoordinatedly. They can interoperate as system-of-systems through local fog nodes, adapting their respective northbound interface and gateway component as necessary.

(iii) Embedded-layer off-loading: the fog layer allows to save entire embedded nodes by directly connecting to the sensors/actuators. Those embedded nodes that remain in the system can implement reduced computational power. Although the additional cost of the fog layer must be considered and compared to the cost savings in the embedded layer, this comparison should address the total cost over the system's lifetime. The fog layer increases the reliability of the embedded layer (as there are fewer or simpler nodes), and maintenance in the fog layer will likely be more cost-efficient than in the embedded layer (as the fog layer is easier to access).

(iv) Optimized multi-core utilization and resource overprovision: scaling up the number of processor cores per chip is the primary factor in today's silicon performance growth. Thus, multi-core SoCs are also readily available for the embedded systems market. However, individual embedded nodes executing only a single application will not necessarily leverage this increase in silicon performance as the applications have a very specialized and often very narrow purpose. Fog nodes enable the efficient usage of multi-core SoCs to consolidate multiple applications from different embedded nodes. Furthermore, fog nodes may implement spare resources to prepare the RT system for future service upgrades (*resource overprovisioning*).

(v) Simplified power and cooling infrastructure: many modern embedded systems have high demands in overall energy consumption resulting in complex energy and cooling infrastructures. These infrastructures can be simplified as the fog computing architecture reduces the number of individual nodes. The fog node may also be located remotely in some use cases, which eases cooling and power supply. However, if a fog node exceeds a certain complexity, the fog node's cooling and power subsystems may become a challenge, and a distributed (fog) solution can be preferable.

(vi) Centralized data gathering and information processing: a fog node can also act as a central entity to collect data from the embedded layer. This data can then be processed in multiple ways, e.g., concentrated and compressed for upload to the cloud layer, used as a basis for intrusion and anomaly detection, application-level processing to generate new information for the embedded layer, simplified redundancy management, and data consistency services for replicas in the embedded layer.

(vii) Centralized configuration and maintenance: modern real-time networks like IEEE 802.1 Time-Sensitive Networking have many configuration and reconfiguration capabilities. The fog layer is well suited to operate as the central network configuration instance for such networks in the embedded layer. Similarly, the fog layer can operate as a centralized instance to receive new software updates for the embedded layer from the cloud layer and manage the update process.

(viii) Security architecture means: the fog layer can operate as one of the multiple lines of defense against security threats. For example, the fog node may implement intrusion detection algorithms or a firewall.

The centralization of functionality from the embedded layer into the fog layer may also pose new dependability concerns and security risks.

Although a system design implements a fog computing architecture, it must still meet application-specific dependability goals. These can be achieved by a sufficient quality of the fog node(s), replication of fog nodes, design of combined dependability measures in the fog and the embedded layer, or a combination of these. In addition, typical system design pitfalls should be avoided: single point of failure, fault and error propagation (in particular between cloud, fog, and embedded layer boundaries), and common-mode failures. Implementing multiple fault-containment units within a single fog node may be justifiable in some system designs. However, such designs must be analyzed with great care, and it must be shown that pitfalls like the ones mentioned above are avoided or at least sufficiently mitigated.

In a fog computing architecture, the fog layer is the only interface toward the cloud. While this has security benefits as discussed above, it will also be the major target for security attacks. Thus, the fog layer itself needs to be designed toward high-security requirements. Depending on the specific use case, it will also be advisable to implement end-to-end security measures down to the embedded layer, for example, by using signatures and secure boot.

14.3.3 General Fog Computing and Comparison to Edge Computing

Shi et al. [Shi16] define "edge as any computing and network resources along the path between data sources and cloud data centers" and argue that "edge computing is interchangeable with fog computing." Although NIST calls out differences between fog and edge, we agree with Shi et al. It seems that the technical realization of edge computing and fog computing are identical. However, the technical realization can be viewed from different perspectives. On the one hand, cloud and Internet service providers differentiate the core and the edge of their infrastructure as the location of computations. Therefore, they tend to use the term *edge computing*, and nowadays, cloud providers also offer edge computing as a service. On the other hand, from the perspective of an architectural style, we prefer *fog computing* with its defining layers of cloud, fog, and embedded. It also seems a better metaphor in the context of cloud computing, not the least because clouds do not have edges.

14.4 Selected Cloud and Fog Technologies

While the fog computing paradigm as an architectural style has directly emerged from cloud computing, the technologies that enable fog computing, particularly technologies necessary to build complex real-time systems, have evolved in parallel to cloud computing technologies. In this section, we discuss enabling technologies for fog computing. The discussion is structured by our previously introduced three fog computing characteristics: resource pooling, connectivity, and configuration.

We also address *system design automation* in fog computing in this section and emphasize the importance of appropriate configuration tooling.

14.4.1   Resource Pooling

A fog node for real-time systems must limit the interference between the different applications it hosts and provide predictable bounds to its southbound user. Only then will the fog node's real-time services adhere to their specifications. Standard software techniques in cloud computing for resource pooling are *virtualization*, *containerization*, and *serverless computing*. Although these techniques limit interference, their actual implementations (i) lack the scrutiny typically required for safety-related and safety-critical systems and (ii) typically only consider the limitation of interference in the value domain but are not concerned with the interference in the time domain that is absolutely necessary for hard real-time systems.

The most potent form of interference limitation is hardware-based: the fog node implements a multitude of chips. As a result, the interference between applications that execute on different chips is limited, and the communication system that connects these chips can exert control.

> **Example: zFAS**—zFAS is a central domain controller for automotive driver-assistance systems. It consists of a single board that implements multiple CPUs and an FPGA. A time-triggered Ethernet switch connects them.

Fog nodes may even implement multiple boards that connect via a backplane. Such fog nodes typically use Ethernet, PCIe, or both as backplane communication.

Besides the limitation of interference, other issues requiring multi-chip fog node implementations are the sheer need for performance, the lack of certified semiconductors, and special-purpose chips as hardware accelerators. However, trends in the semiconductor industry point toward more and more integration into single system-on-chip (SoC) solutions aiming to mitigate the said issues. For example, semiconductors are already available today that realize *safety islands* for critical applications. This safety island part of the chip is developed according to industry-specific safety standards and integrates with other non-safety portions on the same silicon.

Software solutions can also limit interference. For example, in traditional embedded systems, it is the task of the real-time operating system to schedule the real-time tasks. Thus, as long as the operating system meets the tasks' deadlines, temporal interference of tasks on each other is acceptable. Likewise, other operating system services will ensure interference limitation in the value domain (e.g., memory management, I/O).

Virtualization With the growing semiconductor performance, particularly multi-core SoCs, virtualization becomes technically feasible for hard real-time systems, enabling the fog computing benefits discussed in the previous section. Virtualization is a well-established hardware abstraction technology introduced by Goldberg [Gol73], and formal requirements for virtualizable computer architectures are formulated by Popek and Goldberg [Pop74]. Virtualization makes use of a *virtual machine monitor* (VMM) that manages and controls *virtual machines* (VMs). Today, VMMs are commonly referred to as *hypervisors*, and they come in two types. A *type 1 hypervisor* is a software layer that operates directly on top of the hardware compute resource (i.e., the processor); it is also said to operate on *bare metal*. In contrast, a *type 2 hypervisor* operates on top of an operating system. Thus, in the case of type 2 hypervisors, an operating system sits between the hypervisor and the bare metal, called the *host operating system*. Typically, operating systems are also installed inside the VMs—these operating systems are called *guest operating systems*.

Today, modern processors support virtualization not only for computation but also for I/O and graphics. With the growing number of computing cores per processor, virtualization has become a major technology that enables fine-grained resource pooling. For example, VMs may be configured to execute on a fixed set of processor cores called *pinning*. Cores can also be configured to be shared between VMs, including VMs from different consumers.

> **Example:** A specific server-class SoC may implement 64 cores. By using the Xen type 1 hypervisor, many individual cloud consumers can be supported by allocating the appropriate cores to the consumer VMs. In a simple scenario, 63 cloud consumers get a core assigned exclusively, leaving one core for Xen to manage the virtual machines.

In the cloud, the number of VMs per consumer can dynamically scale with the consumer demand, and VMs can be configured to migrate from one machine to another, e.g., in the case of failures. Fog nodes can use hypervisors for hard real-time systems that have been developed for quite some time now. An early open-source hypervisor is RT-Xen [Xi11]. ACRN [Li19] is a more recent one. There also exist a growing number of commercial solutions.

> **Example: Nerve Platform**—Nerve is a fog computing platform for industrial automation. It implements the ACRN 2.0 hypervisor that allows the execution of industrial real-time tasks.

A fundamental problem they must address is the problem of *multilevel scheduling*.

Multilevel Scheduling in Virtualized Systems   Suppose a system that consists of a multi-core SoC virtualized by a type 1 hypervisor that instantiates multiple virtual machines and lets at least one virtual machine implement a real-time operating system. Multilevel scheduling in virtualized systems is the problem of coordinating the hypervisor's scheduling decisions with the guest operating systems' scheduling decisions such that real-time tasks executing in the virtual machines do meet their deadlines.

In the simplest solution, multilevel scheduling reduces to classical real-time scheduling and an appropriate hypervisor configuration: VM-pinning assigns computation cores exclusively to VMs. Thus, whenever the guest operating system schedules a real-time task for execution, it will not be delayed by the hypervisor scheduling another VM on said core. However, VM-pinning may be costly.

Alternatively, a cost-efficient solution is *time-triggered virtual machines* (TTVMs) in which the schedules for the VM activation and real-task execution are jointly generated at design time.

Example: Figure 14.2 depicts a simple example of two TTVMs that execute on a single compute core. One TTVM executes non-real-time tasks (Tasks A–C), while the second TTVM executes real-time tasks (RT Task 1 and RT Task 2). In this example, the activities at points in time *t1–t4* are defined at system design time and repeated cyclically with the activity at t4 repeating the activity of t1. The points in time t1, t2, and t4 are part of the hypervisor configuration and instruct the start times of the respective TTVMs. The points in time t2 and t3 are part of the RTOS configuration and instruct the RTOS when to start the real-time tasks.

Time-triggered virtual machines also require synchronized global time. Ruh et al. [Ruh21] study clock synchronization in virtualized distributed systems using ACRN as hypervisor and IEEE 802.1AS as clock synchronization protocol. They show that clock synchronization quality in virtual environments can come close to non-virtualized systems. Experiments achieve a precision of the clock synchronization well below a microsecond.

Containers Virtualization through a hypervisor provides strong isolation between different VMs, but it comes at a certain cost: each VM needs to implement a guest operating system. Thus, more lightweight forms of virtualization have been developed over the last years. One of these techniques is called *container-based virtualization* [Sol07], with its most prominent implementations LXC (https://linuxcontainers.org/) and Docker [Ber14]. In contrast to VMs, containers are not separated using a hypervisor but directly use an operating system's kernel mechanisms (e.g., in Linux: *namespaces*, *cgroups*, and *capabilities*). For this, containers are managed and controlled by a *container engine*, which sets up the execution of the containers by the appropriate calls to the kernel functions. Containers are, therefore, a form of *OS-level virtualization*.

Containers solve very practical problems in software development: the software image to be executed within the container consists not only of the application code itself but also includes *all* dependencies (e.g., libraries, stacks, and portions of

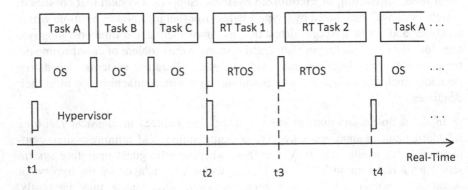

Fig. 14.2 Time-triggered virtual machines (TTVMs)

operating systems user spaces in general). Thus, the image to be executed within a container is *self-contained*. Since images are self-contained, they are also easy to scale. Each physical server that implements a container engine for compatible containers knows how to set up said container's execution.

Container orchestration (e.g., Kubernetes) manages and controls container deployment, communication, and scaling. However, this container orchestration is only concerned with the containers themselves and not with the services executed within. Container orchestration will, e.g., allocate the containers to the appropriate compute resources and set up a virtual networking infrastructure to enable communication between the containers. It can also monitor container execution and restart containers in case of failures. After the successful setup of the containers, the services executing within the containers will coordinate themselves and not depend on container orchestration.

Containers for real-time systems currently receive growing research attention [Cin18]. Analogous to TTVMs, we can also design time-triggered containers (TTCs) that execute at pre-configured points in a synchronized global time, where container orchestration can include the calculation of these instants. The RTOS kernel and the container engine need to ensure the TTCs' timely execution. However, as the isolation between containers is established by the container engine and the underlying operating system kernel, any failure of these can have a severe impact on the system operation. For example, a software failure in the kernel could cause all containers to fail.

Serverless Computation The most recent abstraction technique for compute resources is *serverless computation* introduced by Amazon with AWS Lambda functions [Ama21]. The term *serverless* needs to be interpreted from the consumer perspective: consumers can directly deploy code in the cloud without the need to manage and control a server (i.e., a VM or a container) to execute the said code as a service. Baldini et al. [Bal17] give an overview of the key characteristics of serverless platforms. Among other use cases, serverless platforms typically execute code consisting of a single main function written in one of many supported programming languages (such as Java or Python). Furthermore, the cloud will start serverless functions on-demand and will scale the number of its instances with the actual consumer needs. Therefore, "bursty, compute-intensive workloads" highly benefit from serverless computation [Bal17].

Serverless computation is not yet used for hard real-time systems to the author's best knowledge. However, although speculative at the time of this writing, it is plausible that *serverless real-time computation* will also become a reality in the future as it relieves the application developer from platform concerns entirely. The application designer can focus on solving her domain-specific problem while the platform is responsible for integrating the application in the form of (micro-)services in the system. This integration in the *world of real-time systems* will go way beyond *the world of cloud*, as it must consider characterizing the application in the temporal domain (e.g., by calculating the application's WCET) and their dependability and security attributes.

14.4.2   Connectivity

We distinguish between the northbound communication system that connects the fog layer to the cloud layer and the southbound communication system connecting the fog layer to the embedded layer. These two communication systems are significantly different, particularly regarding their real-time properties. Northbound communication must tie in with the *broad network access* characteristic of cloud computing, while southbound communication follows the requirements of a real-time network.

The lower layers of the northbound communication system of a fog node will typically realize technologies like Ethernet, Wi-Fi, or 5G. On the higher layers, the standard Internet stack of protocols will be realized. For example, a recent survey by Dizdarevic et al. [Diz19] on fog and cloud computing communication systems indicates frequent usage of REST HTTP as an application-level protocol for northbound communication. MQTT is another well-known protocol.

Important properties of the northbound communication system are, for example (not all requirements have to be met, but they will depend on the actual fog computing use case):

(i) Lossless data transfer: as many use cases of cloud computing are concerned with data gathering and analysis, northbound data must not be lost and must be re-transmitted in case of message losses.

(ii) Tolerance of communication brittleness: the northbound communication system may be brittle, both in mobile fog node deployments and in stationary settings. Mobile fog nodes, for example, deployed in a truck, communicate wirelessly on the northbound interface to the cloud, and this wireless link may be subject to outages because of tunnels or other environmental conditions. On the other hand, stationary fog nodes may be deployed in geographic areas with underdeveloped Internet infrastructure. Thus, the northbound interface cannot rely on an always up and running Internet connection.

(iii) Occasional bulky data transfer: bulky data transfers may be required both from the fog to the cloud, e.g., bulky upload of map or video data, and vice versa, e.g., software updates.

The southbound communication will implement real-time networks, as already discussed earlier in this book (Chap. 7). Thus, technologies like CAN or real-time Ethernet variants, including IEEE 802.1 Time-Sensitive Networking (TSN), are used on the lower network layers. Higher-layer protocols on the southbound are often proprietary. However, there are ongoing standardization efforts in various industries, for example, Robot Operating System (ROS) in robotics, SOME/IP in automotive, OPC UA in industrial automation, or Data-Distribution Service (DDS) cross-industry. In addition, special interest groups have been formed to standardize the interaction of higher-layer and lower-layer protocols, e.g., OPC UA over TSN or DDS over TSN.

Example: The Data Distribution Service (DDS) includes a publish-subscribe message-broker system. On the one hand, *DDS nodes* publish their output, called *topics*, as messages. On the other hand, *DDS nodes* may subscribe to topics. DDS establishes the information exchange between publishers and subscribers. DDS is standardized by the OMG [OMG15].

Some industries, especially the automotive industry, differentiate between *signal-based* and *service-oriented communication*. Signal-based typically means message-based communication directly on lower network layers where the communication configuration is also often hard-coded. A signal may, for example, relate to a certain field in the payload of an Ethernet message that has a hard-coded sender and receiver. Service-oriented is also message-based but means richer communication systems that consist of lower and higher network layers. On the higher layers, service-oriented communication may, for example, include a publish-subscribe message broker system or service discovery protocols. Service-oriented communication increases a system's flexibility but also increases its complexity.

Example: Service discovery for fault tolerance—a safety-critical system can be designed to rely on a service discovery process as part of a fail-over routine. For example, if a primary system fails, the service discovery can identify a backup service. In such a setting, the failure of service discovery may lead to a catastrophic event.

As shown by the example, the use of service-oriented communication in hard real-time systems, in particular safety-critical ones, must be carefully examined on a case-by-case basis. For example, one non-critical way to use service-oriented communication is during the development phase of a system when software updates are frequent. However, once the system enters the production phase, the dynamic portions of service-oriented communication are replaced by static configurations.

14.4.3 Configuration

The location of a fog node in a fog computing architecture allows it to execute configuration authority on the embedded layer and be re-configurable by the cloud layer.

The fog node may reconfigure the real-time network toward the embedded layer (the southbound interface). For example, IEEE 802.1 TSN defines the services of *Central Network Configuration* (CNC) and *Central User Configuration* (CUC). Both services target the network configuration and reconfiguration concerning changing application needs. These services can be realized in a fog node. Furthermore, fog nodes can also act as proxies for system-wide software update processes that include the embedded layer: new software images are downloaded to the fog node and distributed to the respective embedded nodes from there.

Example: Over-the-Air (OTA) Update I—a new driver software version for an automotive camera becomes available. This camera is located in a self-driving car that implements a fog computing architecture. The new software version is first loaded into a fog node inside the car, further distributing the software to the embedded node that controls the camera.

The fog node itself can be subject to updates through its northbound interface.

Example: OTA Update II—a new software version of a sensor-fusion algorithm becomes available. This new software version is loaded through the northbound interface to the fog node, which exchanges the old version with the new one and executes it locally.

14.4.4   *System Design Automation*

Fog computing allows realizing quite sophisticated RT systems as, for example, self-driving cars or autonomous factories (Industry 4.0). However, these RT systems incorporate many different technologies, like the ones we discussed in this section, and most of these technologies must be appropriately configured. There will also be many dependencies between the configurations of the different technologies, for example, *multilevel scheduling* discussed above. Other dependencies result from the need to construct *computation chains*, a set of tasks that must execute in a given temporal order within a given deadline, possibly in different compute resources and potentially even in different layers of a fog computing architecture.

Example: MotionWise is an automotive software platform for advanced driver assistance functions and automated driving. The MotionWise Creator is a tool for system design automation. It allows, for example, to specify properties of real-time tasks (e.g., WCET, task dependencies, memory footprint), and MotionWise Creator generates the task and communication schedules.

System design automation differentiates *lower-level configuration* from *higher-level configuration*, where a human produces the higher-level configuration and a tool automatically generates the lower-level configuration. However, with technological progress and the encoding of human expert know-how, the boundary between higher-level and lower-level configuration gradually rises. Thus, more and more configuration tasks become automated.

Example: In the current MotionWise version, a human allocates functions to compute resources (higher-level configuration). In upcoming versions, MotionWise will also automate this function allocation (lower-level configuration).

System design automation in fog computing has enormous growth potential. For example, in the future, it may target (i) to automatically allocate functions to the different fog computing architecture layers (embedded, fog, cloud), (ii) to automate and support the selection of matching technologies for given use cases, and (iii) to guide the selection and placement of fog nodes.

14.5   Example Use Cases

Cloud and fog computing enable many use cases for distributed embedded applications. We discuss some in this final section of this chapter.

14.5.1 Cloud Computing-Enabled Use Cases

This section discusses use cases for architectures in which the embedded layer communicates with the cloud layer, either directly or indirectly via the fog layer. The fog layer is optional.

Distributed Data Collection and Big Data Processing

The data center characteristic of cloud computing allows it to operate as centralized data storage and data processing. This is particularly attractive when not only a single system feeds data into the cloud, but larger collectives of systems as, for example, fleets of cars or a factory floor (or multiple factory floors) of industrial robots do so to achieve a higher common goal.

> **Example:** A fleet of drones is equipped with individual cameras to map an area after an earthquake to aid a rescue mission. The individual camera feeds of the drones are collected in the cloud, which generates a consolidated map. A cloud service can guide the rescue mission based on the consolidated map and continued soft real-time feeds from the drones' cameras.

> **Example:** In the development of self-driving cars, an essential means of validation is street testing. A fleet of self-driving cars provides test data to the cloud as the cars drive on the street. Part of this test data are edge case detections, i.e., situations that the car did not recognize correctly or handle as expected. These edge case detections are collected in the cloud, where they can be used to produce software updates for self-driving cars, including updates of the neuronal networks used in object recognition procedures.

> **Example: Predictive Maintenance** a factory floor of industrial robots delivers diagnostics data to the cloud. The cloud can use pattern recognition methods on this diagnosis data to identify the need to replace components of the industrial robots. In addition, it can learn from past incidents and identify patterns that will likely cause similar issues.

As the previous examples show, the use cases for data collection and processing in the cloud are quite diverse. Indeed, the analysis of *Big Data* (e.g., see Sagrigoglu et al. [Sag13]) has already caused disruptions in established industries like the advertising industry, and it has a similar potential for disruption in traditional real-time control industries, e.g., Industry 4.0.

Digital Twin

The NASA taxonomy defines the digital twin as a means for model-based system engineering (MBSE). Indeed, NASA originally introduced the concept of a *digital twin* in Shafto et al. [Sha10] to describe the NASA digital twin: "A digital twin is an integrated multi-physics, multi-scale, probabilistic simulation of a vehicle or a system [...]" and "the digital twin integrates sensor data from the vehicle's on-board integrated vehicle health management (IVHM) system [...]."

In a more general interpretation, the digital twin concept has two core characteristics:

(i) The digital twin is a *useful* simulation model of the physical system.
(ii) There exists some online synchronization between the physical object and the model realizing the digital twin.

Cloud computing is an ideal use case for the digital twin as the simulation model can become quite computationally intensive, and the synchronization with the physical system can be implemented by the cloud's broad network access characteristic.

Example: Industrial Robot Monitoring—the motion of an industrial robot in response to control commands is simulated in the cloud. This simulation is compared with the actual motion of the physical twin exposed to the same control commands. This continuous monitoring of the physical twin and comparison to the digital twin can detect failures and deteriorations.

The digital twin concept has been adopted in different fields, for example, in the mechatronics domain [Bos16], and it is generally considered a significant element in Industry 4.0.

Cloud Support for Embedded Software Engineering
The traditional way to develop software for embedded systems is to design and code the software on a development computer (a laptop or a workstation), cross-compile, and test directly on a target hardware system close to the production system. Such target hardware typically sits right on the software developer's desk. This setup gives maximum flexibility to the software developer but has severe drawbacks. First, the number of available target hardware systems may be a bottleneck. Especially when working with cutting-edge chipsets, it is rather typical that development boards are scarce. On the other hand, such hardware may be quite expensive, making it difficult to justify the acquisition of large quantities. A second major shortcoming is the lack of inherent collaborative development processes.

Cloud computing techniques can mitigate these issues to a large degree. The cloud data center can be equipped with specialized target hardware shared in a resource pooling fashion between different developers. Alternatively, and indeed the better fit for cloud computing, the target hardware is emulated by a software layer executing on standard cloud hardware or at least less-specific hardware than the concrete target hardware. Indeed, such hardware simulation is done today down to the chip level—it is even possible to develop software for a target chip before this chip is produced. Cloud computing also inherently supports team collaboration. Standard workflow orchestration techniques can be used, for example, to set up a collaborative development workflow.

Although large portions of embedded software developments can be addressed by cloud-computing techniques, at least for critical embedded systems, there will always be a necessity for final verification and validation on the production hardware to meet requirements for a system to be granted regulatory approval.

14.5.2   Fog Computing-Enabled Use Cases

This section discusses use cases in architectures with a mandatory fog layer.

Automotive Domain-Based Architecture and In-Car Compute Platform (ICCP)

Modern cars are sophisticated distributed computers: multitudes of electronic control units (ECUs) connect with in-vehicle (real-time) networks. Altogether the ECUs and networks are often called the *E/E architecture* (for Electrical/Electronic architecture). This E/E architecture continually evolves and is a major source of innovation. The applications a car executes are quite diverse, and car manufacturers typically group them into *domains* like infotainment and digital cockpit, body, energy, advanced driver assistance system (ADAS) and autonomous driving (AD), chassis, powertrain, and gateway/connectivity. In the past, new applications have been introduced into an automobile as hardware/software bundle, i.e., via a new ECU, and typically with the introduction of a new car model or a major model overhaul. This has led to quite a high number of ECUs and complex in-vehicle networks: top-of-the-line cars easily exceed a hundred ECUs. Thus, current E/E architectures are evolving more and more toward fog computing architectures (with benefits and risks discussed in Sect. 14.3.2).

An ongoing step in this E/E architecture evolution is the introduction of *domain-based architectures* in which each domain implements a *domain controller* that can be interpreted as a fog node. Each domain controller is a rather powerful ECU, compared to previous ECUs in the car, and connects to an embedded layer through a southbound interface and to the cloud via a northbound interface. The connection to the cloud is either directly in the gateway/connectivity domain controller or indirectly through the gateway for the other domains.

The hardware and software characteristics of the different domain controllers vary from domain to domain. For example, the infotainment domain controller will be less critical, while the ADAS/AD domain controller will be safety critical. The Audi zFAS is an example of an ADAS/AD domain controller [AUD17].

The E/E architecture continues to evolve toward even more centralization as multiple domain controllers are being integrated into single, more powerful ECUs. Car manufacturers are exploring full centralization into one or two *central car compute platforms,* or *in-car compute platforms,* where the major reason for two ECUs is the isolation of fault-containment units from each other.

In the E/E architecture, innovation also changes the embedded layer. One such innovation is structuring the embedded layer into zones, e.g., into four zones: front-right, front-left, and back-right, back-left. Each zone implements a *zone controller* that connects on the one hand to the sensors and actuators and on the other hand to the central *multi-domain* ECU(s).

Example: Zonal Architecture with In-Car Compute Platform (ICCP)—Figure 14.3 depicts an E/E architecture comprising two ICCP ECUs and four zone controllers (ZC).

Soft Programmable Logic Controller (Soft PLC)

Complex distributed computer systems highly automate the modern production of goods. Traditionally, these distributed computer systems are organized in the so-called automation pyramid that distinguishes different levels (see [Kah79] and the

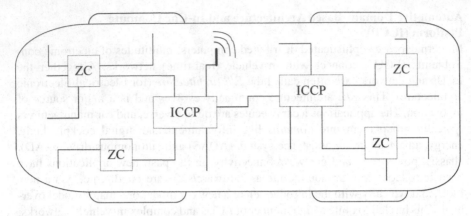

Fig. 14.3 In-car compute platform (ICCP) with four zone controllers (ZC)

Purdue Reference Model [Wil90]). On the lower levels, the pyramid places the physical processes, the sensors and actuators, and the process control. The production planning and supply chain management tasks are allocated on the higher levels in the pyramid. In the context of Industry 4.0, the strict levels of the automation pyramid are relaxed. Fog computing is a means to do so.

> **Example:** Figure 14.4 depicts an example fog computing architecture in which a fog node (FN) controls two industrial robots and a conveyor belt.

Traditional industrial automation systems implement the process control in dedicated hardware, called Programmable Logic Controllers (PLCs). As indicated in Fig. 14.3, when fog nodes are equipped with appropriate real-time hardware, the PLC functionality can also move to the fog node. Indeed, PLC functionality is already migrated to industrial PCs and run in software. We then speak of *Soft PLCs* when the PLC software is executed on a more generic hardware platform. A fog node is ideally positioned to take over the execution of multiple soft PLCs.

14.5.3   Nerve

Nerve [TTT21] is a software platform for industrial fog computing. It consists of the *Nerve Management System* in the cloud layer and the *Nerve Node* that operates in the fog layer. The implementation of the Nerve Node software enables a device to operate as a fog node (also called *edge device* by Nerve). The following table summarizes the fog node characteristics of the *Nerve Node* (Table 14.2).

Nerve use cases are, for example, condition monitoring, remote access to machines, data collection for remote machine learning, or the implementation of a digital twin.

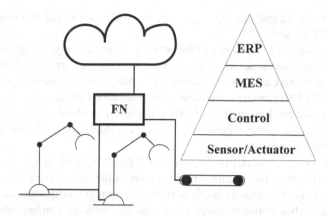

Fig. 14.4 Fog computing architecture in industrial automation and comparison to the automation pyramid with levels according to the Purdue Reference Model [Wil90]: sensor/actuator, process control, manufacturing execution system (MES), and enterprise resource planning (ERP)

Table 14.2 The fog node characteristics provided by *Nerve Node*

Resource pooling	Nerve Node implements a hypervisor (Xen or ACRN) to execute multiple VMs and a Docker Engine to execute containers. Real-time tasks with cycle times of down to 1 ms can be implemented by CODESYS (an appropriate CPU must be used)
Connectivity	Nerve Node connects to the Nerve Management System with MQTT and standard Internet protocols at the northbound interface. The southbound interface depends on the choice of device. Typical interfaces are IEEE 802.1 TSN, Profinet, and EtherCAT
Configuration	VMs and containers can be added and removed on the fly in Nerve Node. Reconfiguration of the embedded layer depends on the concrete southbound interface. For example, in case of IEEE 802.1 TSN, the Nerve Node can implement the *central network configuration*

Points to Remember

- The *world of cloud* is fundamentally different from the *world of RT systems*.
- The *world of cloud* addresses timeliness for the average case, has availability as the top dependability attribute, and aims to provide a consumer the illusion of infinite resources.
- Cloud computing is defined by *five essential characteristics*: resource pooling, broad network access, on-demand self-service, rapid elasticity, and a measured service.
- Cloud computing supports soft real-time services but is unfit for hard real-time systems because it cannot provide timeliness guarantees for the worst case.
- *Fog computing* is an architecture style that differentiates the *cloud*, *fog*, and *embedded* layer.

- *Fog nodes* implement the fog layer and decouple the world of cloud and the world of RT systems.
- Fog nodes have *three essential characteristics*: resource pooling, northbound and southbound connectivity, and configuration (fog and embedded layers).
- Resource pooling can be implemented by virtualization. *Time-triggered virtual machines (TTVMs)* ensure timeliness in virtualized environments.
- Other resource pooling methods heavily deployed in cloud computing, such as *containers* and *serverless computing*, have a certain potential also to become relevant for fog computing.
- The northbound connectivity implements typical Internet protocols while the southbound connectivity realizes real-time communication (see Chap. 7).
- *System design automation* is essential to handle system configuration complexity and distinguishes between *lower-level and higher-level configuration*, where tools automate the lower-level and a human takes care about the higher-level. The boundary between lower- and higher-level continuously rises and, thus, more and more configuration tasks become automated.
- Cloud use cases are, for example, distributed data collection, big data processing, digital twin, and software engineering support.
- Fog use cases are, for example, automotive domain-based architectures, in-car compute platforms (ICCP), and soft programmable logic controls (soft PLCs) in industrial automation.
- *Nerve* is a software platform for industrial use cases. It consists of the *Nerve Management System* for the cloud layer and the *Nerve Node* that enables hardware devices to operate as fog nodes.

Bibliographic Notes

Barroso et al. [Bar18] call data centers for cloud computing warehouse-scale computers (WSCs). These WSCs commonly consist of tens to hundreds of thousands of servers, typically clustered into racks, wherein each rack interconnects its servers with a local Ethernet switch with 40 Gbps or 100 Gbps links and further networks interconnect the racks to each other. Servers are typically standard CPUs, but recently more and more specialized computing accelerators are introduced to WSCs as well [Bar18, p. 56]. These accelerators are, for example, GPUs or neural network accelerators. Further detailed discussions of WSCs, including energy, cooling, and cost aspects can be found in [Bar18] and Hennessy and Patterson [Hen19, Chap. 6].

Cloud applications commonly use software architectures based on services and service interactions. Dragoni et al. [Dra17] discuss the evolution of this software architecture style from *object-oriented design patterns* to *service-oriented computing* (SOC) and *service-oriented architectures* (SOA) and to the more recent software architecture style of microservices.

Cloud providers also offer scalable messaging services, like scalable message queues that operate as an intermediary buffer between cloud services. Message queuing and publish-subscribe communication are well-established techniques. They are reviewed and discussed by Eugster et al. [Eug03], who also establish a common view on publish-subscribe and earlier messaging techniques. However, cloud computing scales these techniques to unprecedented levels. In their webinar, Wojciak and Bray give an overview of the messaging services provided by AWS [Woj17].

Fog computing for industrial automation has been researched by the European Project FORA (Fog Computing for Robotics and Industrial Automation) [Pop18, Pop21].

Review Questions and Problems

14.1 What are the main differentiating characteristics between the *world of cloud* and the *world of RT* systems?

14.2 Why can hard real-time systems not be implemented in the world of cloud?

14.3 What is a service-level agreement?

14.4 What are the three characteristics of the fog layer?

14.5 What are the benefits of fog computing and what are the risks?

14.6 What is a hypervisor and what is the multilevel scheduling problem in virtualized systems?

14.7 What are the main differences between the northbound and the southbound communication system in a fog architecture?

14.8 What is system design automation?

14.9 Give example use cases for cloud and fog computing in real-time systems!

14.10 What is a digital twin?

Annexes

Annex 1: Abbreviations[1]

AD	Automated driving
ADAS	Advanced driver assistance systems
AES	Advanced Encryption Standard
ALARP	As low as reasonably practical
API	Application programming interface
ASIC	Application-specific integrated circuit
AVB	Audio-Video Bridging
BMTS	Basic message transport service
CAN	Control area network
CCF	Concurrency control field
EDF	Earliest-deadline-first
EMI	Electro-magnetic interference
EPC	Electronic Product Code
ET	Event-triggered
FRU	Field-replaceable unit
FTU	Fault-tolerant unit
GPS	Global positioning system
IaaS	Infrastructure as a service
ICCP	In-car compute platform
IoT	Internet of Things
LIF	Linking interface
LL	Least-laxity
MARS	Maintainable real-time system
MPSoC	Multiprocessor system-on-chip

---

[1] This annex contains a list of frequently used abbreviations.

© The Editor(s) (if applicable) and The Author(s), under exclusive license
to Springer Nature Switzerland AG 2022
H. Kopetz, W. Steiner, *Real-Time Systems*,
https://doi.org/10.1007/978-3-031-11992-7

MSD	Message structure declaration
NBW	Non-blocking write
NDDC	Non-deterministic design construct
NoC	Network-on-chip
NTP	Network time protocol
OTA	Over the air (update)
PaaS	Platform as a service
PAR	Positive-acknowledgment-or-retransmission
PFSM	Periodic finite state machine
PIM	Platform independent model
PSM	Platform specific model
RFID	Radio frequency identification
RT	Real time
SaaS	Software as a service
SLA	Service-level agreement
SOC	Sphere of control
SoC	System-on-chip
SRU	Smallest replaceable unit
TADL	Task descriptor list
TAI	International Atomic Time
TDMA	Time-division multiple access
TMR	Triple-modular redundancy
TSN	Time-sensitive networking
TT	Time-triggered
TTA	Time-triggered architecture
TTEthernet	Time-triggered Ethernet
TTP	Time-triggered protocol
TTVM	Time-triggered virtual machine
UID	Unique identifier
UTC	Universal Time Coordinated
WCAO	Worst-case administrative overhead
WCCOM	Worst-case communication delay
WCET	Worst-case execution time
WSN	Wireless sensor network

Annex 2: Glossary[2]

Absolute Time-stamp: An *absolute time-stamp* of an *event e* is the *time-stamp* of this *event* that is generated by the *reference clock* (Sect. 3.1.2).

Accuracy Interval: The maximum permitted time interval between the *point of observation* of a *real-time entity* and the *point of use* of the corresponding *real-time image* (Sect. 1.2.1).

Accuracy of a Clock: The *accuracy* of a clock denotes the maximum offset of a given clock from the external time reference during the time interval of interest (Sect. 3.1.3).

Action: An *action* is the execution of a (part of a) program or a (part of a) communication protocol (Sect. 1.3.1).

Action Delay: The *action delay* is the maximum time interval between the start of sending a message and the instant when this message becomes *permanent* at the receiver (Sect. 5.5.1).

Actuator: A *transducer* that accepts data and *trigger* information from a *gateway component* and realizes the intended physical effect in the *controlled object* (Sect. 9.5.4).

Advanced Encryption Standard (AES): An international standard for the encryption of data (Sect. 6.2.2).

Audio Video Bridging (AVB): The IEEE 802.1 audio/video bridging (AVB) task force has developed a set of protocols based on the Ethernet standard that meets the requirements of multimedia systems (Sect. 7.4.2).

Agreed Data: An *agreed data element* is a *measured data element* that has been checked for plausibility and related to other measured data elements, e.g., by the use of model of the *controlled object*. An agreed data element has been judged to be a correct image of the corresponding real-time entity (→*raw data, measured data*) (Sect. 9.6.1).

Agreement Protocol: An *agreement protocol* is a protocol that is executed among a set of *components* of a distributed system to come to a common (agreed) view about the state of the world, both in the discrete value domain and in the sparse time domain (Sect. 9.6).

Alarm Monitoring: *Alarm monitoring* refers to the continuous observation of the *RT entities* to detect an abnormal behavior of the *controlled object* (Sect. 1.2.1).

Alarm Shower: An *alarm shower* is a correlated set of alarms that is caused by a single *primary event* (Sect. 1.2.1).

Analytic Rational Subsystem: A conscious human problem-solving subsystem that operates according to the laws of causality and logic (Sect. 2.1.1).

Anytime Algorithm: An anytime algorithms consist of a *root segment* that calculates a first approximation of the result of sufficient quality and a *periodic segment*

---

[2]All terms that are defined in this glossary are put in *italics*. At the end of each entry the section of the book that introduces or discusses the term is mentioned in the parenthesis.

that improves the quality of the previously calculated result. The periodic segment is executed repeatedly until the deadline is reached (Sect. 10.2.3).

Aperiodic Task: An *aperiodic task* is a *task* where neither the *task request times* nor the minimum time interval between successive requests for activation are known (→*periodic task,* →*sporadic task)* (Sect. 10.1.2).

Application Programming Interface (API): The interface between an application program and the operating system within a component (Sect. 9.1.4).

A Priori Knowledge: Knowledge about the future behavior of a system that is available ahead of time (Sect. 1.5.5).

Assumption Coverage: *Assumption coverage* is the probability that assumptions that are made in the model building process hold in reality. The *assumption coverage* limits the probability that conclusions derived from a perfect model will be valid in the real world (Sect. 1.5.3).

Atomic Action: An *atomic action* is an action that has the all-or-nothing property. It either completes and delivers the intended result or does not have any effect on its environment (Sect. 4.2.3).

Atomic Data Structure: An *atomic data structure* is a data structure that has to be interpreted as a whole (Sect. 5.2.).

Availability: *Availability* is a measure of the correct service delivery regarding the alternation of correct and incorrect service, measured by the fraction of time that the system is ready to provide the service (Sect. 1.4.4).

Babbling Idiot: A *component* of a distributed computer system that sends messages outside the specified time interval is called a *babbling idiot* (Sect. 4.7.1).

Back-Pressure Flow Control: In *back-pressure flow control* the receiver of a sequence of messages exerts back pressure on the sender so that the sender will not outpace the receiver (Sect. 7.2.3).

Basic Message Transport Service (BMTS): The *basic message transport service* transports a *message* from a sending component to one or more receiving components (Sect. 7.2).

Benign Failure: A *failure* is *benign* if the worst-case failure costs are of the same order of magnitude as the loss of the normal utility of the system (Sect. 6.1.3).

Best Effort: A *real-time system* is a *best-effort* system if it is not possible to establish the temporal properties by analytical methods, even if the *load- and fault hypothesis* holds (→guaranteed timeliness) (Sect. 1.5.3).

Bit-length of a Channel: The *bit length of a channel* denotes the number of bits that can traverse the channel within one *propagation delay* (Sect. 7.2.4).

Bus Guardian: The independent hardware unit of a TTP controller that ensures *fail silence* in the temporal domain (Sect. 7.5.1).

Byzantine Error: A *Byzantine error* occurs if a set of receivers observes different (conflicting) values of a *RT entity*. Some or all of these values are incorrect (synonym: malicious error, two-faced error, inconsistent error) (Sect. 3.4.1).

Causal Order: A *causal order* among a set of *events* is an order that reflects the cause-effect relationships between the *events* (Sect. 3.1.1).

Causality: The *causality* relationship between a cause C and an event E is defined as follows: If C happens, then E is always produced by it (Sect. 2.1.1).

Clock: A *clock* is a device for time measurement that contains a counter and a physical oscillation mechanism that periodically generates an *event,* the →*tick* or →*microtick* of the clock, to increase the counter (Sect. 3.1.2).

Cluster: A *cluster* is a subsystem of a real-time system. Examples of clusters are the *real-time computer system,* the operator, or the *controlled object* (Sect. 1.1).

Cognitive Complexity: The elapsed time needed to →*understand* a model by a given observer is a measure for the cognitive effort and thus for the cognitive complexity of a model relative to the observer. We assume that the given observer is representative for the intended user group of the model. (Sect. 2.1.3)

Complex Task (C-task): A *complex task (C-task)* is a *task* that contains a blocking synchronization statement (e.g., a semaphore operation *wait*) within the *task* body (Sect. 9.2.3).

Component: A *component* is a hardware-software unit, i.e., a self-contained computer including system and application software that performs a well-defined function within a distributed computer system (Sect. 4.1.1).

Composability: An architecture is *composable* regarding a specified property if the system integration will not invalidate this property, provided it has been established at the subsystem level (Sect. 4.7.1).

Computational Cluster: A subsystem of a real-time system that consists of a set of *components* interconnected by a *real-time communication network* (Sect. 1.1).

Concept: A *concept* is a category that is augmented by a *set of beliefs* about its relations to other categories. The set of beliefs relates a *new concept* to already *existing concepts* and provides for an *implicit theory* (Sect. 2.1.2).

Conceptual Landscape: The *conceptual landscape* refers to the *personal knowledge base* that has been built up and maintained by an individual in the *experiential* and *rational* subsystem of the mind (Sect. 2.2).

Concrete World Interface: The *concrete world interface* is the physical I/O *interface* between an *interface component* and an external device or another external component (Sect. 4.5).

Concurrency Control Field (CCF): The *concurrency control field (CCF)* is a single-word data field that is used in the *NBW protocol* (Sect. 9.4.2).

Consistent Failure: A *consistent failure* occurs if all users see the same erroneous result in a multi-user system (Sect. 6.1.3).

Contact Bounce: The random oscillation of a mechanical contact immediately after closing (Sect. 9.5.2).

Control Area Network (CAN): The *control area network (CAN)* is a low-cost *event-triggered* communication network that is based on the carrier-sense multiple-access collision-avoidance technology (Sect. 7.3.1).

Controlled Object: The *controlled object* is the industrial plant, the process, or the device that is to be controlled by the *real-time computer system* (Sect. 1.1).

Convergence Function: The *convergence function* denotes the maximum *offset* of the local representations of the global time within an ensemble of *clocks* (Sect. 3.4.1).

Deadline: A *deadline* is the instant when a result should/must be produced (→*soft deadline, firm deadline, and hard deadline*) (Sect. 1.1).

Deadline Interval: The *deadline interval* is the interval between the *task request time* and the *deadline* (Sect. 10.1).

Determinism: A physical system behaves deterministically if given an initial state at instant t and a set of future timed inputs, then the future states and the values and times of future outputs are entailed. In a deterministic distributed computer system, we must assume that all events, e.g., the observation of the initial state at instant t and the timed inputs, are *sparse events* on a *sparse* global time base (Sect. 5.6.1).

Drift: The *drift* of a physical *clock k* between *microtick i* and *microtick i+1* is the frequency ratio between this *clock k* and the *reference clock* at the time of *microtick i*. (Sect. 3.1.2).

Drift Offset: The *drift offset* denotes the maximum deviation between any two good *clocks* if they are free running during the resynchronization interval (Sect. 3.1.3).

Duration: A *duration* is a section of the timeline (Sect. 3.1.1).

Dynamic Scheduler: A *dynamic scheduler* is a *scheduler* that decides at run time after the occurrence of a significant *event* which *task* is to be executed next (Sect. 10.4).

Earliest-Deadline-First (EDF) Algorithm: An optimal dynamic preemptive scheduling algorithm for scheduling a set of independent *tasks* (Sect. 10.4.1).

Electro-Magnetic Interference (EMI): The disturbance of an electronic system by electromagnetic radiation (Sect. 11.3.4).

Electronic Product Code (EPC): A code designed by the RFID community that can be used to uniquely identify every product on the globe (Sect. 13.4.2).

Embedded System: A *real-time computer* that is embedded in a well-specified larger system, consisting in addition to the embedded computer of a mechanical subsystem and, often, a man-machine *interface* ($\rightarrow$*intelligent product*) (Sect. 1.6.1).

Emergence: We speak of *emergence* when the interactions of subsystems give rise to unique global properties at the system level that are not present at the level of the subsystems (Sect. 2.4).

End-to-End Protocol: An *end-to-end protocol* is a protocol between the users (machines or humans) residing at the end points of a communication channel (Sect. 1.7).

Environment of a Computational Cluster: The *environment* of a given *computational cluster* is the set of all *clusters* that interact with this *cluster*, either directly or indirectly (Sect. 1.1).

Error: An *error* is that part of the state of a system that deviates from the intended specification (Sect. 6.1.2).

Event: An *event* is a happening at a cut of the time-line. Every change of state is an *event* (Sect. 1.1).

Event Message: A message is an *event message* if it contains information about events and if every new version of the message is queued at the receiver and consumed on reading ($\rightarrow$*state message*) (Sect. 4.3.3).

Event-triggered (ET) Observation: An *observation* is *event-triggered* if the *point of observation* is determined by the occurrence of an *event* other than a *tick* of a *clock* (Sect. 5.2).

Event-Triggered (ET) System: A *real-time computer system* is *event-triggered* (ET) if all communication and processing activities are triggered by *events* other than a clock *tick* (Sect. 1.5.5).

Exact Voting: A *voter that* considers two messages the same if they contain the exactly same sequence of bits (→inexact voter) (Sect. 6.4.2).

Execution Time: The *execution time* is the *duration* it takes to execute an *action* by a computer. If the speed of the oscillator that drives a computer is increased, the execution time is decreased. The *worst-case execution time* is called →*WCET* (Sect. 10.2).

Explicit Flow Control: When a sender or multiple senders aim to exceed the resource capacities of the network or the receiver, messages will be lost. *Explicit flow control* prevents these situations by (the receiver or the communication system) signaling all or some senders to pause and resume their message transmissions. (→*flow control,* →*implicit flow control*) (Sect. 7.2.3).

External Clock Synchronization: The process of synchronization of a *clock* with a *reference clock* (Sect. 3.1.3).

Fail-Operational System: A *fail-operational system* is a *real-time system* where a safe state cannot be reached immediately after the occurrence of a *failure* (Sect. 1.5.2).

Fail-Safe System: A *fail-safe system* is a *real-time system* where a safe state can be identified and quickly reached after the occurrence of a *failure* (Sect. 1.5.2).

Fail-Silence: A subsystem is *fail-silent* if it either produces correct results or no results at all, i.e., it is quiet in case it cannot deliver the correct service (Sect. 6.1.3).

Failure: A *failure* is an *event* that denotes a deviation of the actual service from the intended service (Sect. 6.1.3).

Fault: A *fault* is the cause of an *error* (Sect. 6.1.1).

Fault Hypothesis: The *fault hypothesis* identifies the assumptions that relate to the type and frequency of faults that a fault-tolerant computer system is supposed to handle (Sect. 6.4).

Fault-Tolerant Average Algorithm (FTA): A distributed clock synchronization algorithm that handles *Byzantine* failures of *clocks* (Sect. 3.4.3).

Fault-Containment Unit (FCU): A unit that contains the direct consequences of a fault. Different FCUs must fail independently. A *component* should be an FCU (Sect. 6.4.1).

Fault-Tolerant Unit (FTU): A unit consisting of a number of replica deterministic →FCUs that provides the specified service even if some of its constituent FCUs *(components)* fail (Sect. 6.4.2).

Field Replaceable Unit (FRU): An FRU is a subsystem that is considered atomic from the point of view of a repair action (Sect. 1.4.3).

Firm Deadline: A *deadline* for a result is *firm* if the result has no utility after the deadline has passed (Sect. 1.1).

FIT: A FIT is a unit for expressing the failure rate. 1 FIT is 1 failure in 10^9 h (Sect. 1.4.1).

Flow Control: *Flow control* assures that the speed of the information flow between a sender and a receiver is such that the receiver can keep up with the sender ($\rightarrow$*explicit flow control*, $\rightarrow$*implicit flow control*) (Sect. 7.2.3).

Gateway component: A *component* of a distributed real-time system that is a member of two *clusters* and implements the relative views of these two interacting *clusters* (Sect. 4.5).

Global Time: The *global time* is an abstract notion that is approximated by a properly selected subset *of the microticks* of each synchronized local clock of an ensemble. The selected *microticks* of a local *clock* are called the *ticks* of the *global time* (Sect. 3.2.1).

Granularity of a Clock: The *granularity* of a *clock* is the nominal number of *microticks* of the *reference clock* between two *microticks* of the *clock* (Sect. 3.1.2).

Ground (g) State: The *ground state* of a *component* of a distributed system at a given level of abstraction is a *state* at an instant where there is a minimal dependency of future behavior on past behavior. At the ground state instant all information of the past that is considered relevant for the future behavior is contained in a declared ground state data structure. At the ground state instant no *task* is active and all communication channels are flushed. The instants of the ground state are ideal for reintegrating components (Sect. 4.2.3).

Guaranteed Timeliness: A *real-time system* is a *guaranteed timeliness* system if it is possible to reason about the temporal adequacy of the design without reference to probabilistic arguments, provided the assumptions about the *load- and fault hypothesis* hold ($\rightarrow$*best effort*) (Sect. 1.5.3).

Hamming Distance: The *Hamming distance* is one plus the maximum number of bit errors in a codeword that can be detected by syntactic means (Sect. 6.3.3).

Hard Deadline: A *deadline* for a result is *hard* if a catastrophe can occur in case the deadline is missed (Sect. 1.1).

Hard Real-Time Computer System: A *real-time computer system* that must meet at least one *hard deadline* (Synonym: *safety-critical real-time computer system*) (Sect. 1.1).

Hazard: A *hazard* is an undesirable condition that has the potential to cause or contribute to an accident (Sect. 11.4.2).

Hidden Channel: A communication channel outside the given *computational cluster* (Sect. 5.5.1).

Idempotency: *Idempotency* is a relation between a set of replicated messages arriving at the same receiver. A set of replicated messages is *idempotent* if the effect of receiving more than one copy of a message is the same as receiving only a single copy (Sect. 5.5.4).

Implicit Flow Control: In *implicit flow control*, the sender and receiver agree a priori, i.e., before the start of a communication session, about the instants when messages will be sent, or the message send rate. The sender commits to send only messages at the agreed instants or rate, and the receiver commits to accept all messages sent by the sender, as long as the sender fulfills its obligation ($\rightarrow$*explicit flow control*, $\rightarrow$*flow control*) (Sect. 7.2.3).

Inexact Voting: A *voter* that considers two messages the "same" if both of them conform to some application-specific "sameness" criterion (→*exact voter*) (Sect. 6.4.2).

Instant: An *instant* is a cut of the timeline (Sect. 1.1).

Instrumentation Interface: The *instrumentation interface* is the *interface* between the *real-time computer system* and the *controlled object* (Sect. 1.1).

Intelligent Actuator: An *intelligent actuator* consists of an actuator and a microcontroller, both mounted together in a single housing (Sect. 9.5.5).

Intelligent Product: An *intelligent product* is a self-contained system that consists of a mechanical subsystem, a user *interface*, and a controlling embedded *real-time computer system* (→*embedded system*) (Sect. 1.6.1).

Intelligent Sensor: An *intelligent sensor* consists of a sensor and a microcontroller such that *measured data* is produced at the output *interface*. If the *intelligent sensor* is fault-tolerant, *agreed data* is produced at the output *interface* (Sect. 9.5.5).

Interface: An *interface* is a common boundary between two subsystems (Sect. 4.4).

Interface Component: A *component* with an *interface* to the external environment of a component. An *interface component* is a *gateway* (Sect. 4.5).

Internal Clock Synchronization: The process of mutual synchronization of an ensemble of *clocks* in order to establish a *global time* with a bounded *precision* (Sect. 3.1.3).

International Atomic Time (TAI): An international time standard, where the second is defined as 9 192 631 770 periods of oscillation of a specified transition of the Cesium atom 133 (Sect. 3.1.4).

Intrusion: The successful exploitation of a *vulnerability* (Sect. 6.2).

Intuitive Experiential Problem-Solving System: A human preconscious emotionally based problem-solving subsystem that operates holistically, automatically, and rapidly and demands minimal cognitive resources for its execution (Sect. 2.1.1).

Internet of Things (IoT): The direct connection of physical things to the Internet such that remote access and control of physical devices is enabled (13).

Irrevocable action: An action that cannot be undone, e.g., drilling a hole, activation of the firing mechanism of a firearm (Sect. 1.5.1).

Jitter: The *jitter* is the difference between the maximum and the minimum duration of an *action* (processing action, communication action) (Sect. 1.3.1).

Laxity: The *laxity* of a *task* is the difference between the *deadline interval* minus the *execution time* (the *WCET*) of the *task* (Sect. 9.2.2).

Least-Laxity (LL) Algorithm: An optimal dynamic preemptive *scheduling* algorithm for scheduling a set of independent *tasks* (Sect. 10.4.1).

Logical Control: *Logical control* is concerned with the control flow *within* a *task*. The *logical control* is determined by the given program structure and the particular input data to achieve the desired data transformation (→*temporal control*) (Sect. 4.1.3).

Maintainability: The *Maintainability M(d)* is the probability that the system is restored to its operational state and restarted within a time interval *d* after a failure (Sect. 1.4.3).

Malicious Code Attack: A malicious code attack is an attack where an adversary inserts malicious code, e.g., a virus, a worm, or a Trojan horse, into the software in order that the attacker gets partial or full control over the system (Sect. 6.2.2).

Measured Data: A *measured data element* is a *raw data element* that has been pre-processed and converted to standard technical units. A sensor that delivers *measured data* is called an *intelligent sensor* (→*raw data, agreed data*) (Sect. 9.6.1).

Membership Service: A *membership service* is a service in a distributed system that generates consistent information about the operational state (operating or failed) of all *components* at agreed instants (membership points). The length of the interval between a membership point and the moment when the consistent membership information is available at the other *components* is a quality of service parameter of the membership service (Sect. 5.3.2).

Message Structure Declaration (MSD): A specification that explains how the data field of a message is structured into syntactic units and assigns names to these syntactic units. The names identify the *concepts* that explain the meaning of the data (Sect. 4.6.2).

Microtick: A *microtick* of a physical clock is a periodic *event* generated by this *clock* (→*tick*) (Sect. 3.1.2).

Non-Blocking Write Protocol (NBW): The *non-blocking write protocol* (*NBW*) is a synchronization protocol between a single writing task and many reading tasks that achieves data consistency without blocking the writer (Sect. 9.4.2).

Non-Deterministic Design Construct (NDDC): A non-deterministic design construct is a design construct that produces unpredictable results either in the value domain or the temporal domain (Sect. 5.6.3).

Observation: An observation of a real-time entity is an atomic triple consisting of the name of the real-time entity, the instant of the observation, and the value of the real-time entity (Sect. 5.2).

Offset: The offset between two *events* denotes the time difference between these *events* (Sect. 3.1.3).

Periodic Finite State Machine (PFSM): A PFSM is an extension of the finite state machine model to include the progression of real time (Sect. 4.1.3).

Periodic Task: A *periodic task* is a *task* that has a constant time interval between successive *task* request times (→*aperiodic task*, →*sporadic task*) (Sect. 10.1.2).

Permanence: *Permanence* is a relation between a given message and all related messages that have been sent to the same receiver before this given message has been sent. A particular message becomes *permanent* at a given *component* at the moment when it is known that all earlier sent related messages have arrived (or will never arrive) (Sect. 5.5.1).

Phase-Aligned Transaction: A *phase-aligned transaction* is a *real-time transaction* where the constituting processing and communication *actions* are synchronized (Sect. 5.4.1).

Point of Observation: The instant when a *real-time entity* is observed (Sect. 1.2.1).

Precision: The *precision* of an ensemble of clocks denotes the maximum *offset* of respective ticks of any two clocks of the ensemble over the period of interest. The *precision* is expressed in the number of *ticks* of the *reference clock* (Sect. 3.1.3).

Primary Event: A *primary event* is the cause of an *alarm shower* (Sect. 1.2.1).

Priority Ceiling Protocol: A *scheduling* algorithm for *scheduling* a set of dependent periodic *tasks* (Sect. 10.4.2).

Process Lag: The delay between applying a step function to an input of a *controlled object* and the start of response of the *controlled object* (Sect. 1.3.1).

Propagation Delay: The *propagation delay* of a communication channel denotes the time interval it takes for a single bit to traverse the channel (Sect. 7.2.4).

Protocol: A *protocol* is a set of rules that governs the communication among partners (Sect. 2.2.3).

Radio Frequency Identification (RFID): A technology for the identification of objects by electronic means (Sect. 13.4)

Rare Event: A *rare event* is a seldomly occurring event that is of critical importance. In a number of applications the predictable performance of a *real-time computer system* in *rare event* situations is of overriding concern (Sect. 1.2.1).

Rate-Monotonic Algorithm: A dynamic preemptive *scheduling* algorithm for *scheduling* a set of independent periodic *tasks* (Sect. 10.4.1).

Raw Data: A *raw data element* is an analog or digital data element as it is delivered by an unintelligent sensor (→*measured data, agreed data*) (Sect. 9.6.1).

Real-Time (RT) Entity: A *real-time (RT) entity* is a state variable, either in the *environment* of the *computational cluster*, or in the *computational cluster* itself, that is relevant for the given purpose. Examples of *RT entities* are the temperature of a vessel, the position of a switch, the set point selected by an operator, or the intended valve position calculated by the computer (Sect. 5.1).

Real-Time (RT) Image: A *real-time (RT) image* is a current picture of a *real-time entity* (Sect. 5.3).

Real-Time Computer System: A *real-time computer system* is a computer system, in which the correctness of the system behavior depends not only on the logical results of the computations, but also on the physical time when these results are produced. A real-time computer system can consist of one or more *computational clusters* (Sect. 1.1).

Real-time Data Base: The *real-time database* is formed by the set of all *temporally accurate real-time images* (Sect. 1.2.1).

Real-Time Object: A *real-time (RT) object* is a container inside a computer for a *RT entity* or a *RT image*. A *clock* with a granularity that is in agreement with the dynamics of the *RT object* is associated with every RT object (Sect. 5.3.2).

Real-Time Transaction: A *real-time (RT) transaction* is a sequence of computational and communication *actions* between a stimulus from the environment and a response to the environment of a *computational cluster* (Sect. 1.7.3).

Reasonableness Condition: The *reasonableness condition* of clock synchronization states that the *granularity* of the *global time* must be larger than the *precision* of the ensemble of *clocks* (Sect. 3.2.1).

Reference Clock: The *reference clock* is an ideal *clock* that ticks always in perfect agreement with the international standard of time (Sect. 3.1.2).

Reliability: The *reliability R* (*t*) of a system is the probability that a system will provide the specified service until time *t*, given that the system was operational at $t = t_0$. (Sect. 1.4.1).

Replica Determinism: *Replica Determinism* is a desired relation between replicated *RT objects*. A set of replicated *RT objects* is *replica determinate* if all objects of this set have the same visible *state* and produce the same output messages at instants that are at most an interval of *d* time units apart (Sect. 5.6).

Resource Adequacy: A *real-time computer system* is *resource adequate* if there are enough computing resources available to handle the specified *peak load* and the *faults* specified in the *fault hypothesis*. Guaranteed response systems must be based on *resource adequacy* (→*guaranteed timeliness*) (Sect. 1.5.4).

Rise Time: The *rise time* is the time required for the output of a system to rise to a specific percentage of its final equilibrium value as a result of step change on the input (Sect. 1.3.1).

Risk: *Risk* is the product of *hazard* severity and *hazard* probability. The severity of a *hazard* is the worst-case damage of a potential accident related to the *hazard* (Sect. 11.4.2).

Safety: *Safety* is *reliability* regarding *critical failure* modes (Sect. 1.4.2).

Safety Case: A *safety case* is a combination of a sound set of arguments supported by analytical and experimental evidence substantiating the *safety* of a given system (Sect. 11.4.3).

Safety-Critical Real-Time Computer System: Synonym to *hard real-time computer system* (Sect. 1.1).

Sampling: In *sampling*, the state of a RT entity is periodically interrogated by the computer system at instants that are in the *sphere of control* of the computer system. If a memory element is required to store the effect of an *event*, the memory element is outside the *sphere of control* of the computer system (Sect. 1.3.1).

Schedulability Test: A *schedulability test* determines whether there exists a schedule such that all *tasks* of a given set will meet their deadlines (Sect. 10.1.2).

Semantic Agreement: An agreement among measured variables is called *semantic agreement* if the meanings of the different *measured values* are related to each other by a process model that is based on a priori knowledge about the physical characteristics and the dynamics of the *controlled object* (Sect. 9.6.3).

Semantic Content: The essential meaning of a statement or variable as understood by an end-user. The same *semantic content* can be represented in different syntactic forms (Sect. 2.2.4).

Signal Conditioning: *Signal conditioning* refers to all processing steps that are required to generate a *measured data element* from a *raw data element* (Sect. 1.2.1).

Soft Deadline: A *deadline* for a result is *soft* if the result has utility even after the *deadline* has passed (Sect. 1.1).

Soft Real-Time Computer System: A *real-time computer system* that is only concerned with *soft deadlines* (Sect. 1.1).

Sparse Event: An event that occurs in the active interval of a →*sparse time base* (Sect. 3.3).

Sparse Time Base: A time base in a distributed computer systems where the physical time is partitioned into an infinite sequence of active and silent intervals and where *sparse events* may be generated only in the active intervals (Sect. 3.3).

Sphere of Control (SOC): The *sphere of control* of a subsystem is defined by the set of *RT entities* the values of which are established within this subsystem (Sect. 5.1.1).

Sporadic Task: A *sporadic task* is a *task* where the *task* request times are not known but where it is known that a minimum time interval exists between successive requests for execution (→*periodic task*, →*aperiodic task*) (Sect. 10.1.2).

Spoofing Attack: A security attack where an adversary masquerades as a legitimate user in order to gain unauthorized access to a system (Sect. 6.2.2).

State: The state of a component at a given instant is a data structure that contains all information about the past that is considered relevant for the future operation of the component (Sect. 4.2).

State Estimation: *State estimation* is the technique of building a model of a *RT entity* inside a *RT object* to compute the probable state of a *RT entity* at a selected future instant and to update the related *RT image* accordingly (Sect. 5.4.3).

State Message: A message is a *state message* if it contains information about states, if a new version of the message replaces the previous version, and the message is not consumed on reading (→*event message*) (Sect. 4.3.4).

Synchronization Condition: The *synchronization condition* is a necessary condition for the synchronization of clocks. It relates the *convergence function*, the *drift offset*, and the *precision* (Sect. 3.4.1).

System of Systems (SoS): A system consisting of a set of nearly autonomous constituent systems that decide to cooperate in order to achieve a common objective (Sect. 4.7.3).

Task Descriptor List (TADL): The *task* descriptor list (TADL) is a static data structure in a time-triggered operating system that contains the instants when the *tasks* have to be dispatched (Sect. 9.2.1).

Task Request Time: The *task request time* is the instant when a *task* becomes ready for execution (Sect. 10.1.2).

Task: A *task* is the execution of a program (→*simple task*, →*complex task*) (Sect. 9.2).

Temporal Accuracy: A *real-time image* is *temporally accurate* if the time interval between the moment "now" and instant when the current value of the real-time image was the value of the corresponding *RT entity* is smaller than an application-specific bound (Sect. 5.4).

Temporal Control: *Temporal control* is concerned with the determination of the real-time instants when a *task* must be activated or when a *task* must be blocked (→*logical control*) (Sect. 4.1.3).

Temporal Failure: A *temporal failure* occurs when a value is presented at the system-user *interface* outside the intended interval of real time. Temporal failures can only exist if the system specification contains information about the expected temporal behavior of the system (Synonym timing failure) (Sect. 6.1.3).

Temporal Order: The *temporal order* of a set of *events* is the order of *events* as they occurred on the time line (Sect. 3.1.1).

Thrashing: The phenomenon that a system's throughput decreases abruptly with increasing load is called *thrashing* (Sect. 7.2.5).

Tick: A *tick* (synonym: macrotick) of the global time is a selected *microtick* of the local clock. The *offset* between any two respective global ticks of an ensemble of synchronized *clocks* must always be less than the *precision* of the ensemble (→*microtick, reasonableness condition*) (Sect. 3.2.1).

Time-Stamp: A *time-stamp* of an *event* with respect to a given clock is the state of the clock at the instant of occurrence of the *event* (Sect. 3.1.2).

Time-Division Multiple Access (TDMA): *Time-division multiple access* is a time-triggered communication technology where the time axis is statically partitioned into slots. Each slot is statically assigned to a *component*. A *component* is only allowed to send a message during its slot (Sect. 7.5).

Time-Triggered Architecture (TTA): A distributed computer architecture for real-time applications, where all components are aware of the progression of the global time and where most actions are triggered by the progression of this global time (Sect. 11.7).

Time-Triggered Ethernet (TTEthernet): An extension of standard Ethernet that supports deterministic message transport (Sect. 7.5.2).

Time-Triggered Protocol (TTP): A communication protocol where the instant of starting a message transmission is derived from the progression of the global time (Sect. 7.5.1).

Time-Sensitive Networking (TSN): A set of protocols and mechanisms for real-time communication, standardized within the IEEE 802.1, including time-triggered communication (Sect. 7.5.3).

Timed Message: A *timed message* is a message that contains the time-stamp of an *event* (e.g., point of observation) in the data field of the message (Sect. 9.3.1).

Timing Failure: →Temporal failure

Transducer: A device converting energy from one domain into another. The device can either be a *sensor* or an *actuator* (Sect. 9.5)

Transient Fault: A *transient fault* is a fault that exists only for a short period of time after which it disappears. The hardware is not permanently affected by a transient fault (Sect. 6.1.1).

Trigger: A *trigger* is an *event* that causes the start of some action (Sect. 1.5.5).

Trigger Task: A *trigger task* is a time-triggered *task* that evaluates a condition on a set of temporally accurate variables and generates a *trigger* for an application *task* (Sect. 9.2.2).

Triple-Modular Redundancy (TMR): A fault-tolerant system configuration where a *fault-tolerant unit (FTU)* consists of three synchronized replica deterministic *components*. A value or timing failure of one *component* can be masked by the majority (→*voting*) (Sect. 6.4.2).

Understanding: *Understanding* develops if the concepts and relationships that are employed in the representation a model have been adequately linked with the →*conceptual landscape* and the methods of reasoning of the observer (Sect. 2.1.3).

Universal Time Coordinated (UTC): An international time standard that is based on astronomical phenomena (→*International Atomic Time*) (Sect. 3.1.4).

Value Failure: A *value failure* occurs if an incorrect value is presented at the system-user *interface* (Sect. 6.1.3).

Voter: A *voter* is a unit that detects and masks errors by comparing a number of independently computed input messages and delivers an output message that is based on the analysis of the inputs (→*exact voting*, →*inexact voting*) (Sect. 6.4.2).

Vulnerability: A *deficiency* in the design or operation of a computer system that can lead to a security incident, such as an *intrusion* (Sect. 6.2).

Watchdog: A *watchdog* is an independent external device that monitors the operation of a computer. The computer must send a periodic signal (*life sign*) to the *watchdog*. If this life sign fails to arrive at the *watchdog* within the specified time interval, the *watchdog* assumes that the computer has failed and takes some action (e.g., the *watchdog* forces the *controlled object* into the safe state) (Sect. 9.7.4).

Worst-Case Administrative Overhead (WCAO): The *worst-case execution time* of the administrative services provided by an operating system (Sect. 10.2).

Worst-Case Communication Delay (WCCOM): The *worst-case communication delay* is the maximum duration it may take to complete a communication action under the stated *load- and fault hypothesis* (Sect. 5.4.1).

Worst-Case Execution Time (WCET): The *worst-case execution time (WCET)* is the maximum duration it may take to complete an *action* under the stated *load and fault hypothesis*, quantified over all possible input data (Sect. 10.2).

References

[Ahu90] Ahuja, M., Kshemkalyani, A. D., & Carlson, T. (1990). A basic unit of computation in a distributed system. In *Proceedings of the 10th IEEE distributed computer systems conference* (pp. 12–19). IEEE Press.

[Ale77] Alexander, C. S., et al. (1977). *A pattern language.* Oxford University Press.

[All21] Allende, I., et al. (2021). Towards Linux based safety systems – A statistical approach for software execution path coverage. *Elsevier Journal of Systems Architecture.*

[Alu15] Alur, R. (2015). *Principles of cyber-physical systems.* MIT Press.

[Ama21] Amazon. (2021). *AWS Lambda.* https://aws.amazon.com/lambda/. Accessed 3 Aug 2021.

[Ami01] Amidzic, O., Riehle, H. J., Fehr, T., Wienbruch, C., & Elbert, T. (2001). Pattern of focal y-bursts in chess players. *Nature, 412,* 603.

[And01] Anderson, D. L. (2001). Occam's razor; simplicity, complexity, and global geodynamics. *Proceedings of the American Philosophical Society, 14*(1), 56–76.

[And95] Anderson, J., Ramamurthy, S., & Jeffay, K. (1995). Real-time computing with lock-free shared objects. In *Proceedings RTSS 1995* (pp. 28–37). IEEE Press.

[ANS20] ANSI/UL. (2020). *Standard for evaluation of autonomous products* (ANSI/UL 4600). UL.

[ARI05] ARINC. (2005). *Design assurance guidance for airborne electronic hardware* (RTCA/ DO-254). ARINC.

[ARI11] ARINC. (2011). *Software considerations in airborne systems and equipment certification* (ARINC DO-178C). ARINC.

[ARI91] ARINC. (1991). *Multi-transmitter data bus ARINC 629 – Part 1: Technical description.* ARINC.

[Arl03] Arlat, J., et al. (2003). Comparison of physical and software-implemented fault injection techniques. *IEEE Transactions on Computers, 52*(9), 1115–1133.

[Arm10] Armbrust, M., et al. (2010). A view of cloud computing. *Communications of the ACM, 53*(4), 50–58.

[Art06] ARTEMIS. (2006). *Strategic research agenda. Reference designs and architectures.* https://www.artemisia-association.org/downloads/RAPPORT_RDA.pdf

[Art94] Arthur, W. B. (1994). On the evolution of complexity. In G. Cowan, D. Pines, & D. Meltzer (Eds.), *Complexity: Metaphors, models, and reality.* Westview Press.

[Att09] Attaway, S. (2009). *Matlab, a practical introduction to programming and problem solving.* Elsevier.

© The Editor(s) (if applicable) and The Author(s), under exclusive license to Springer Nature Switzerland AG 2022
H. Kopetz, W. Steiner, *Real-Time Systems,*
https://doi.org/10.1007/978-3-031-11992-7

[AUD17] Audi. (2017). *Audi A8 – Zentrales Fahrerassistenzsteuergerät (zFAS)*. https://www.audi-technology-portal.de/de/elektrik-elektronik/fahrerassistenzsysteme/audi-a8-zentrales-fahrerassistenzsteuergeraet-zfas. Accessed 27 Jan 2022.

[Avi04] Avizienis, A., et al. (2004). Basic concepts and taxonomy of dependable and secure computing. *IEEE Transactions on Dependable and Secure Computing, 1*(1), 11–33.

[Avi82] Avizienis, A. (1982). The four-universe information system model for the study of fault tolerance. In *Proceedings of the 12th FTCS symposium* (pp. 6–13). IEEE Press.

[Avi85] Avizienis, A. (1985). The N-version approach to fault-tolerant systems. *IEEE Transactions on Software Engineering, 11*(12), 1491–1501.

[Avr92] Aversky, D., Arlat, J., Crouzet, Y., & Laprie, J. C. (1992). Fault injection for the formal testing of fault tolerance. In *Proceedings of FTCS 22* (pp. 345–354). IEEE Press.

[Bag10] Baggerman, C. (2010). *Avionics system architecture for NASA Orion vehicle* (Presentation, 2010). NASA JSC.

[Bal17] Baldini, I., et al. (2017). Serverless computing: Current trends and open problems. In *Research advances in cloud computing* (pp. 1–20). Springer.

[Bar01] Baresi, L., & Young, M. (2001). *Test Oracles*. University of Oregon, Department of Computer Science.

[Bar07] Baronti, P., et al. (2007). Wireless sensor networks: A survey on the State of the Art and the 802.15.4 and Zigbee Standards. *Computer Communication, 30*, 1655–1695. Elsevier.

[Bar10] Barreno, M., et al. (2010). The security of machine learning. *Machine Learning, 81*, 121–148. Springer.

[Bar18] Barroso, L. A., Hölzle, U., & Ranganathan, P. (2018). The datacenter as a computer: Designing warehousescale machines. *Synthesis Lectures on Computer Architecture, 13*(3), i–189. Morgan&Claypool Publishers.

[Bar93] Barborak, M., & Malek, M. (1993). The consensus problem in fault-tolerant computing. *ACM Computing Surveys, 25*(2), 171–218.

[Bau09] Bauer, H., Scharbarg, J.-L., & Fraboul, C. (2009). Applying and optimizing trajectory approach for performance evaluation of AFDX avionics network. In *2009 IEEE conference on emerging technologies & factory automation* (pp. 1–8). IEEE.

[Bea08] Beautement, A., Sasse, M. A., & Wonham, M. (2009). The compliance budget: Managing security behavior in organizations. In *Proceedings of NSPW 08* (pp. 47–58). ACM Press.

[Bed08] Bedau, M. A., & Humphrey, P. (2008). *Emergence*. MIT Press.

[Ben00] Benini, L., & DeMicheli, G. (2000). System level power estimation: Techniques and tools. *ACM Transactions on Design Automation of Electronic Systems, 5*(2), 115–192.

[Ber01] Berwanger, J., et al. (2001). FlexRay – The communication system for advanced automotive control systems. In *Proceedings of the SAE world congress 2001* (pp. 2001–0676). SAE Press.

[Ber07] Bertolino, A. (2007). Software testing research: Achievements, challenges, dreams. In *Proceedings of FOSE 07* (pp. 85–103). IEEE Press.

[Ber14] Bernstein, D. (2014). Containers and cloud: From LXC to Docker to Kubernetes. *IEEE Cloud Computing, 1*(3), 81–84.

[Ber85] Berry, G., & Cosserat, L. (1985). The synchronous programming language ESTEREL and its mathematical semantics. In *Proceedings of the seminar on concurrency* (LNCS 197). Springer.

[Bha10] Bhattacharayya, R. et al. (2010). Low-cost, ubiquitous RFID-tag-antenna-based sensing. Proceedings of the IEEE. 98 (10). 1593-1600.

[Bla09] Black, D. C., Donovan, J., & Bunton, B. (2009). *System C: From the ground up*. Springer.

[Boe01] Boehm, B., & Basili, V. (2001). *Software defect reduction top 10 list* (pp. 135–137). IEEE Computer.

[Bor07]   Borkar, S. (2007). Thousand core chips – A technology perspective. In *Proceedings of DAC 2007* (pp. 746–749). ACM Press.

[Bou61]   Boulding, K. E. (1961). *The image*. Ann Arbor Paperbacks.

[Bou96]   Boussinot, F., & Simone, R. (1996). The SL synchronous language. *IEEE Transactions on Software Engineering, 22*(4), 256–266.

[Bos16]   Boschert, S., & Rosen, R. (2016). Digital twin - the simulation aspect. In Mechatronic Futures. Springer. (pp. 59–74).

[Bro00]   Brown, S. (2000). Overview of IEC 61508 – Design of electrical/electronic/programmable electronic safety-related systems. *Computing and Control Engineering Journal, 11*(1), 6–12.

[Bro10]   Brooks, F. P. (2010). *The design of design: Essays from a computer scientist.* Addison Wesley.

[Bun08]   Bunge, M. (2008). *Causality and modern science.* Transaction Publishers.

[Bur09]   Burns, A., & Wellings, A. (2009). *Real-time systems and programming languages: Ada, real-time Java and C/real-time POSIX.* Addison-Wesley.

[But04]   Buttazzo, G. (2004). *Hard real-time computing systems: Predictable scheduling algorithms and applications.* Springer.

[Cad22]   Cadavid, H., et al. (2022). *System and software architecting harmonization practices in ultra-large-scale Systems of Systems.* Preprint at: https://arxiv.org/pdf/2201.03275.pdf. Accessed 22 Feb 2022.

[CAN90]   CAN. (1992). *Controller area network CAN, an in-vehicle serial communication protocol* (SAE Handbook 1992) (pp. 20.341–20.355). SAE Press.

[Car08]   Cardenas, A., Amin, S., & Shastry, S. (2008). Research challenges for the security of control systems. In *Proceedings of the workshop on Hot topics in security.* Usenix Association. http://portal.acm.org/citation.cfm?id=1496671.1496677

[Cha09]   Chandola, V., Banerjee, A., & Kumar, V. (2009). Anomaly detection: A survey. *ACM Computing Surveys, 41*(3), 15.1–15.58.

[Che87]   Cheng, S. C. (1987). Scheduling algorithms for hard real-time systems – A brief survey. In *Hard real-time systems.* IEEE Press.

[Cin18]   Cinque, M., & Cotroneo, D. (2018). Towards lightweight temporal and fault isolation in mixed-criticality systems with real-time containers. In *48th annual IEEE/IFIP international conference on dependable systems and networks workshops* (pp. 59–60).

[Cla03]   Clark, E., et al. (2003). Counterexample-guided abstraction refinement for symbolic model checking. *Journal of the ACM, 50*(5), 752–794.

[Cla14]   Clavier, R., Sautereau, P., & Dufour, J.-F. (2014). TTEthernet, a promising candidate for Ariane 6. In *Proceedings of DASIA 2014 DAta systems in aerospace* (Vol. 725, p. 34 f). ESA SP.

[Cla88]   Clark, D. (1988). The design philosophy of the DARPA internet protocols. *Computer Communication Review, 18*(4), 106–114.

[Cle22]   Clerckx, B., Popovic, Z., & Murch, R. (2022). Future networks with wireless power transfer and energy harvesting. *Proceedings of the IEEE, 110*(1), 3–7.

[Con02]   Constantinescu, C. (2002). Impact of deep submicron technology on dependability of VLSI circuits. In *Proceedings of DSN 2002* (pp. 205–209). IEEE Press.

[Cra16]   Craciunas, S., et al. (2016). Scheduling real-time communication in IEEE 802.1 Qbv time sensitive networks. In *24th international conference on real-time networks and systems* (pp. 183–192). ACM.

[Cri89]   Cristian, F. (1989). Probabilistic clock synchronization. *Distributed Computing, 3*, 146–185. Springer.

[Cru91a] Cruz, R. L. (1991). A calculus for network delay. I. Network elements in isolation. *IEEE Transactions on Information Theory, 37*(1), 114–131.

[Cru91b] Cruz, R. L. (1991). A calculus for network delay. II. Network analysis. *IEEE Transactions on Information Theory, 37*(1), 132–141.

[Cum10]   Cumming, D. M. (2010, March 11). Haven't found that software glitch, Toyota? Keep trying. *Los Angeles Times*.

[Dav79]   Davies, C. T. (1979). Data processing integrity. In *Computing systems reliability* (pp. 288–354). Cambridge University Press.

[Dea02]   Deavours, D., et al. (2002). The Mobius framework and its implementation. *IEEE Transactions on Software Engineering, 28*(10), 1–15.

[Deg95]   Degani, A., Shafto, M., & Kirlik, A. (1995). Mode usage in automated cockpits: Some initial observations. In *Proceedings of IFAC 1995* (pp. 1–13). IFAC Press.

[Diz19]   Dizdarevic, J., et al. (2019). A survey of communication protocols for internet of things and related challenges of fog and cloud computing integration. *ACM Computing Surveys, 51*(6), 1–29.

[Dra17]   Dragoni, N., et al. (2017). Microservices: Yesterday, today, and tomorrow. In *Present and ulterior software engineering*. Springer.

[Dri03]   Driscoll, K., et al. (2003). Byzantine fault-tolerance: From theory to reality. In *Proceedings of SAFECOMP 2003* (LNCS 2788) (pp. 235–248). Springer.

[Dys98]   Dyson, G. B. (1998). *Darwin among the machines – The evolution of global intelligence*. Basic Books.

[EAR14]   European Association of Research & Technology Organisations. (2014). *The TRL scale as a research & innovation policy tool, EARTO recommendations*.

[Edm00]   Edmonds, B. (2000). Complexity and scientific modeling. In *Foundations of science* (pp. 379–390). Springer.

[Eid06]   Eidson, J. (2006). *Measurement, control and communication using IEEE 1588*. Springer.

[Eps08]   Epstein, S. (2008). Intuition from the perspective of cognitive experiential self-theory. In *Intuition in judgment and decision making* (pp. 23–38). Lawrence Erlbaum.

[ETS19]   ETSI. (2019). *Analysis of the collective perception service (CPS)* (ETSI TR 103 562). ETSI.

[Eug03]   Eugster, P. T., et al. (2003). The many faces of publish/subscribe. *ACM Computing Surveys, 35*(2), 114–131.

[Eve10]   Evensen, K., & Weiss, K. (2010). *A comparison and evaluation of real-time software systems modeling languages* (p. 3504). AIAA Infotech@ Aerospace.

[Fei06]   Feiler, P., Gluch, D., & Hudak, J. (2006). *The architecture analysis and design language (AADL): An introduction* (Report CMU-SEI 2006-TN-011). Software Engineering Institute.

[Fel04]   Feltovich, P. J., et al. (2004). Keeping it too simple: How the reductive tendency effects cognitive engineering. In *IEEE intelligent systems* (pp. 90–94). IEEE Press.

[Fel04a] Feldhofer, M., Dominikus, S., & Wokerstorfer, J. (2004). *Strong authentication for RFID systems using the AES algorithms* (LCNS 3156) (pp. 357–370). Springer.

[Fin03]   Finkenzeller, K. (2003). *RFID handbook*. Wiley.

[Fis06]   Fisher, D. A. (2006). *An emergent perspective on the operation of system-of-systems* (CMU/SEI-2006-TR-003). Carnegie Mellon Software Engineering Institute. http://www.dtic.mil/cgi-bin/GetTRDoc?Location=U2&doc=GetTRDoc.pdf&AD=ADA449020

[Foh94]   Fohler, G. (1994). *Flexibility in statically scheduled hard real-time systems*. PhD thesis, Institut für Technische Informatik. Technical University of Vienna.

[Fra01]   Frank, D., et al. (2001). Device scaling limits of Si MOSFETs and their application dependencies. *Proceedings of the IEEE, 89*(3), 259–288.

[Fri20]   Freitag, C., et al. (2020). *The climate impact of ICT: A review of estimates, trends and regulations*. Lancester University. https://arxiv.org/abs/2102.02622. Accessed 11 Mar 2022

[Gad10]   Gadh, R., et al. (2010). RFID: A unique radio innovation for the 21st century. *Proceedings of the IEEE, 98*(2), 1541–1680.

[Gaj09]   Gajski, D. D., et al. (2009). *Embedded system design*. Springer.

[Gar75]   Garey, M. R., & Johnson, D. S. (1975). Complexity results for multiprocessor scheduling under resource constraints. *SIAM Journal of Computing, 4*(4), 397–411.

[Gau05] Gaudel, M. C. (2005). Formal methods and testing: Hypotheses, and correctness approximations. In *Proceedings of formal methods 2005* (LNCS 3582). Springer.

[Gol73] Goldberg, R. P. (1973). *Architectural principles for virtual computer systems* (Technical Report). Harvard University.

[Gra86] Gray, J. (1985). Why do computers stop and what can be done about it? Tandem Technical Report TR85.7.

[Hal05] Halford, G. S., et al. (2005). How many variables can humans process? *Psychological Science, 16*(1), 70–76.

[Hal92] Halbwachs, N. (1992). *Synchronous programming of reactive systems*. Springer.

[Hal96] Halford, G. S., Wilson, W. H., & Phillips, S. (1996). *Abstraction, nature, costs, and benefits*. Department of Psychology, University of Queensland.

[Hau06] Hause, M. (2006). The SysML modelling language. In *Fifteenth European systems engineering conference* (Vol. 9, pp. 1–12).

[Hay90] Hayakawa, S. I. (1990). *Language in thought and action*. Harvest Original.

[Hei00] Heinzelman, W. R., Chandrakasan, A., & Balakrishan, H. (2000). Energy-efficient communication protocol for wireless microsensor networks. In *Proceedings of the 33rd Hawaii international conference on system science* (pp. 3722–3725). IEEE Press.

[Hen03] Henzinger, T., Horowitz, B., & Kirsch, C. M. (2003). Giotto: A time-triggered language for embedded programming. *Proceedings of the IEEE, 91*(1), 84–99.

[Hen19] Hennessy, J. L., & Patterson, D. A. (2019). *Computer architecture: A quantitative approach* (6th ed.). Elsevier.

[Her09] Herault, L. (2009). Holistic approach for future energy-efficient cellular networks. In *Proceedings of the second Japan-EU symposium on the future internet* (pp. 212–220). European Communities Brussels.

[HER15] HERE. (2015). *Sensor ingestion interface specification*. https://360.here.com/2015/06/23/here-sensor-data-ingestion/. https://developer.here.com/documentation/sdii-data-spec/. Accessed 27 Jan 2022.

[Hme04] Hmelo-Silver, C. E., & Pfeffer, M. G. (2004). Comparing expert and novice understanding of a complex system from the perspective of structures, behaviors, and functions. *Cognitive Science, 28*, 127–138. Elsevier.

[Hoe10] Hoefer, C. (2010). Causal determinism. In *Stanford encyclopedia of philosophy* (pp. 1–24) http://plato.stanford.edu/entries/determinism-causal/

[Hop78] Hopkins, A. L., Smith, T. B., & Lala, J. H. (1978). FTMP: A highly reliable fault-tolerant multiprocessor for aircraft control. *Proceedings IEEE, 66*(10), 1221–1239.

[Hue08] Huebscher, M. C., & McCann, J. A. (2008). A survey of autonomic computing–degrees, models and applications. *ACM Computing Surveys, 40*(3), 7.1–7.28.

[IEA21] *Data Centres and Data Transmission Networks*. (2022). IEA. https://www.iea.org/reports/data-centres-and-data-transmission-networks. Accessed 11 Mar 2022.

[IEC21] IEC. (2021). *Industrial communication networks – Network and system security* (IEC 62443). IEC.

[Int09] Intel. (2009). *Teraflop research chip*. http://techresearch.intel.com/articles/TeraScale/1449.htm

[Ior18] Iorga, M., et al. (2018). *Fog computing conceptual model* (NIST SP) (pp. 500–325). NIST.

[ISO21] ISO. (2021). *Road vehicles – Cybersecurity engineering* (ISO/SAE 21434:2021). ISO.

[ITR09] ITRS Roadmap. (2009). *International technology roadmap for semiconductors, 2009 ed. Executive summary*. Semiconductor Industry Association.

[Jac07] Jackson, D., Thomas, M., & Millet, L. I. (2007). *Software for dependable systems: Sufficient evidence?* National Academic Press.

[Jer77] Jerri, A. J. (1977). The Shannon sampling theorem – Its various extensions and applications: A tutorial review. *Proceedings of the IEEE, 65*(11), 1565–1596.

[Joh92] Johnson, S. C., & Butler, R. W. (1992). Design for validation. *IEEE Aerospace and Electronic Systems Magazine, 7*(1), 38–43.

[Jon78] Jones, J. C. (1978). *Design methods, seeds of human futures*. Wiley.

[Jon97]   Jones, M. (1997). *What really happened on Mars Rover Pathfinder*. http://catless.ncl.
           ac.uk/Risks/19.49.html#subj1

[Jou18]   Jouppi, N., et al. (2018). Motivation for and evaluation of the first tensor processing
           unit. *IEEE Micro, 38*(3), 10–19.

[Jue05]   Juels, A., & Weis, S. A. (2005). Authenticating pervasive devices with human protocols.
           In *Proceedings of CRYPTO 2005* (pp. 293–308). Springer.

[Jur04]   Juristo, N., Moreno, A. M., & Vegas, S. (2004). Reviewing 25 years of testing technique
           experiments. *Empirical Software Engineering, 9*, 7–44. Springer.

[Kah79]   Kahne, S., Lefkowitz, I., & Rose, C. (1979). Automatic control by distributed intel-
           ligence. *Scientific American, 240*(6), 78–92.

[Kan95]   Kantz, H., & Koza, C. (1995). The ELECTRA railway signaling-system: field experi-
           ence with an actively replicated system with diversity. *Proceedings FTCS, 25*, 453–458.

[Kar95]   Karlson, J., et al. (1995). Integration and comparison of three physical fault-injection
           experiments. In *Predictably dependable computing systems*. Springer.

[Kau93]   Kauffman, S. (1993). *The origins of order: Self-organization and selection in evolution.*
           Oxford University Press.

[Kea07]   Keating, M., et al. (2007). *Low power methodology manual for chip design.* Springer.

[Kim94]   Kim, K. H., & Kopetz, H. (1994). A real-time object model RTO.k and an experimental
           investigation of its potential. In *Proceedings COMPSAC 94*. IEEE Press.

[Kni86]   Knight, J. C., & Leveson, N. G. (1986). An experimental evaluation of the assump-
           tion of independence in multiversion programming. *IEEE Transactions on Software
           Engineering, SE-12*(1), 96–109.

[Koo04]   Koopman, P. (2004). Embedded system security. *IEEE Computer*, 95–97.

[Koo20]   Koopman, P., & Wagner, M. (2020). Positive trust balance for self-driving car deploy-
           ment. In *Proceedings of international conference on computer safety, reliability, and
           security* (pp. 351–357). Springer.

[Kop03]   Kopetz, H., & Suri, N. (2003). Compositional design of RT systems: A conceptual basis
           for the specification of linking interfaces. In *Proceedings of 6th ISORC* (pp. 51–59).
           IEEE Press.

[Kop03a]   Kopetz, H., & Bauer, G. (2003). The time-triggered architecture. *Proceedings of the
           IEEE, 91*(1), 112–126.

[Kop04]   Kopetz, H., Ademai, A., & Hanzlik, A. (2004). Integration of internal and external clock
           synchronization by the combination of clock state and clock rate correction in fault tol-
           erant distributed systems. In *Proceedings of the RTSS 2004* (pp. 415–425). IEEE Press.

[Kop06]   Kopetz, H. (2006). Pulsed data streams. In *From model-driven design to resource
           management for distributed embedded systems* (IFIP Series 225/2006) (pp. 105–114).
           Springer.

[Kop07]   Kopetz, H., et al. (2007). Periodic finite-state machines. In *Proceedings of ISORC 2007*
           (pp. 10–20). IEEE Press.

[Kop08]   Kopetz, H. (2008). *The rationale for time-triggered ethernet* (RTSS 2008) (pp. 3–11).
           IEEE Press.

[Kop08a]   Kopetz, H. (2008). The complexity challenge in embedded system design. In
           Proceedings of ISORC 2008 (pp. 3–12). IEEE Press.

[Kop09]   Kopetz, H. (2009). *Temporal uncertainties in cyber-physical systems* (Report 1/2009).
           Institute für Technische Informatik, TU Vienna.

[Kop10]   Kopetz, H. (2010). Energy-savings mechanism in the time-triggered architecture. In
           Proceedings of the 13th ISORC (pp. 28–33). IEEE Press.

[Kop21]   Kopetz, H. (2021). *An architecture for driving automation.* https://www.the-autonomous.
           com/news/an-architecture-for-driving-automation. Accessed 13 Dec 2021.

[Kop22]   Kopetz, H. (2022). *Data, information and time: The DIT model* (Springer Briefs in
           Computer Science). Springer.

[Kop85]   Kopetz, H., & Merker, W. (1985). The architecture of MARS. In *Proceedings of
           FTCS-15* (pp. 274–279). IEEE Press.

[Kop87] Kopetz, H., & Ochsenreiter, W. (1987). Clock synchronization in distributed real-time systems. *IEEE Transactions on Computers, 36*(8), 933–940.

[Kop90] Kopetz, H., & Kim, K. (1990). Temporal uncertainties in interactions among real-time objects. In *Proceedings 9th IEEE symposium on reliable distributed systems* (pp. 165–174). IEEE Press.

[Kop91] Kopetz, H., Grünsteidl, G., & Reisinger, J. (1991). Fault-tolerant membership service in a synchronous distributed real-time system. In *Dependable computing for critical applications* (pp. 411–429). Springer.

[Kop92] Kopetz, H. (1992). Sparse time versus dense time in distributed real-time systems. In *Proceedings 14th international conference on distributed computing systems* (pp. 460–467). IEEE Press.

[Kop93] Kopetz, H., & Gruensteidl, G. (1993). TTP – A time-triggered protocol for fault-tolerant real-time systems. In *Proceedings FTCS-23* (pp. 524–532). IEEE Press.

[Kop93a] Kopetz, H., & Reisinger, J. (1993). The non-blocking write protocol NBW: A solution to a real-time synchronisation problem. In *Proceedings of RTSS 1993* (pp. 131–137). IEEE Press.

[Kop98] Kopetz, H. (1998). The time-triggered model of computation. In *Proceedings of RTSS 1998* (pp. 168–176). IEEE Press.

[Kop99] Kopetz, H. (1999). Elementary versus composite interfaces in distributed real-time systems. In *Proceedings of ISADS 99* (pp. 26–34). IEEE Press.

[Kor10] Kortuem, G., et al. (2010). Smart objects as the building block for the internet of things. *IEEE Internet Computing, 2010*, 44–50.

[Kum10] Kumar, K., & Lu, Y. H. (2020). Cloud computing for mobile users: Can offloading computation save energy? *IEEE Computer, 2010*, 51–56.

[Lal94] Lala, J. H., & Harper, R. E. (1994). Architectural principles for safety critical real-time applications. *Proceedings of the IEEE, 82*(1), 25–40.

[Lam74] Lamport, L. (1994). A new solution of Dijkstra's concurrent programming problem. *Communications of the ACM, 8*(7), 453–455.

[Lam78] Lamport, L. (1978). Time, clocks, and the ordering of events. *Communications of the ACM, 21*(7), 558–565.

[Lam82] Lamport, L., Shostak, R., & Pease, M. (1982). The byzantine generals problem. *Communications of the ACM TOPLAS, 4*(3), 382–401.

[Lam85] Lamport, L., & Melliar Smith, P. M. (1985). Synchronizing clocks in the presence of faults. *Journal of the ACM, 32*(1), 52–58.

[Lan09] Langley, P., Laird, J. E., & Rogers, S. (2009). Cognitive architectures: Research issues and challenges. *Cognitive System Research, 10*(2), 141–160.

[Lan81] Landwehr, C. E. (1981). Formal models for computer security. *ACM Computing Suverys, 13*(3), 248–278.

[Lan97] Landwehr, C. E., & Goldschlag, D. M. (1997). Security issues in networks with internet access. *Proceedings of the IEEE, 85*(12), 2034–2051.

[Lap08] Laprie, J.-C. (2008). From dependability to resilience. In *Proceedings of 38th IEEE/ IFIP international conference on dependable systems and networks* (pp. G8–G9). IEEE Press.

[Lau06] Lauwereins, R. (2006). *Multi-core platforms are a reality...but where is the software support?* (Visual Presentation). IMEC. http://www.mpsoc-forum.org/2006/slides/Lauwereins.pdf

[Lee02] Lee, E. A. (2002). Embedded software. In *Advances in computers* (Vol. 56). Academic.

[Lee10] Lee, E. A., & Seshia, S. A. (2010). *Introduction to embedded systems – A cyber-physical systems approach.* http://LeeSeshia.org

[Lee90] Lee, P. A., & Anderson, T. (1990). *Fault tolerance: Principles and practice.* Springer.

[Leh85] Lehmann, M. M., & Belady, L. (1985). *Program evolution: Processes of software change.* Academic.

[Lev08]   Leverich, J., et al. (2008). Comparative evaluation of memory models of chip multiprocessors. *ACM Transactions on Architecture and Code Optimization, 5*(3), 12.1–12.30.

[Lev95]   Leveson, N. G. (1995). *Safeware: System safety and computers.* Addison Wesley Company.

[Li19]   Li, H., et al. (2019). ACRN: A big little hypervisor for IoT development. In *15th ACM SIGPLAN/SIGOPS international conference on virtual execution environments* (pp. 31–44).

[Lie10]   Lienert, D., & Kriso, S. (2010). Assessing criticality in automotive systems. *IEEE Computer, 43*(5), 30.

[Lit93]   Littlewood, B., & Strigini, L. (1993). Validation of ultra-high dependability for software-based systems. *Communications of the ACM, 36*(11), 69–80.

[Liu00]   Liu, J. W. S. (2000). *Real-time systems.* Prentice Hall.

[Liu73]   Liu, C. L., & Layland, J. W. (1973). Scheduling algorithms for multiprogramming in a hard-real-time environment. *Journal of the ACM, 20*(1), 46–61.

[LoB19]   Lo Bello, L., & Steiner, W. (2019). A perspective on IEEE time-sensitive networking for industrial communication and automation systems. *Proceedings of the IEEE, 107*(6), 1094–1120.

[Lun84]   Lundelius, L., & Lynch, N. (1984). An upper and lower bound for clock synchronization. *Information and Control, 62*, 199–204.

[Lv09]   Lv, M., et al. (2009). *A survey of WCET analysis of real-time operating systems.* http://www.neu-rtes.org/publications/lv_ICESS09.pdf

[Mai98]   Maier, M. W. (1998). Architecting principles for system of systems. *Systems Engineering, 1*(4), 267–284.

[Mar91]   MARS. (1991). *The Mars video.* TU Vienna. http://pan.vmars.tuwien.ac.at/mars/video

[Mar99]   Martin, T. L., & Siewiorek, D. P. (1999). The impact of battery capacity and memory bandwidth on CPU speed setting: A case study. In *Proceedings of ISLPED99* (pp. 200–205). IEEE Press.

[McC09]   McCabe, M., et al. (2009). Avionics architecture interface considerations between constellation vehicles. In *Proceedings DASC'09* (pp. 1.E.2.1–1.E.2.10). IEEE Press.

[Mel11]   Mell, P., & Grance, T. (2011). *The NIST definition of cloud computing* (NIST SP 800-145). NIST.

[Mes04]   Mesarovic, M. D., Screenath, S. N., & Keene, J. D. (2004). Search for organizing principles: Understanding in systems biology. *Systems Biology On Line, 1*(1), 19–27.

[Mes89]   Mesarovic, M. D. (1989). *Abstract system theory* (Lecture Notes in Control and Information Science) (Vol. 116). Springer.

[Met76]   Metcalfe, R. M., & Ethernet. (1976). Distributed packet switching for local computer networks. *Communications of the ACM*, 395–404.

[Mil04]   Miller, D. (2004, October 27). *AFDX determinism.* Visual Presentation at ARINC general session, Rockwell Collins.

[Mil56]   Miller, G. A. (1956). The magical number seven, plus or minus two: Some limits on our capacity for processing information. *The Psychological Review, 63*, 81–97.

[Mil91]   Mills, D. L. (1991). Internet time synchronization: The network time protocol. *IEEE Transactions on Communications, 39*(10), 1482–1493.

[Min02]   Miner, P. S., Malekpour, M., & Torres, W. (2002). A conceptual design for a reliable optical bus (ROBUS). In *Proceedings of the 21st digital avionics systems conference* (p. 13D3). IEEE Press.

[Mog06]   Mogul, C. (2006). Emergent (mis)behavior vs. complex software systems. In *Proceedings of EuroSys 2006.* ACM Press.

[Mok83]   Mok, A. (1983). *Fundamental design problems of distributed systems for the hard real-time environment.* PhD thesis, Massachusetts Institute of Technology.

[Mor07]   Morin, E. (2007). *Restricted complexity, general complexity.* World Scientific Publishing Corporation.

[Mos]   Moses, J. *Complexity and flexibility* (Working Paper).

[NAS99]   NASA. (1999). *Mars climate orbiter Mishap investigation report*. Washington, DC. ftp://ftp.hq.nasa.gov/pub/pao/reports/2000/MCO_MIB_Report.pdf

[Neu56]   Neumann, J. (1956). Probabilistic logic and the synthesis of reliable organisms from unreliable components. In C. E. Shannon & J. McCarthy (Eds.), *Automata studies* (Annals of Mathematics Studies, No 34) (pp. 43–98). Princeton University Press.

[Neu94]   Neuman, B. C., & Ts'o, T. (1994). Kerberos – An authentication service for computer networks. *IEEE Communication Magazine, 32*(9), 33–38.

[Neu95]   Neumann, P. G. (1995). *Computer related risks*. Addison Wesley/ACM Press.

[Neu96]   Neumann, P. G. (1996). Risks to the public in computers and related systems. *Software Engineering Notes, 21*(5), 18. ACM Press.

[NIS21]   Ross, R., et al. (2021). *Developing cyber-resilient systems: A systems security engineering approach* (NIST SP 800-160) (Vol. 2, Revision 1). NIST.

[Obe09]   Obermaisser, R., & Kopetz, H. (2009). *GENSYS – An ARTEMIS cross-domain reference architecture for embedded systems*. Südwestdeutscher Verlag für Hochschulschriften (SVH).

[OMG08]   OMG, MARTE. (2008). *Modeling and analysis of real-time and embedded systems*. Object Management Group.

[OMG15]   OMG. (2015). *OMG data distribution service (DDS). Version 1.4*. Object Management Group.

[Pap02]   Pappu, R., et al. (2002). Physical one-way functions. *Science, 297*, 2026–2030.

[Par97]   Parunak, H. V. D., et al. (1997). Managing emergent behavior in distributed control systems. *Proceedings of ISA Tech, 97*, 1–8.

[Pau08]   Paukovits, C., & Kopetz, H. (2008). Concepts of switching in the time-triggered network-on-chip. In *14th IEEE conference on embedded and real-time computing systems and applications* (pp. 120–129). IEEE Press.

[Pau98]   Pauli, B., A. Meyna, & P. Heitmann. (1998). Reliability of electronic components and control units in motor vehicle applications. Verein Deutscher Ingenieure (VDI). 1009-1024.

[Pea80]   Pease, M., Shostak, R., & Lamport, L. (1980). Reaching agreement in the presence of faults. *Journal of the ACM, 27*(2), 228–234.

[Ped06]   Pedram, M., & Nazarian, S. (2006). Thermal modeling, analysis and management in VLSI circuits: Principles and methods. *Proceedings of the IEEE, 94*(8), 1487–1501.

[Per10]   Perez, J. M. (2010). *Executable time-triggered model (E-TTM) for the development of safety-critical embedded systems* (pp. 1–168). PhD. thesis, Institut für Technische Informatik, TU Wien, Austria.

[Per99]   Perrow, C. (1999). *Normal accidents: Living with high risk technologies*. Princeton University Press.

[Pet79]   Peters, L. (1979). Software design: Current methods and techniques. In *Infotech state of the art report on structured software development*. Infotech International.

[Pet96]   Peterson, I. (1996). Comment on time on Jan 1, 1996. *Software Engineering Notes, 19*(3), 16.

[Pol07]   Polleti, F., et al. (2007). Energy-efficient multiprocessor systems-on-chip for embedded computing: Exploring programming models and their architectural support. *Proceedings of the IEEE, 56*(5), 606–620.

[Pol95]   Poledna, S. (1995). *Fault-tolerant real-time systems, the problem of replica determinism*. Springer.

[Pol95b] Poledna, S. (1995). Tolerating sensor timing faults in highly responsive hard real-time systems. *IEEE Transactions on Computers, 44*(2), 181–191.

[Pop18]   Pop, P., et al. (2018). Enabling fog computing for industrial automation through time-sensitive networking (TSN). *IEEE Communications Standards Magazine, 2*(2), 55–61.

[Pop21]   Pop, P., et al. (2021). The FORA fog computing platform for industrial IoT. *Elsevier Information Systems, 98*, 101727.

[Pop68]   Popper, K. R. (1968). *The logic of scientific discovery*. Hutchinson.

[Pop74]   Popek, G. J., & Goldberg, R. P. (1974). Formal requirements for virtualizable third generation architectures. *Communications of the ACM, 17*(7), 412–421.

[Pot21]   Pott, C. (2021). Firmware security module. *Springer Journal of Hardware and Systems Security, 2021*(5), 103–113.

[Pow91]   Powell, D. (1991). *Delta-4, A generic architecture for dependable distributed computing* (pp. 1–484). Springer.

[Pow95]   Powell, D. (1995). Failure mode assumptions and assumption coverage. In B. Randell, J. C. Laprie, H. Kopetz, & B. Littlewood (Eds.), *Predictably dependable computing systems* (pp. 123–140). Springer.

[Pus89]   Puschner, P., & Koza, C. (1989). Calculating the maximum execution time of real-time programs. *Real-Time Systems, 1*(2), 159–176. Springer.

[Qas19]   Qasaimeh, M., et al. (2019). Comparing energy efficiency of CPU, GPU and FPGA implementations for vision kernels. In *Proceedings of IEEE international conference on embedded software and systems* (pp. 1–8). IEEE Press.

[Ran75]   Randell, B. (1975). System structure for software fault tolerance. *IEEE Transactions on Software Engineering, SE-1*(2), 220–232.

[Ray10]   Ray, K. (2010). *Introduction to service-oriented architectures.* http://anengineersper-spective.com/wp-content/uploads/2010/03/Introduction-to-SOA.pdf

[Rec02]   Rechtin, E., & Maier, M. W. (2002). *The art of systems architecting.* CRC Press.

[Rec91]   Rechtin, E. (1991). *Systems architecting, creating and building complex systems.* Prentice Hall.

[Rei10]   Reisberg, D. (2010). *Cognition.* W.W. Norton.

[Rei57]   Reichenbach, H. (1957). *The philosophy of space and time.* Dover.

[Ric20]   Ritchie, H. (2020). *Climate change and flying: What share of global CO_2 emissions come from aviation?* Our World in Data. https://ourworldindata.org/co2-emissions-from-aviation. Accessed 11 Mar 2022

[Riv78]   Rivest, R. L., Shamir, A., & Adleman, L. (1978). A method for obtaining signatures and public-key cryptosystems. *Communications of the ACM, 21*(2), 120–126.

[Rob16]   Robati, T., Gherbi, A., & Mullins, J. (2016). A modeling and verification approach to the design of distributed IMA architectures using TTEthernet. *Elsevier Procedia Computer Science, 83*, 229–236.

[Rom07]   Roman, R., Alcarez, C., & Lopez, J. (2007). A survey of cryptographic primitives and implementations for hardware-constrained sensor network nodes. *Mobile Network Applications, 12*, 231–244. Springer.

[Ruh21]   Ruh, J., Steiner, W., & Fohler, G. (2021). Clock synchronization in virtualized distributed real-time systems using IEEE 802.1AS and ACRN. *IEEE Access, 9*, 126075–126094.

[Rui14]   Ruiz De Azua, J. A., & Boyer, M. (2014). Complete modelling of AVB in network calculus framework. In *22nd international conference on real-time networks and systems* (pp. 55–64). ACM.

[Rum08]   Rumpler, B. (2008). *Design comprehension of embedded rea-time systems.* PhD thesis, Institut für Technische Informatik, TU Wien.

[Rus02]   Rushby, J. (1999). Systematic formal verification for fault-tolerant time-triggered algorithms. *IEEE Transactions on Software Engineering, 25*(5), 651–660.

[Rus03]   Rushby, J. (2003). A comparison of bus architectures for safety critical embedded systems. Report NASA/CR-2003–212161.

[Rus93]   Rushby, J. (1993). *Formal methods and the certification of critical systems* (Research Report No. SRI-CSL-93-07). Computer Science Lab, SRI.

[SAE95]   SAE. (1995). Class C application requirements, survey of known protocols, J20056. In *SAE handbook* (pp. 23.437–23.461). SAE Press.

[Sag13]   Sagiroglu, S., & Sinanc, D. (2013). Big Data: A review. In *International conference on collaboration technologies and systems* (pp. 42–47). IEEE Press.

[Sah20]   Sah, K. D., & Amgoth, T. (2020). Reliable energy harvesting schemes in wireless sensor networks: A survey. *Elsevier Information Fusion, 63*, 223–247.

[Sal10] Salfner, F., Lenk, M., & Malek, M. (2010). A survey of online failure prediction methods. *ACM Computing Surveys, 42*(3), 10.10–10.42.

[Sal84] Saltzer, J., Reed, D. P., & Clark, D. D. (1984). End-to-end arguments in system design. *ACM Transactions on Computer Systems, 2*(4), 277–288.

[Sch88] Schwabl, W. (1988). *The effect of random and systematic errors on clock synchronization in distributed systems*. PhD thesis, Technical University of Vienna, Institut für Technische Informatik.

[Sel08] Selberg, S. A., & Austin, M. A. (2008). *Towards an evolutionary system of systems architecture*. Institute for Systems Research. http://ajcisr.eng.umd.edu/~austin/reports.d/INCOSE2008-Paper378.pdf

[Ses08] Sessions, R. (2008). *Simple architectures for complex enterprises*. Microsoft Press.

[Sha04] Sha, L., et al. (2004). Real-time scheduling theory: A historical perspective. *Real-Time Systems Journal, 28*(3/4), 101–155. Springer.

[Sha10] Shafto, M., et al. (2010). *Draft: NASA technology roadmap: Modeling, simulation, information technology & processing roadmap – Technology area 11*. NASA.

[Sha89] Shaw, A. C. (1989). Reasoning about time in higher-level language software. *IEEE Transactions on Software Engineering, SE-15*, 875–889.

[Sha90] Sha, L., Rajkumar, R., & Lehoczky, J. P. (1990). Priority inheritance protocols: An approach to real-time synchronization. *IEEE Transactions on Computers, 39*(9), 1175–1185.

[Sha94] Sha, L., Rajkumar, R., & Sathaye, S. S. (1994). Generalized rate-monotonic scheduling theory: A framework for developing real-time systems. *Proceedings of the IEEE, 82*(1), 68–82.

[Shi16] Shi, W., et al. (2016). Edge computing: Vision and challenges. *IEEE Internet of Things Journal, 3*(5), 637–646.

[Sie00] Siegel, J. (2000). *CORBA 3 – Fundamentals and programming*. OMG Press/Wiley.

[Sim11] Siminiceanu, R., Miner, P. S., & Person, S. (2011). *A methodology for evaluating artifacts produced by a formal verification process* (Technical Memorandum 20110022654). NASA.

[Sim81] Simon, H. A. (1981). *Science of the artificial*. MIT Press.

[Smi97] Smith, J. S. S. (1997). *Application specific integrated circuits*. Addision Wesley.

[Soe01] Soeleman, H., Roy, K., & Paul, B. C. (2001). Robust subtreshold logic for ultra-low power operation. *IEEE Transactions on VLSI Systems, 9*(1), 90–99.

[Sol07] Soltesz, S., et al. (2007). Container-based operating system virtualization: A scalable, high-performance alternative to hypervisors. In *2nd ACM SIGOPS/EuroSys European conference on computer systems* (pp. 275–287).

[Spr89] Sprunt, B., Sha, L., & Lehoczky, J. (1989). Aperiodic task scheduling for hard real-time systems. *Real-Time Systems, 1*(1), 27–60.

[Sta03] Stamatis, D. H. (2003). *Failure mode and effect analysis: FMEA from theory to execution*. ASQ Quality Press.

[Sta18] Stallings, W. (2018). *Operating systems: Internals and design principles*. Pearson.

[Ste10] Steiner, W. (2010). An evaluation of SMT-based schedule synthesis for time-triggered multi-hop networks. In *31st IEEE real-time systems symposium* (pp. 375–384). IEEE.

[Ste11a] Steiner, W., & Dutertre, B. (2011). Automated formal verification of the TTEthernet synchronization quality. In *NASA formal methods symposium* (LNCS) (Vol. 6617, pp. 375–390). Springer.

[Ste11b] Steiner, W., & Rushby, J. (2011). TTA and PALS: Formally verified design patterns for distributed cyber-physical systems. In *2011 IEEE/AIAA 30th digital avionics systems conference* (pp. 7B5-1–7B5-15). IEEE.

[Sus00] Sussman, J. (2000). *Introduction to Transportation Systems*. Artech House Publishers.

[Sus03] Sussman, J. (2003). *Collected views on complexity in systems* (Working Paper Series. ESD-WP-2003-01.06-ESD). MIT Engineering Systems Division.

[Szy99] Szyperski, C. (1999). *Component software – Beyond object-oriented programming*. Addision Wesley.

[Tai03]   Taiani, F., Fabre, J. C., & Killijian, M. O. (2003). Towards implementing multi-layer reflection for fault-tolerance. In *Proceedings of the DSN 2003* (pp. 435–444). IEEE Press.

[Tal08]   Taleb, N. N. (2008). *The black swan: The impact of the highly improbable*. Penguin.

[Tas03]   Force, T. (2004). *Final report on the August 14, 2003 Blackout in the United States and Canada*. US Department of Energy.

[Tel09]   Japan, T. (2009). Cyber-Clean Center (CCC) project for anti-bot countermeasures in Japan. In *Proceedings of the second Japan-EU symposium on the future internet* (pp. 212–220). European Communities Brussels.

[Tin95]   Tindell, K. (1995). Analysis of hard real-time communications. *Real-Time Systems, 9*(2), 147–171.

[Tra88]   Traverse, P. (1988). AIRBUS and ATR system architecture and specification. In *Software diversity in computerized control systems*. Springer.

[TTT21]   TTTech. (2021). *Nerve*. https://www.tttech-industrial.com/products/nerve. Accessed 28 Dec 2021.

[Ver09]   Vermesan, O., et al. (2009). *Internet of things – Strategic research roadmap*. European Commission-Information Society and Media DG.

[Ver94]   Verissimo, P. (1994). Ordering and timeliness requirements of dependable real-time programs. *Real-Time Systems, 7*(3), 105–128.

[Vig10]   Vigras, W. J. (2010). *Calculation of semiconductor failure data*. http://rel.intersil.com/docs/rel/calculation_of_semiconductor_failure_rates.pdf

[Vig62]   Vigotsky, L. S. (1962). *Thought and language*. MIT Press.

[Wen78]   Wensley, J. H., et al. (1978). SIFT: The design and analysis of a fault-tolerant computer for aircraft control. *Proceedings IEEE, 66*(10), 1240–1255.

[Wik10]   Washington's Axe. (2010). *Wikipedia*. http://en.wikipedia.org/wiki/George_Washington%27s_axe#George_Washington.27s_axe

[Wil08]   Wilhelm, R., et al. (2008). The worst-case execution time problem – Overview of methods and survey of tools. *ACM Transactions on Embedded Computing Systems, 7*(3), 1–53.

[Wil21]   Williamsson, D., & Sellgren, U. (2022). *Integrated modularization methodology and process of heavy-duty trucks*. https://www.diva-portal.org/smash/record.jsf?pid=diva2%3A1532165&dswid=-5065. Accessed 22 Feb 2022.

[Wil90]   Williams, T. J. (1990). A reference model for computer integrated manufacturing from the viewpoint of industrial automation. *Elsevier IFAC Proceedings Volumes, 23*(8), 281–291.

[Wil98]   Wilson, E. O. (1998). *Consilience – The unity of knowledge*. Vintage Books.

[Win01]   Winfree, A. T. (2001). *The geometry of biological time*. Springer.

[Wit90]   Withrow, G. J. (1990). *The natural philosophy of time*. Oxford Science Publications/Clarendon Press.

[Woj17]   Wojciak, J., & Bray, T. (2017). *Decouple and scale applications using Amazon SQS and Amazon SNS*. https://aws.amazon.com/pub-sub-messaging/. Accessed 3 Aug 2021.

[Xi11]   Xi, S., et al. (2011). RT-Xen: Towards real-time hypervisor scheduling in Xen. In *International conference on embedded software (EMSOFT)* (pp. 39–48). IEEE Press.

[Xin08]   Xing, L., & Amari, S. V. (2008). *Handbook of performability engineering*. Springer.

[Xu91]   Xu, J., & Parnas, D. (1990). Scheduling processes with release times, deadlines, precedence, and exclusion relations. *IEEE Transactions on Software Engineering, 16*(3), 360–369.

[Zil96]   Zilberstein, S. (1996). Using anytime algorithms in intelligent systems. *AI Magazine, 16*(3), 73–83.

Index

Printed in the United States
by Baker & Taylor Publisher Services